DARK
MISSION

Lu~

Thanks for all your help while I am on my own "Dark Mission". Thanks for being there & loving me unconditionally.

I Love You

~Me

DARK
MISSION

The Secret History
of the
National Aeronautics
and Space Administration

by
Richard C. Hoagland
and
Michael Bara

A Feral House Book
ISBN: 978-1-932595-26-0

10 9 8 7 6 5 4 3

Feral House
PO Box 39910
Los Angeles, CA 90039

www.FeralHouse.com
info@FeralHouse.com

Design by Sean Tejaratchi

Table of Contents

Introduction

The NASA that we've known for over 50 years has been a lie.

My name is Richard C. Hoagland. As the brief authors' biography page states, I was indeed a NASA consultant to the Goddard Spaceflight Center in the post-Apollo era, and Science Advisor to Walter Cronkite and CBS News, Special Events, advising CBS on the science of the NASA missions to the Moon and Mars, during the Apollo Program. I currently run an independent NASA watchdog and research group, the Enterprise Mission, attempting to figure out how much of what NASA has found in the solar system over the past 50 years has actually been silently filed out of sight as classified material, and therefore totally unknown to the American people.

My friend and colleague Mike Bara and I are going to attempt the impossible in the next few hundred pages: we're going to try to describe, and then carefully document, exactly what's been going with NASA in terms of that classified data and information. It won't be an easy task.

The predisposition of most Americans—even after the Challenger and Columbia disasters and a host of other "missing" spacecraft—is to place NASA somewhere on par with Mother Teresa in terms of public confidence and credibility. This is, in major part, due to the average American's (to say nothing of the media's) inability to figure out a reason why NASA—ostensibly a purely scientific Agency—would actually lie. NASA is, after all, holding high the beacon of our last true heroes, the astronauts. I mean, what's to hide regarding moon rocks, craters and space radiation?

If we're right, a lot.

However, even a hint that NASA—or, more precisely, its leadership—has been carrying out any kind of hidden agenda for over 50 years is, at best, met with disbelief. The vast majority of NASA's nearly 18,000 full-time employees are, in our analysis, innocent of the wrongdoing of the few that we are going to describe.

To even begin to understand the extraordinary case we are presenting in this book, to fully appreciate what NASA has been quite consciously, deliberately and methodically concealing from the American people and the world for all these years, you have to begin with NASA's turbulent past—specifically an account of its origins in the increasingly dangerous geopolitical environment Americans were thrust into in the wake of World War II.

The governmental institution known as NASA is a department of the Executive Branch, ultimately answerable solely to the President of the United States, an Agency created through the National Aeronautics and Space Act of 1958. NASA ostensibly is "a *civilian* agency exercising control over aeronautical and space activities sponsored by the United States."[Emphasis added.][1]

But contrary to common public and media perception that NASA is an open, strictly civilian scientific institution, is the legal fact that the Space Agency was quietly founded as a direct adjunct to the Department of Defense, tasked with specifically assisting the national security of the United States in the midst of a deepening Cold War with its major geopolitical adversary, the Soviet Union. It says so right in the original NASA Charter:

"Sec. 305... (i) The [National Aeronautics and Space] Administration shall be considered *a defense agency of the United States* for the purpose of Chapter 17, Title 35 of the United States Code..." [Emphasis added.]

In another section[2] of the act, this seldom-discussed *defense* responsibility—the ultimate undercutting of NASA's continuing public façade as a strictly civilian, scientific agency—is blatantly spelled out:

"Sec. 205... (d) No [NASA] information which *has been classified for reasons of national security* shall be included in *any report* made under this section [of the Act]..." [Emphasis added.]

Clearly, from this and the other security provisions[3] incorporated in the Act, what the Congress, the press and the American taxpayers get to see of NASA's ultimate activities—including untouched images and data regarding what's *really* on the Moon, on Mars or anywhere else across the solar system—is totally dependent on whether the President of the United States (and/or his legal surrogates in the Department of Defense and the "intelligence community") has already secretly classified that data. This is directly contrary to everything we've been led to believe regarding NASA for over 50 years now.

After NASA was formed, almost before the ink was dry on the Bill that brought it into being (which, among many other detailed objectives, called for "the establishment of long-range studies of the potential benefits to be gained from, the opportunities for and the problems involved in the use of aeronautical and space activities for peaceful and scientific purposes"), NASA commissioned a formal "futures study" into the projected effects on American society of its many planned activities (including covert ones).

Carried out as a formal NASA contract to the Brookings Institution—a well-known Washington, D.C.-based think tank—the 1959 study was officially titled "Proposed Studies on the Implications of Peaceful Space Activities for Human Affairs."[4] The results of this multi-disciplinary investigation were officially submitted to the administrator of NASA in late 1960, and after the Kennedy Administration was elected, to Congress in April 1961.

One area of unusual interest covered in the report—easily overlooked amid mountains of interminable statistics and analyses —was a quiet assessment of the near-certainty of a NASA discovery of intelligent extraterrestrial life:[5]

"While face-to-face meetings with it [extra-terrestrial life] *will not* occur within the next 20 years (unless its technology is more advanced than ours, qualifying it to visit Earth), *artifacts left at some point in time by these life forms might possibly be discovered through our* [NASA's] *space activities* on the Moon, Mars, or Venus." [Emphasis added.]

This quietly inserted sub-section of Brookings is revealing on many levels, and it forms the documented basis of our case—that the NASA "you thought you knew" doesn't actually exist, and that NASA has been deliberately concealing and classifying its most significant discoveries because of "national security" rationales.

Brookings officially affirmed NASA's expectations that the Agency would fly to nearby planets in the solar system, and would thus be physically capable, for the first time, of confronting "extraterrestrials" *right in their backyard.*

Did any skeptics even know this official document existed, before we made it public in 1996? As you shall see documented later in this book, beginning in the mid-1960s with unmanned spacecraft, NASA actually discovered its projected extraterrestrial artifacts—but then, the Agency never got around to telling the rest of us!

NASA would clandestinely confirm with these earliest robotic probes, and then proceed to cover up, the first awesome remains of a once-extraordinary, solar-system-wide, ancient technological civilization on the Moon—precisely as Brookings had predicted. Four years later, the Apollo Program would come to full fruition, and the lunar astronauts themselves would personally

witness and extensively document, with tens of thousands of high quality photographs, from both lunar orbit and the surface, extraordinary "glass-like" structures on the Moon! The Apollo crews would also bring back to NASA laboratories not just rocks, but actual *samples* of the ancient technologies they found—for highly classified efforts at "back engineering."

Above is just one example of the ancient, glass-like lunar ruins photographed in person by the Apollo astronauts, and hidden away (by a former NASA employee) in a private archive for more than 30 years. Later in this volume, far more details on those historic in-situ astronaut lunar ruins observations will be dealt with—and additional, uncensored Apollo photographs of these extensive structures, as well as photographs of some of the actual artifacts brought back to Earth—presented, and analyzed in depth.

A skeptic might well ask at this point, how can we be presenting valid, official NASA images of suppressed ruins and technology if the Agency has spent so much time and energy over the last 40 years covering them up?

The answer is that after two generations, leaked images like the one above—displaying stunning details of ancient lunar structures arching overhead, as well as key alien artifacts that have been brought back—have also suddenly begun appearing on the internet, on official NASA websites!

A small cadre of loyal NASA employees were witnesses to what actually went on, and agreed at that time to keep the secret in the interest of national security. Some of these NASA employees, apparently, have finally "seen the light"—that this continued deception, no matter what the legal rationale or national security implications, was fundamentally extra-Constitutional. Because of these true NASA heroes, real space history as about to officially begin, again.

Based on our analysis as presented in this book, it is also our opinion that NASA's entire lunar exploration program—culminating with the incredibly successful manned Apollo Project—was carefully conceived, from the beginning, as a kind of "alien reconnaissance" followed by an "alien artifacts retrieval" program.

Again, the intention to do just that was blatantly laid out in Brookings. We now believe this is the reason by which President John F. Kennedy—reported to be "totally disinterested in space"[6]—was quietly convinced to announce his historic decision to "send men to the Moon... and return them safely to the Earth... within a decade" in May of 1961. This was widely believed, then and now, to be Kennedy's effort to demonstrate to the world the superiority of the American system, as opposed to Soviet communism.

However, at the United Nations on September 20, 1963, the President suddenly issued a public invitation to the Soviets[7] only two years into the Apollo "race" to the Moon: an offer of a "cooperative, joint U.S./USSR lunar expedition."

Of course, if there were a "hidden agenda" to Apollo, this move would have revealed that the prime objective was not to beat the Soviet Union, but to covertly find and return samples of the incredibly advanced lunar technology that had been waiting on the Moon for eons ... and then to share them with the Soviets! Curiously enough, a mere two months following Kennedy's startling U.N. proposition, the President was killed.

The enthusiastic architects of the continuing NASA Brookings cover-up, in part, are the same heroes we have been encouraged to worship as some of the leading pioneers of our technological era. Their names are synonymous with America's achievements in space science and rocket engineering. In many cases, they are also men with secret pasts—Germans, Egyptians, Englishmen and Americans, men at the very fringes of rational thought and conventional wisdom. These literal "fringe elements," then, are divided into three main groups inside the Agency, as best as we can tell at present. For the purposes of this volume, we shall call them the "Magicians," the "Masons" and the "Nazis"—and deal with each group separately.

Each "sect" is led by prominent individuals, and supported by lesser-known players. Each has stamped their own agenda on our space program, in indelible but traceable ways. And each, remarkably, is dominated by a secret or "occult" doctrine, that is far more closely aligned with "ancient religion and mysticism" than it is with the rational science and cool empiricism these men promote to the general public as NASA's overriding mantra.

Using commercially available celestial mechanics/astronomical software—programs like the popular "Red Shift" series (which uses the official JPL ephemeris as its database)—we have been able to establish a pattern of behavior on NASA's part that points to something truly as inexplicable as it is exotic: a bizarre internal obsession by the Agency with three "gods" and "goddesses," reaching across the millennia directly from ancient Egypt—Isis, Osiris and Horus.

It is these same three Egyptian gods (whose mythic story has been documented by many Egyptologists and authors, including Christopher Knight and Robert Lomas, in *The Hiram Key*) that are also key to understanding the history of the Masonic Order. As we shall show, it is this same mythology that is also at the heart of the v systems of the NASA "Magicians" and Nazis as well. This ritual Egyptian symbolism, secretly practiced by NASA throughout these past five decades, publicly shows up only in its repeating, blatant choices of simple mission patch designs.

For instance, if one looks at the official patch for the Apollo Program (below), armed with our preceding "heads-up" regarding the bizarre NASA focus on all things "Egyptian," it becomes elemental to match the "A" (for "Apollo") as an actual stand-in for "Asar"—the Egyptian designation for "Osiris." This successful decoding of the hidden Egyptian meaning of the Apollo patch is redundantly confirmed—because "Asar/Osiris" is none other than the familiar Greek constellation of "Orion"—which is, of course, the background stellar constellation on the patch itself.

In case you think such ritual symbolism is some kind of temporary historical aberration, confined only to the Apollo Program and the 1960s, think again; when NASA recently selected a patch design for its new "CEV" spacecraft, which will eventually replace the Shuttle—and ultimately take American astronauts back to the Moon, look what NASA curiously picked again (below).

Later in this book, we shall present a documented history of these continuing, inexplicable "secret society" manipulations inside NASA—not only of its personnel, but also of its major policies—which, in fact, has been going on since its Congressional formation, and all with this inexplicable "Egyptian focus."

We will also investigate the purposes behind these apparently repeating rituals and identify the key players who have quietly made them happen.

You will also read more, in due course, regarding the accumulated evidence of widespread corruption, usurpation of the law and petty rivalries within the Agency—that have allowed these irrational religious practices to continue unabated.

The evidence that NASA is something other than the benevolent civilian science institution it pretends to be, is as overwhelming now as it is disturbing.

In the years following JFK's assassination, when Apollo finally became an engineering reality, only nine successful Apollo missions to and from the Moon were carried out; only six of those were actually lunar landings.

Then (apparently), a critical number and type of lunar artifacts was successfully identified, and returned to Earth by the Apollo crews—at which point the entire Apollo Program was abruptly terminated with Apollo 17.

In our model, it was this successful completion of Apollo's secret mission and agenda, and not Congressional budget cuts, which was the real reason for this abrupt cessation of America's historic journeys to the Moon, and the primary reason no one has gone near the lunar surface for over 30 years.

All of which makes the sudden announcement by President George W. Bush of a new White House/NASA program "to return to the Moon by 2020," made at NASA Headquarters on January 14, 2004, so incredibly intriguing. What does the current Bush Administration know, 30 years after the termination of Apollo, regarding what is waiting on the Moon for human beings to return? And is this why this Administration has mounted an Apollo-style program "on steroids"—as new NASA Administrator Mike Griffin, specifically appointed by President Bush to head the new lunar return program, wryly terms it?

Is NASA's sudden interest in returning to the Moon actually an effort to get back as quickly as its current budgets will allow, before a host of other countries do the same? Countries that have, independently, suddenly announced their plans for going to the Moon—countries like China, India, Japan and Russia, and even the European Space Agency? Are we seeing a second space race being born? A race not for mere propaganda victories this time, but a much more important race, among a much wider field of players, for sole access to the scientific secrets the set of surviving lunar structures surveyed by Apollo must inevitably contain—which, to those who successfully decode what they discover this time could mean the ultimate domination of the Earth?

It was just a few years after the start of the Apollo missions that even more extraordinary solar system ruins were first observed on Mars—beginning with images and other instrumented scans sent back by NASA's first Mars orbiter, the unmanned Mariner 9 spacecraft, in late 1971.

This earliest robotic confirmation that there is also something "anomalous" on Mars paved the way for far more extensive observations when the first Viking orbiters and landers arrived half a decade later. Again, the details of these critical observations and discoveries—and their documentation—will be laid out later in this volume.

The critical thing here is that, in direct contradiction to everything the press and the American people were being led to believe that NASA stood for—program transparency, open scientific inquiry, freedom of publication—the Agency quietly and methodically covered-up up the most astonishing wonders it had found.

Many of NASA's consulting sociologists and anthropologists to Brookings (like Dr. Margaret Meade, whom I had the privilege of actually working with in later years, at New York's Hayden Planetarium) had been warning NASA, even as Brookings was being researched and assembled, of "the enormous potential for social instability" if the existence of bona fide extraterrestrials—or even their ruins—was officially revealed in the socially repressive and heavily religious environment of the late 1950s.[8]

With those first Lunar Orbiter images taken of the Moon, *everything*—the reality of ruins, their extraordinary scale, their obvious presence on more than one world in the solar system, how their builders vanished—suddenly was all too real.

There *had* been a powerful, enormously encompassing, extraordinary solar-system-wide civilization that had simply disappeared, only to be rediscovered by NASA's primitive initial probes. A civilization that, it would turn out later, had been wiped out through a series of all-encompassing, solar-system-wide cataclysms.[9]

The most disturbing part of "Brookings" to policy makers, however—even before these shattering discoveries were verified—was its thinly-veiled, authoritative warnings regarding what could happen to our civilization if NASA's 1950s-style "ET predictions" were confirmed.

"Anthropological files contain many examples of societies, sure of their place in the universe, which have *disintegrated* when they had to associate with previously unfamiliar societies espousing different ideas and different life ways; others that survived such an experience usually did so by paying the price of changes in values and attitudes and behavior..." [Emphasis added.]

The literal *disintegration of society*—simply from knowing that "we're not alone."[10]

The Brookings discussion of the implications of such a crucial discovery also encompassed a critical second-level problem: What to do if the Agency, at some point in the future, actually made such a momentous, world-changing confirmation of extraterrestrial intelligence next door? Or even of their surviving ruins and artifacts?

NASA's discussion of these problems before they occurred—and the draconian measures it was seriously considering—is critically revealing:

"Studies might help to provide programs for meeting and adjusting to the implications of such a discovery. Questions one might wish to answer by such studies would include: How might such information, under what circumstances, be presented to *or withheld from* the public, for what ends? *What might be the role of the discovering scientists* and other decision makers regarding release of the fact of discovery?" [Emphasis added.]

Following the political tumult and excitement of the first successful Apollo Lunar Landings, the White House and NASA dramatically changed the direction of the entire space program—under the excuse of a lack of public interest and insufficient funding.

The Agency quickly dropped any pretense of following up on the Apollo Program with permanent bases on the Moon, as well as indefinitely postponing all discussion and plans for going on to Mars.

Instead, under the now-proven lie of developing an economical, reliable, reusable space transportation system, and a "world-class" space research laboratory for it to re-supply—i.e., the Shuttle, and the International Space Station—NASA collaborated with the White House in a fateful set of decisions in the early 1970s that would consign American astronauts to endlessly circle the Earth for decades, while the Moon—with stunning ruins and bits and pieces of a miraculous, preserved technology orbiting just a quarter of a million miles away—was totally ignored.[11]

On February 15, 2001, Fox Television aired a widely-advertised show titled *Conspiracy Theory: Did We Land On the Moon?* With this program, Fox removed the last weak link in NASA's ongoing, 40-year-old chain of overlapping cover-ups.

It is our assertion that not only was this "Moon hoax" tale carefully constructed as an elegant piece of professional disinformation—as a desperately-required distraction from the *real* lunar conspiracy documented here, which was beginning to seriously unravel as early as 1996. For, I can personally testify that I was a first-hand witness to "the Moon hoax" true beginnings far, far earlier than the 2001 Fox Special—back in 1969, and in the heart of NASA itself!

The occasion was the unforgettable Apollo summer of Neil Armstrong and Buzz Aldrin's epic journey to the Moon—the amazing July landing of Apollo 11. I, of course, had been deeply immersed in all aspects of our CBS coverage of the upcoming Apollo 11 mission for months as official science advisor to CBS News Special Events and chief correspondent Walter Cronkite.

For the actual flight of Apollo 11, I was assigned (at my own request) to the Downey, California facility of the prime contractor for the Apollo Command and Service Modules, North American Rockwell. I was there to personally oversee construction and special effects use of my pet project for our nonstop CBS coverage of "Lunar Landing Day"—a "walk-through solar system" constructed by North American technicians specifically for myself and CBS in a huge, drafty aircraft hanger. It was in this miniature, recreated version of the solar system that I had successfully proposed that Walter Cronkite interview via satellite key engineers, project managers and "special guests"—those who had built the Apollo spacecraft at North American or had special knowledge in the realm of history and space—to comment on the historic legacy of the Apollo 11 flight.

One luminary I was proud to bring before the cameras, to chat with Walter in New York regarding the extraordinary nature of events occurring that historic night, was Robert A. Heinlein, the dean of American science fiction. Decades earlier, Bob had co-written the screenplay for *Destination Moon*, one of the first technically accurate film depictions of the lunar journey then unfolding on live television before a billion people all over planet Earth. As the successful author of a pioneering series of "juvenile" SF novels that, for the first time, introduced *realistic* space travel and engineering concepts to an entire generation of future NASA scientists and engineers, Robert Heinlein had, almost single handedly, "inspired the workforce" for the entire space program.

I must admit, I had a certain smug satisfaction that night, watching Bob Heinlein stroll through "the solar system," emphatically predicting to Walter and literally the world, via satellite, that "henceforth, this night—July 20, 1969—will be known as 'the Beginning of the True History of Mankind.'"

After the heady events of that unforgettable 32 hours—the landing; the eerie EVA, complete with ghostly television shots "live from the Moon"; and then, after the crew had slept for a few hours for the first time on the Moon, the successful liftoff of the Lunar Module "Eagle" and rendezvous with the Command Module "Columbia," still in lunar orbit—CBS moved our unit up the street, to the Jet Propulsion Laboratory (JPL) in Pasadena. There we would cover the remainder of the flight, arriving at JPL right after the three Apollo 11 astronauts blasted home toward Earth and "splashdown" in the South Pacific, three days later.

The reason was that NASA had another mission underway during "the Epic Journey of Apollo 11"—a fly-by of two unmanned Mariner spacecraft past Mars, for only the second time in NASA's history.

With only one "CBS Special Events Unit" in California, to cover *all* of NASA's space activities on the West Coast in those years, it was up to our small group in Los Angeles—a producer, a correspondent, a couple of camera guys, maybe a couple of technicians, a make-up person and me—to overlap our continuing coverage of Apollo 11, now originating from the Von Karman Auditorium at JPL, with new commentary covering the second unmanned NASA mission to fly by Mars in history.

Mariner 6 was to cruise past Mars on July 31—recording television images, making spectral scans, conducting remote atmospheric measurements, etc.— just ten days after "Columbia" left lunar orbit, heading for the Pacific Ocean.

Our arrival at JPL on the afternoon of July 22, in preparation for this Mariner 6 fly-by was heady stuff for a 23-year-old network science consultant, as this was my first "in-person" tour to cover an actual live mission.

The circumstances of my first fly-by live from JPL are etched indelibly in my brain, if for no other reason than it was the moment when television lightning struck. One morning our Executive Producer, Bob Wussler, suddenly decided to put *me* on the air across the entire CBS television network to explain the upcoming Mariner fly-by to the nation!

How could one ever forget their first professional network television appearance—and their first official network commentary for a NASA mission flying by Mars, no less? For the life of me, I can't remember a thing I said that morning. I do remember that I had to literally borrow a sportcoat and tie from one of the floor crew for my first appearance on network television.

And, I vividly remember a bizarre scene that happened only a couple days before at JPL, as we arrived.

It was controlled bedlam. Close to a thousand print reporters, television correspondents, technicians, special VIPs, as well as half the staff at JPL itself, were all attempting to register for the limited seating in the (relatively) small Von Karman Auditorium—that had been the scene for all live network coverage of JPL's previous extravaganzas ever since Explorer 1 had been placed in orbit by an Army/JPL team one January night in 1958.

This warm July afternoon only eleven years later, it seemed that everyone was in a mad scramble—simultaneously—to register at the lobby desks specifically set up for members of the press, trying to grab the limited number of press kits on the Mission, and then nail down a seat in the Auditorium beyond.

It was at this point, as I was drifting around Von Karman, trying to spot where the CBS anchor desk was positioned, that I noticed something strange.

Even to my untrained eye, it looked out of place: a man, wearing jeans and a long, light-colored raincoat (it was typical L.A. weather outside—so, why the coat?). This man, wearing one of those floppy "great coats" that cowpunchers used to wear in old Westerns, complete with a dark leather bag slung over one shoulder, was slowly, methodically, placing "something" on each chair in Von Karman.

As he got closer, I suddenly realized he was accompanied by a more conventionally dressed representative from JPL itself: coatless, in white shirt and black tie—the second figure was, in fact, none other than the head of the JPL press office, Frank Bristow.

In the midst of all the commotion, why was Bristow—again, the head of the JPL press office—*personally* squiring this very out-of-place individual around the Auditorium?

Then, as if that wasn't mystery enough, Bristow began moving "great coat guy" back out to the cramped "press room area" beyond the glassed-in foyer of the Auditorium. There, in an office where space correspondents, like Walter Sullivan (*New York Times*), Frank Pearlman (*San Francisco Chronicle*), Jules Bergmann (ABC), and Bill Stout (our local guy from CBS) hung out, and wrote their leads and copy after each formal press briefing held in Von Karman itself, a handful of reporters were now being introduced, again by Bristow, to "great coat guy." Why was the official head of the JPL press office doing this?

I soon had my answer.

As Bristow watched approvingly, his "guest" proceeded to hand each available reporter a copy of whatever he'd been putting on the seats back in the Auditorium.

As I opened up the handout, something yellow and silvery fell on the tile floor. It was a shiny American flag, maybe four inches lengthwise, made of aluminized mylar. I turned to the couple of mimeographed pages and began to read—and couldn't believe my eyes.

The date was July 22, 1969. The three Apollo 11 astronauts—Neil Armstrong, Buzz Aldrin and Mike Collins, two of whom had just successfully walked on "the frigging Moon." and wouldn't splash down in the South Pacific Ocean for two more days, were still halfway between Earth and the "Sea of Tranquility." Yet here, someone with an obvious "in" to JPL was handing out a mimeographed broadsheet to all the *real* reporters ... claiming that "NASA has just faked the entire Apollo 11 Lunar Landing... on a soundstage in Nevada!"

And, if that wasn't weird enough, this individual was being *personally* escorted around Von Karman by none other than the head of the JPL press office himself!

I did what I saw the other veterans do: I casually threw the two pages in the trash and tucked the shiny flag into my notebook. But the seed had been planted.

Looking back, based on all our hard-won knowledge of what is really "out there" in the solar system, and experiencing the outrageous lengths NASA will go to keep "the secret," I can now put the pieces together.

This was an official "Op"—Bristow's job was to make sure that all the national reporters covering NASA at least saw what was handed out that afternoon, complete with shiny flag to act as a mnemonic device to trigger the memory of what was in the pamphlet long after it was history. Sooner or later, a percentage of those who read it that afternoon at JPL would write it up—as a quirky angle on the far-too-dry official tale of Apollo 11.

In this way, it would become a naturally-reproducing meme—"a unit of cultural information, such as a cultural practice or idea, that is transmitted verbally or by repeated action from one mind to another"—which is exactly what NASA apparently intended to plant at JPL that afternoon. To deliberately "infect" the American culture with the story that "the Moon landing was all a fake!"

Was this all some far-seeing "back-up plan" if, in some point in the future, it started to emerge why the astronauts had *really* gone to the Moon?

Fox, the "fair and balanced" network, activated the meme in 2001—with the *Did We Land on the Moon?* special. There, waiting in the wings was a neatly-packaged 30-year-old "conspiracy theory" perfectly gift-wrapped for those finally beginning to "disbelieve" in NASA. An officially concocted "inoculation" against troublemakers who would one day place before many of those same national reporters a set of embarrassing official Apollo photographs, asking the crucial question: "What did NASA *really* find during its Apollo missions to the Moon?"

Read on...

Chapter One

The Monuments of Mars

One of the core problems most readers have with the question of extraterrestrial artifacts is that the story starts not at the beginning, when the artifacts may have been built, or even the middle, when they may have been abandoned, but very near the end. The possible existence of alien artifacts didn't get its initial push into mass consciousness until July 25, 1976, when a project scientist at NASA's Jet Propulsion Laboratory named Toby Owen put a magnifying glass over Viking Orbiter 1 frame 35A72 and exclaimed "Hey, look at this!" [Fig. 1-1]

After the initial splash created by what came to be known as "The Face on Mars," NASA held a daily press briefing in which the Face was the unquestioned highlight. Gerald Soffen, a Viking project scientist, addressed the assembled press, including at the time one Richard C. Hoagland. Soffen introduced the Face image with the statement "Isn't it peculiar what tricks of light and shadow can do...? When we took another picture a few hours later, it all went away; it was just a trick, just the way the light fell on it."[12] That last statement was later proven to be an outright falsehood, and it eventually became the first chink in the armor of the previously un-assailed integrity of the space agency. Although the Face made newspaper headlines all over the world the next day, no journalist, including Richard C. Hoagland, took it seriously. They all accepted NASA's explanation that there were disconfirming photos taken later that same Martian day.

Yet the image of the Face apparently caused quite a bit of consternation at JPL. The Viking missions actually consisted of four vehicles—two Landers and two Orbiters grouped together and called Viking 1 and Viking 2, respectively. The Landers would separate from the Orbiters and descend to the planet's surface to test for signs of life and take pictures from the Martian surface. The first Viking Lander put down on July 20, 1976 in the Chryse Planitia region of Mars, sending back photograph after photograph of the planet's surface. Cydonia was the selected landing site of the second of the Viking Landers, but within a few days of the first "Face" image, 35A72, rumblings began about changing the Viking 2 landing site.

Cydonia (designated landing site B.1, 44.3°N, 10°W) had been chosen as the Viking 2 prime site because it was low, about five to six kilometers below the mean Martian surface, and because it was near the southernmost extremity of the wintertime north polar hood. B-1 also had the advantage of being in line with the first landing site, so the Viking 1 Orbiter could relay data from the second Lander while the second Orbiter mapped the poles and other parts of Mars during the proposed extended mission. While this was considered a good spot to find water, Viking project scientist Hal Masursky was worried about the geology of the region. He asked David Scott, who had prepared the geology maps, to work up a special hazard map for B-1. After studying the map, Masursky came to the conclusion that the area was not "landable." This analysis, of course, was made with maps based on Mariner 9 photographs. He told Tom Young and Jim Martin, however, that there was one hope; wind-borne material may have mantled the rough terrain and covered "up all those nasties we see."

So the ostensible reason for changing the targeted landing site was that Cydonia was suddenly considered "too rocky" for the Viking Lander to risk a touchdown. It was further claimed that the "northern latitude" of Cydonia was partly to blame for this rough surface, and a more suitable landing sight would be sought farther south. In the end, Viking 2 set down in a region known as Utopia Planitia, an even more northerly and rocky site than Cydonia.[13]

Nobody thought much of the venue change at the time, but since their new choice for a landing site contradicted their reasons for scuttling Cydonia, it seemed that somebody at JPL was nervous enough about the Face to make sure Viking stayed well away from it. One NASA scientist, nonplussed by the odd flip-flop on the landing site, compared the choice to landing in the Sahara desert on Earth to look for life, rather than a more hospitable climate.[14] In an even more bizarre decision, NASA took two more high resolution images of Cydonia—70A11 and 70A13—in mid-August, well after they decided the region was unsuitable for a landing. In doing so, they sacrificed precious Orbiter resources that could have been used to photograph another presumably more suitable region of Mars. Had they seen something in 35A72 that made them curious?

Things were pretty quiet on the Cydonia front after that until 1979, when a couple of imaging specialists at NASA's Goddard Space Flight Center, Vince Dipietro and Greg Molenaar, decided to look up the Face. They quickly found 35A72 (labeled simply "Head" in the Viking image files) and their early enhancements seemed to argue against the "trick of light and shadow" explanation. They then decided to look for other possible images of the Face taken on other orbits. They were surprised to find both that potentially interesting images of the Face taken on subsequent orbits seemed to have disappeared, and there seemed to be no trace of the "disconfirming photographs" that Gerald Soffen had alluded to 5 years earlier. After an exhaustive search of the Viking archives, they discovered a second *misfiled* Face image, 70A13, taken 35 orbits later at a 17° higher sun angle. They never did find the supposed "disconfirming" image and subsequently established that since the next Viking orbit took it nowhere near Cydonia and was at Martian nighttime, no such image could conceivably exist.

They then began to seek out other input. Although stymied in their attempts to get articles on the Face published in the peer-reviewed journals, Dipietro and Molenaar eventually managed to get some of their enhancements of the Face into the hands of Richard C. Hoagland. Although Hoagland had requested the prints in order to study the image enhancement technique (called S.P.I.T.) being used by Dipietro and Molenaar rather than the Face itself, he was intrigued by what he saw. After some discussions with them Hoagland was able to secure funding for the first Independent Mars Investigation under the auspices of the Stanford Research Institute.

From the beginning, Hoagland realized that the question of the Face required special consideration. As far as any members of the IMI knew, no one had ever attempted such an investigation before, and there were therefore no set rules as to how the "Face problem" should be approached. Working from the idea that if the Face were indeed artificial it would likely be beyond the experience of geologists and planetary scientists, Hoagland determined that the research required a group with a broad cross section of skills from the various "hard" and "soft" sciences. This "multidisciplinary" approach allowed the original members of the IMI to examine the Face from every possible scientific perspective, and to cross-reference their results with a ready-made peer review panel.

What they found only deepened the mystery. After close scrutiny of both 35A72 and 70A13, some initial conclusions were immediately evident.

Since the Face was not a profile view as seen in terrestrial rock faces like the Old Man in the Mountain in New Jersey, but rather a direct, overhead view more akin to the presidential monument at Mount Rushmore, they quickly decided against the idea they were just "seeing things." The Face seemed to have specific characteristics of human visages, including a brow ridge, eye sockets, a full mouth and a nasal protuberance. The higher sun angle image 70A13 showed that the beveled platform upon which the Face was seated appeared to be in the range of 90% symmetrical, despite the presence of a data error in the image that distorted the area around the eastern "jaw."

This second image also confirmed the presence of a second "eye socket," and that the level upon which the facial features rested was uniform in height and symmetrical in layout at least as far down as the "mouth." This image also eventually revealed (under enhancements by Dr. Mark Carlotto) what appeared to be teeth in the mouth, bilaterally crossed lines on the forehead and lateral striping on the western half. Both images also showed a mark of some kind, dubbed the "teardrop," on the western side of the face just below the eye socket.

Later, using a "bit slice" imaging technique, Dipietro found what he claimed was a spherical "pupil" in the western eye socket. It will be important later to remember that the critics of the investigation, among them Dr. Michael Malin of Malin Space Science Systems (who controls the camera for the current Mars Global Surveyor probe), claimed that the "pupil" was not really there and was beyond the resolution limits of the data.

But the most controversial features by far were Carlotto's "teeth."

Dr. Mark Carlotto had been brought in to the second Mars research group organized by Hoagland, the Mars Investigation Group, in 1985. He used new imaging techniques to bring out more detail than Dipietro and Molenaar's earlier method had from the two Viking images. In both of the original Viking images (35A72 and 70A13), there were fine but obvious structures in the mouth that seemed to represent teeth [Fig. 1-2]. One of the key tests for the artificiality of the Face has centered on this issue. In fact, it's hard to imagine a more decisive test of artificiality then the representation of teeth in the mouth. Dr. Malin apparently realized this as well, because he made a special effort to debunk the presence of teeth in the Viking data by placing fake images of the "teeth" on his website. He also went well out of his way to misrepresent the arguments made by the researchers advocating the presence of "Teeth."[15]

One of the first contentions made against the presence of actual teeth was the claim that they were merely artifacts of the image enhancement process. However, the possibility that a teeth-like set of artifacts could appear on two very different (although covering much of the same area) images in precisely the same location are non-existent. It is even less likely when you consider that there are no other appearances of teeth-like "artifacts" anywhere else in either image, and the features are well beyond the range of any individual data errors. Finally, the two images are oriented at differing perspectives relative to the pixel grids. In spite of this, Malin and others have persisted in their mischaracterization of the issue.

The City and Other Anomalies at Cydonia

Hoagland was the first to realize that all of this detail was ultimately meaningless if it turned out that the Face was an isolated landform. No matter how much it looked like a Face, if it was all by itself, with no evidence of any civilization around it to have constructed the monument, then it *could* simply be a marvelous trick of erosion and shadow after all.

So Hoagland and the members of the investigation began to look in the immediate vicinity of the Face to see if there was any other evidence of anomalous objects nearby. Dipietro and Molenaar had previously noted a cluster of "pyramidal" mountains to the west of the Face, and they had also pointed out a massive object (1.5 km in height) to the south that appeared to be a four-sided pyramidal mountain. Hoagland dubbed this cluster of mountains the "City," and the massive pyramidal mountain the "D&M Pyramid" [Fig. 1-3], in honor of Dipietro and Molenaar. Enhancements by Carlotto revealed that the "D&M" seemed to be a *five-sided pentagonal* object, rather than four-sided, as Dipietro had argued, and the "City" objects displayed a number of unusual geomorphic characteristics as well. In time, features like the "City Square" (an arrangement of equally spaced mounds with a direct sightline to the Face) the "Fortress" (a object just outside the "City" which seemed to have a triangular shape and two straight walls) the "Tholus" (a rounded mound which closely matched man-made earthen mounds in England in shape and layout—complete with a "trench" around it) the "Cliff" (a long, almost perfectly straight ridge atop what appeared to be a platform built over the ejecta from a nearby impact crater) and the "Crater Pyramid" (a tetrahedral pyramidal mound somehow perched on the rim of the impact crater) formed what became known as the "Cydonia Complex" [Fig. 1-4].

Further examination provided additional details. There was evidence of digging next to the Cliff, implying that the platform upon which it rested had been built up from this material. The Tholus turned out to have an "entrance" of sorts at the top, a walkway that went from the base to this entrance and a pointed, almost pyramidal cap on it. The D&M had what appeared to be almost a bottomless crater next to it, and the right side of the object seemed to bulge out slightly, as if from an internal blast (caused by whatever made the crater?). The City turned out to have a degree of organization to it, and architect Robert Fiertek did an extensive reconstruction of the original layout.[16]

By the mid 1980s, the various members of the investigation were ready to present their findings to the scientific community and call for more analysis and better pictures to determine the validity of their observations. They met with a chilly reception.

Efforts to get their work published in peer-reviewed journals were quickly rebuked. Members later found out that in most cases, the papers were rejected without even having been read, much less "reviewed by peers." Behind-the-scenes efforts to get assistance from prominent members of the scientific community met with a bit more success, as Carl Sagan helped Carlotto get a couple of papers published in computer optics journals. Oddly, at the same time he was doing this, Sagan was attacking the whole issue publicly with an infamous disinformation piece in *Parade* magazine.[17] This would not be the last time that Sagan contradicted himself on Cydonia.

Attempts to present their data to peers directly, via scientific conferences and the like, were also met with resistance. When members of the Mars Investigation Group presented a poster session and a paper at the 1984 "Case for Mars" conference, they were surprised to find out that their presentation and their paper had been expunged from the officially published record of the conference, as if they had never been there.[18]

Undaunted, Hoagland and the others continued their research. Yet, as documented by Dr. Stanley V. McDaniel of Sonoma State University in his voluminous *McDaniel Report*,[19] NASA seemed to have an aversion to investigating what seemed to be an ideal subject for the agency's agenda. In fact, they vociferously refused to even consider making the imaging of Cydonia a priority for any new Mars missions. Beyond that, they continued to insist, in response to inquiries from congressional leaders and the public, that the non-existent "disconfirming photos" proved that the Face was just an illusion. Only after many years (17) of repeatedly pointing out to NASA that no such images existed did they finally cease making this claim.

Dr. Carlotto moved the research in a new direction when he developed a fractal analysis technique, to discern which objects in an image were the least consistent with the "natural" background, to be used on the Cydonia images. After an initial study of about 3,000 square kilometers around the Face, Carlotto and his partner, Michael C. Stein, determined that the Face and the Fortress were the two most "non-fractal" objects in that terrain. Pressed to go even further, they eventually used the program on images covering some 15,000 square kilometers around the Face. The results were consistent with the earlier run-through. The Face was by far the most non-natural object in the surveyed terrain. NASA responded through Dr. Malin to the effect that Carlotto had not measured anything other than the fact that the Face was different, rather than artificial, and suggested that if he applied the technique to a broader area, he would find that the curve would smooth out, and that the Face was not all that unusual.

This response ignored the fact that Carlotto had already done just that by expanding the survey from 3,000 square kilometers to 15,000 and that, contrary to Malin's assertions, the Face's uniqueness was even more pronounced. Lacking the funds to expand the research even further, Carlotto offered to turn the program over to NASA so that the agency could continue the survey over the entire Martian surface. NASA's response was a polite "thanks, but no thanks."

Up to this point, a lot of the behavior of NASA and the planetary science community could be viewed through the tint of simple prejudice or ignorance. No one wanted to be the next Percival Lowell, sticking their chin out on the issue of life on Mars only to have their reputation forever soiled if the data turned out to be wrong. Other members of the broader scientific community simply refused to even consider the possibility. Their models and training had taught them that Mars was a cold, dead world, and had been for billions of years. The notion that someone had been there, built these monuments and then left sometime in the distant past was just too destabilizing to their way of thinking.

The next step in the investigation was even more radical however, and it is here that NASA's resistance turned to active disinformation and suppression.

Mathematical Message?

Early on in the Cydonia investigations, Hoagland had proposed that there might be a broader, contextual relationship between the various landforms identified as anomalous. By themselves, the Face, Fort, City, Tholus, Cliff, Crater Pyramid and the D&M Pyramid were anomalous geomorphic objects that were incongruous with the existing geologic model of Cydonia.

But Hoagland had also noted several "interesting" relationships between the potential monuments on the Cydonia plane. He noticed, for instance, that the three Northward edges of the pentagonal D&M seemed to point to other key features of the complex. Using orthographically rectified images provided by the U.S. Geological Survey and the Rand Corporation, he drew lines defined by these faceted edges across the images of Cydonia. One edge passed right through the center of the City Square, the next right between the eyes of the Face, and the next straight across the apex of the Tholus. He also noted several "mounds" in and around the City. They were consistent in terms of size (about the scale of the Great Pyramid at Giza) and shape, and also seemed to form a perfect equilateral triangle [Fig. 1-5].

It is important to appreciate the sequence in which these observations were made. Hoagland has often been accused of "circular reasoning," of just drawing lines on the photos until they "hit" something and then declaring that object to be a "monument." This is not, in fact, the case.

As we have seen, and has been well documented by Hoagland, Carlotto, Pozos, McDaniel and others who were *there*, the anomalous geomorphic characteristics came first. It was only later, when some thought was given to how the unusual landforms might have a contextual relationship that the measurements were made. Even then, the methodology could have become "circular" if certain precautions were not taken. Hoagland carefully used only techniques that had been previously established by archeologists in their surveys of ancient ruins.

Taking a page out of the SETI manual, Hoagland decided that any intended message would almost certainly have been inscribed more than once. If an architect were seeking to send a clear mathematical signal to a civilization that would happen upon his creation, he would surely have *reinforced* the message, since a single mathematical relationship could not be distinguished from random "noise." So a cornerstone of the whole process was that any "significant" mathematical relationship must occur redundantly. He also made certain not to include any object that was not significant in some other way to the model. If an object was not anomalous in any way, but stood at a significant location in the alignment model, it was rejected. Each and every relationship that would be considered significant had to be a candidate for inclusion on at least two grounds.

A prime example of this is the City Square. It was originally considered a potential candidate for artificiality because of the way the four mounds were equally spaced around a central nexus. Additionally, the four mounds seemed to be almost identical in height, scale and volume. So the fact that the center of the City Square was later found by Hoagland to lie along a direct line marked by the northwest facet of the D&M was only significant because of these previous observations. Without the initial geomorphic issues calling the features into question, the later determined alignment would have been meaningless in Hoagland's methodology.

Yet he still faced a significant degree of criticism from "reductionists" inside NASA. The reductionist method seeks to isolate each and every data point in a given argument and break it down without reference to the greater context. Hoagland argued that this isolationist approach could not be valid in an investigation such as this one, since there would likely have been some form of *intent* in the mind of any "Martian architect," just as there was in any earthly monumental architecture.

This was not the first time that someone had faced this sort of criticism from NASA.

On November 22, 1966, three years to the day from the date President Kennedy had been killed, NASA released a Lunar Orbiter 2 image from the Moon in the vicinity of the crater Cayley B in the Sea of Tranquility. In it, there were objects casting extremely long shadows that seemed to imply that the objects themselves were "towers" of seventy feet or more [Fig. 1-6]. Such objects, if they really were present on the lunar surface, would almost by definition be artificial. Eons of meteoric bombardment would have long since blasted any such naturally occurring objects into dust.

William Blair, a Boeing anthropologist, noted that the "spires" had a series of contextual, geometric relationships to each other. "If such a complex of structures were photographed on Earth, the archeologist's first order of business would be to inspect and excavate test trenches and thus validate whether the prospective site has archeological significance," he was quoted in the *L.A. Times*.[20] Blair had extensive experience analyzing aerial survey maps to look for possible prehistoric archeological sites in the Southwest United States.

The response from Dr. Richard V. Shorthill of the Boeing Scientific Research Laboratory was swift and eerily reminiscent of the criticism aimed at Hoagland. "There are many of these rocks on the Moon's surface. Pick some at random and you eventually will find a group that seems to conform to some kind of pattern." He went on to claim that the long shadows were caused by the fact that ground was sloping away from relatively short objects, thereby elongating the shadows. Subsequent analysis has proven Shorthill wrong on all counts.[21] The objects are indeed very tall, and the shadows are not caused by a sloping hill. Beyond that, the geometric relationships cited by Blair turned out to be based on *tetrahedral* geometry, which will become very significant as you read on.

Blair's rebuttal would later put the reductionist arguments in their appropriate context: "If this same axiom were applied to the origin of such surface features on Earth, more than half of the present known Aztec and Mayan architecture would still be under tree- and bush-studded depressions— the result of natural geophysical processes. The science of archeology would have never been developed, and most of the present knowledge of man's physical evolution would still be a mystery."

In 1988, Hoagland was approached by Erol Torun, a cartographer and satellite imagery interpreter for the Defense Mapping Agency. Torun was probably the most uniquely qualified person on the planet to render a judgment on the potential artificiality of the Cydonia enigmas. After attaining a degree in geology with a specialty in geomorphology, he had spent more than ten years of his professional life looking at remote imagery just like the original Viking data and distinguishing artificial structures from naturally occurring landforms.

After reading *Monuments*, he had written Hoagland expressing his surprise that his initial assumptions about the subject were not supported by his subsequent analysis. He was particularly impressed with the geometry and geology of the D&M Pyramid. "I have a good background in geomorphology and know of no mechanism to explain its formation," he wrote Hoagland.[22]

Torun had come to the Mars investigation as a skeptic; relatively certain he would find that the geomorphic interpretations and the early contextual alignments cited by Hoagland would turn out to be "false positives" in the search for answers to the riddle of Cydonia.

Yet once he had a chance to study the Cydonia images in detail, Torun concluded that the D&M itself was nothing less than the "Rosetta Stone" of Cydonia, finding a series of "significant" mathematical constants expressed in the internal geometry of the D&M. Being careful to avoid projecting his own biases on the measurements, Torun decided beforehand that he would restrict his analysis to just a few possible relationships. As it turned out, not only did the D&M *have* a consistent internal geometry, it was also one full of rich geometric clues that spoke to him of a specific mathematical *message*. He found numerous repetitive references to specific mathematical constants, like e/π, √2, √3, √5 and references to ideal hexagonal and pentagonal forms. He also found geometry linking the shape of the D&M to other ideal geometric figures, like the Golden Ratio (φ) and the *Vesica Piscis*, which is the root symbol of the Christian church—and the five basic "Platonic Solids"—the tetrahedron, cube, octahedron, dodecahedron and icosahedron. Further studies found that the reconstructed shape of the D&M, as determined by Torun *before he took any of these measurements*, is the only one that could produce this specific set of constants and ratios.[23] More than that, these same constants showed up redundantly in all the different methods of measurement, and were not dependant on terrestrial methods of measurement (i.e. a radial measurement system based on a 360° circle). As Torun put it: "All of this geometry is 'dimensionless'; i.e. it is not dependent on such cultural conventions as counting by tens, or measuring angles in the 360 system. This geometry will 'work' in any number system" [Fig. 1-7].

After receiving Torun's study, Hoagland quickly realized that they were on the verge of a potentially important discovery. If Torun's numbers were repeatable throughout the Cydonia Complex, if the same angles and ratios appeared in the larger relationships between the *already* established potential "monuments," then they would have a very strong argument that Torun's model was valid. Again being careful to only take measurements between obvious features, the apex of the Tholus and D&M, the straight line defined by the Cliff, the center of the City Square, the apex of the tetrahedral Crater Pyramid, Hoagland found that many of the same angles, ratios and trig functions applied *all over the Cydonia Complex* [Fig. 1-8].

Somewhat stunned by what they had found, Hoagland and Torun had come to the realization that there *was* a message on the ground at Cydonia. The problem was that they didn't know what that message was trying to say.

In the Message itself was the key to decoding it. One of the angles noted by Torun within the D&M was 19.5°, which occurred twice. Hoagland also found the same "19.5°" encoded in the broader Cydonia complex *three more times*. Searching for the significance of this number, they eventually determined that it related to the geometry of the tetrahedron. The simplest of the five so-called "Platonic Solids" (because it is the most fundamental three-dimensional form that can exist); it made a certain kind of sense to use this "lowest order" geometric shape as a basis for establishing communication across the eons.

If a tetrahedron is circumscribed by a sphere, with its apex anchored at either the North Pole or South Pole, then the three vertices of the base will "touch" the sphere at 19.5° in the hemisphere opposite the polar apex alignment. In addition, the value of e (as in the e/π ratio which is encoded at least ten times throughout the Cydonia complex) is 2.718282, a near exact match for the ratio of the surface area of a sphere to the surface area of a tetrahedron (2.720699).

This whole "tetrahedral" motif was reinforced when they went back to the original Cydonia images. Some of the small mounds Hoagland had noted earlier had the look of tetrahedral pyramids, and the Crater Pyramid, which is involved with one of the key 19.5 measurements, is also clearly tetrahedral. The mounds themselves were also arranged into a couple of sets of equilateral triangles, the 2D base figure for a tetrahedral pyramid.

Later, Dr. Horace Crater, an expert in probabilities and statistics, did a study of the mounds at Cydonia with Dr. Stanley McDaniel.[24] What Crater found was that not only was there a non-random pattern in the distribution of nearly identical mounds at Cydonia, but that the pattern of distribution was overwhelmingly *tetrahedral*—and to a factor of 200 million to one against a natural origin.

The Message of Cydonia

In 1989, Hoagland and Torun proceeded to publish their results in a new paper titled, appropriately, "The Message of Cydonia."[25] Based on the barrage of ad-hominem criticism Hoagland had experienced after "Monuments" was published two years before, they assumed that it would be pointless to try to have their paper published in the NASA controlled peer review press. Instead, they decided to go "straight to the people" and uploaded the paper to CompuServe, the largest online message board of its time. The paper contained a number of predictions based on their evolving theory of the tetrahedral Message of Cydonia and also the even more radical new idea that within the tetrahedral mathematics was nothing less than an entirely new physics model. Hoagland then found that there was a long-abandoned line of thought by some of the masters of early physics, including James Clerk Maxwell, which included the idea that certain problems in electromagnetics could be solved by the imposition of higher spatial dimensions into the equations. The energies coming from these higher dimensions would then be "reflected" in our lower three dimensional universe through tetrahedral geometries. It was this crucial insight, they decided, that the builders of Cydonia were ultimately trying to impart.

The reductionists were quick to attack Hoagland and Torun's model. The critics argued against the validity of the model on one of two basic counts—that the measurements were either inaccurate, or if they were accurate, they did not mean what Hoagland and Torun implied they meant.

Anonymous memos from within NASA in the late 1980s used the same sorts of tactics they had leveled at Dr. Blair years earlier. They argued that Torun's measurements were not reliable because of the amount of error built into the ortho-rectified images. They frequently disputed the measurements themselves, but did not actually bother to try and reproduce them. Dr. Ralph Greenberg, a University of Washington mathematics professor, has more recently taken up this view. Greenberg has written several documents critical of Hoagland and Torun's model, and has also made something of a mini-career for himself accusing Hoagland of lying about his contributions to the idea of life under the oceans of Europa.

Dr. Michael Malin of Malin Space Science Systems (who controlled the camera for the then planned Mars Observer and the current Mars Global Surveyor) took a slightly different tack, agreeing that the measurements made by Hoagland and Torun "are not wholly in dispute"[26] but arguing that even if the numbers were right, it did not necessarily follow that they *meant* something significant.

Most of these critiques are typical of the type of reaction you get from scientists when their established paradigms are threatened or when experts in particular fields try to apply the standards they are familiar with to a problem that is outside their experience. The issue of margins of error, especially, is one that (even today) is simply misunderstood, even by experienced mathematicians. Put simply, Greenberg argues—as many have before him—that the margin of error built into the measurements of the Cydonia Complex renders them useless, because they are large enough to make almost "any" mathematical constants and ratios possible. Greenberg, who has become pretty much the point man for attacks on the Cydonia Geometric Relationship Model, also claimed that Hoagland and Torun "selected"[27] the angles they found, implying that they were looking for specific relationships before they ever started.

For the record, Greenberg also argues that the frequently cited mathematical and astronomical alignments of the Pyramids in Egypt are fallacious, even though few Egyptologists doubt them. It is by now well established that the base of the Great Pyramid is a square with right angle corners accurate to 1/20th of a degree. The side faces are all perfect equilateral triangles which align precisely with true north, south, east and west. The length of each side of the base is 365.2422 Hebrew cubits, which is the exact length of the solar year. The slope angle of the sides results in the pyramid having a height of 232.52 cubits. Dividing two by the side length by this height gives a figure of 3.14159. This figure gives the circumference of a circle when multiplied by its diameter. The perimeter of the base of the pyramid is exactly equal to the circumference of a circle with a diameter twice the height of the pyramid itself.[28]

Because of the angle of the slope sides, for every ten feet you ascend on the pyramid, your altitude is raised by nine feet. Multiplying the true altitude of the Great Pyramid by ten to the power of nine, you get 91,840,000, which is the exact distance from the sun to the earth in miles.[29] In addition, the builders also apparently knew the tilt of the earth's axis (23.5°), how to accurately calculate degrees of latitude (which vary as an observer ventures farther from the equator) and the length of the earth's precessionary cycle.

And all of this, according to the brilliant Dr. Greenberg, is just a coincidence. Just an example of "the power of randomness."

Greenberg's arguments are pure reductionism. Forgetting for a moment the sheer unlikelihood of finding consistent and redundant mathematical linkages among a very few objects pre-selected *only for their anomalous geology,* (which Greenberg does not address in any of his arguments) *not* for their possible mathematical relationships to one another, and using only clear and obvious structural points on these objects from which to measure, Greenberg completely fails to grasp the issue—Hoagland and Torun's measurements are *nominal*, meaning that they are valid to the closest fit of the methodology employed. They are not saying, "these are the numbers within a loose tolerance range," they are saying flatly "these *are* the numbers." The tolerances are just what we have to live with pending higher resolution images. Further, having stated that the measurements reflect a specific tetrahedral geometry—not just any set of "significant" mathematical numbers, as Greenberg implies—and that they encode a predictable *physics*, it becomes very easy to simply test their contextual model vs. his reductionist view. Greenberg seeks to isolate the numbers themselves, and argue only his view of the "power of randomness," rather than simply test the alignments in the greater context of the physics they imply.

Fortunately, "The Message of Cydonia" contained three predictions that would provide the ideal opportunity to do just that. At that time, Voyager 2 was approaching Neptune but had yet to image the planet up close. At the end of their paper, Hoagland and Torun put in three specific predictions about what Voyager would see. They predicted a storm or disturbance within a few degrees of the tetrahedral 19.5° latitude. Based on their further interpretation of the hyperdimensional physics they were developing, they also predicted that this disturbance would be in the southern hemisphere of the planet, and that the magnetic dipole polarity of Neptune's magnetic field would be anchored at the Northern Pole.

All three predictions—remember, based on the supposedly "fallacious" numbers derived from a supposedly "meaningless" set of alignments of possible *ruins* on Mars—turned out to be...

Absolutely correct.

Greenberg and the reductionists then argued "a single prediction, no matter what it is based on, cannot be relied upon as proof of anything." This tactic, combining the predictions into a single one instead of three, is a common means of dismissing the frequency of Hoagland and Torun's successes. As Harvard astronomer Halton Arp put it in his excellent book, *Seeing Red*, "The game here is to lump all the previous observations into one 'hypothesis' and then claim there is no second, confirming observation."

There is, flatly, no way that Hoagland and Torun could use a set of "meaningless" or "fallacious" data to make three such accurate predictions about features on a planet the human race had never seen up close before. These features have no explanation in the conventional models, at least as far as providing a mechanism for the storm, the location of the storm, and its relationship to the magnetic pole of the planet. In other words, there is no way they could have just "gotten lucky" by using established models of the solar system. Their predictions come solely from the Cydonia Geometric Alignment model. This is not only a ringing endorsement of the validity of both the measurements and the physics model deduced from them, but also a harsh indictment of the methods and motives of both Greenberg and Malin (Greenberg at one point challenged Hoagland to a "debate" on the mathematics of Cydonia, but only if he could exclude Crater's tetrahedral mound data, which he acknowledged he could not explain away).

Armed with the suspicion that they had found the intent of the builders of the "Monuments of Mars," Hoagland and Torun now turned their attention to the possible application of the geometry they had discovered.

Chapter One Images

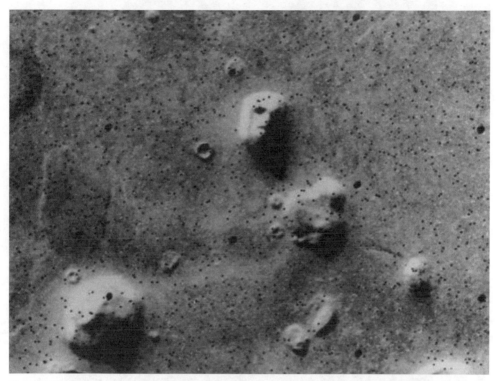

Fig. 1-1 – The "Face on Mars" (Viking frame 35A72) as it initially appeared in the press, July 25, 1976. Black dots are data transmission dropouts. (NASA)

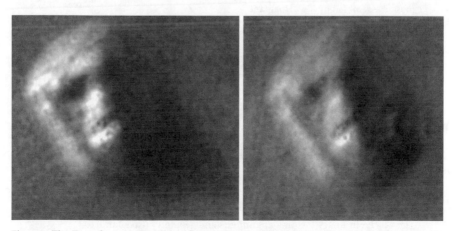

Fig. 1-2 - The Face, from NASA Viking frames 35A72 (L), and 70A13 (R) processed by Dr. Mark Carlotto.

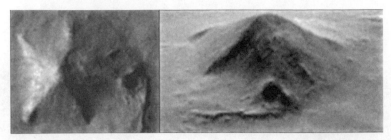

Fig. 1-3 – Two views of the Pentagonal D&M Pyramid from Viking frame 35A72. Plan view (L) and 3D perspective view (R). (Carlotto)

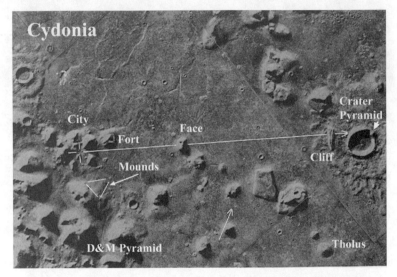

Fig. 1-4 - The original "Cydonia Complex," with key features identified. Hardcopy mosaic assembled by R. C. Hoagland.

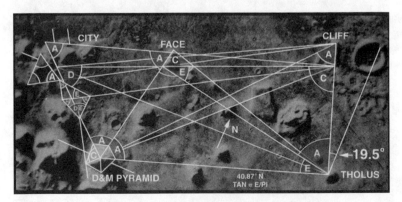

Fig. 1-5 - Preliminary Cydonia alignments and geometry (Hoagland).

Fig. 1-6 – The "Blair Cuspids" in Mare Tranquillitatis, from Lunar Orbiter frame LO2-61H3 (NASA)

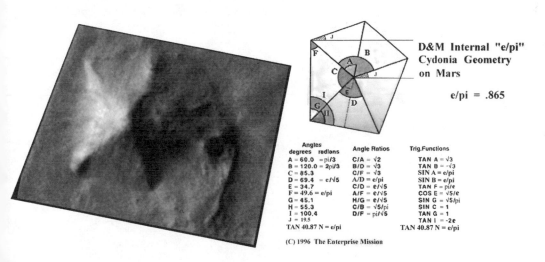

D&M Internal "e/pi" Cydonia Geometry on Mars

e/pi = .865

Angles degrees radians	Angle Ratios	Trig.Functions
A = 60.0 = pi/3	C/A = √2	TAN A = √3
B = 120.0 = 2pi/3	B/D = √3	TAN B = -√3
C = 85.3	C/F = √3	SIN A = e/pi
D = 69.4 = e/√5	A/D = e/pi	SIN B = e/pi
E = 34.7	C/D = e/√5	TAN F = pi/e
F = 49.6 = e/pi	A/F = e/√5	COS E = √5/e
G = 45.1	H/G = e/√5	SIN G = √5/pi
H = 55.3	C/B = √5/pi	SIN C = 1
I = 100.4	D/F = pi/√5	TAN G = 1
J = 19.5		TAN I = -2e
TAN 40.87 N = e/pi		TAN 40.87 N = e/pi

(C) 1996 The Enterprise Mission

Fig. 1-7 – D&M Pyramid internal geometric relationships as defined by Erol Torun in 1988 (Hoagland and Torun).

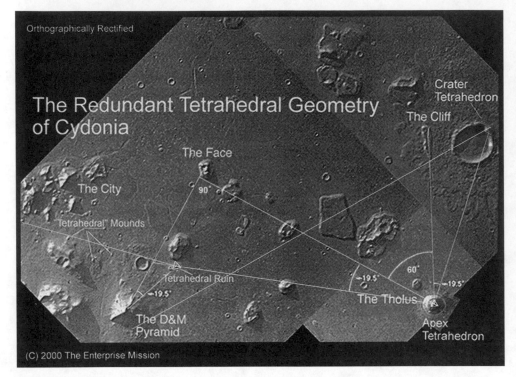

Fig. 1-8 - Hoagland's Cydonia "Geometric Relationship Model" – (Carlotto Enhanced Digital Mosaic – Hoagland overlay)

Chapter Two

Hyperdimensional Physics

One of the first things Hoagland and Torun noticed was that throughout the observed solar system, planetary disturbances and upwellings of energy seemed to preferentially cluster around this key 19.5º latitude. In addition to Neptune's Great Dark Spot (as it came to be known), the Great Red Spot of Jupiter, the erupting volcanoes of Jupiter's moon Io, Olympus Mons on Mars (the largest shield volcano in the solar system—below) and Earth's own Mauna Kea volcano in Hawaii all were at, or very near, 19.5º latitude [Fig. 2-1].

Beyond that, clusters of sunspots occurring from the heightened energy output of the sun during the peak of the solar cycle also centered around 19.5º. And interestingly, whether the upwelling event was located in the north or south hemisphere depended on the alignment of the bodies' magnetic field. If the field were anchored at the South Pole, the disturbance appeared at or around 19.5º in the northern hemisphere. Conversely, if it were anchored at the North Pole, the disturbance appeared in the south. It was as if there really were "giant tetrahedrons" inside the planets, driving the physics of these energy upwellings and forcing them to comply with some mysterious, unseen rules of behavior.

Excess Heat

Another key observation Hoagland and Torun made in these early days was the role that this theorized "tetrahedral physics" might play in another ongoing mystery of the outer solar system.

Beginning in the mid 1960s, ground-based telescopic observations of the solar system began to turn up the startling detection of anomalous internal infrared radiation coming from the planet Jupiter [Fig. 2-2]. Later Pioneer and Voyager spacecraft observations across the 1970s and '80s added the other "gas giant planets," Saturn, Uranus and Neptune, to the list of solar system worlds that somehow, without internal nuclear fusion processes (like stars) still managed to radiate more energy out into space than they receive directly from the sun (below).[30]

After much initial debate, the conventional understanding of these anomalous "infrared excesses" eventually settled on three possible internal sources:

1. Primordial Heat. Leftover "fossilized thermal echoes" from the enormous energy associated with the accretion and collapse of the planet during its formation. Under this scenario, this energy was still stored inside the planet after literally *billions* of years, but slowly being radiated into space.

2. The Helium Drip Model. Heating caused by eventual internal separation of light elements in so-called "gas giant" planets (helium from hydrogen), releasing potential energy as the helium falls further toward the center of the planet (a form of ultra-slow, continued gravitational contraction).

3. Radioactive Decay. Anomalous energy release due to excess radioactive decay of heavy element concentrations located within massive gas giant "rocky cores."

Of the three conventional explanations for these "energy anomalies," only the first applies to Jupiter. Because of its mass (318 "Earths"), Jupiter just barely falls into a category of worlds that could retain such primordial heating for the lifetime of the solar system (almost five billion years) and still be able to radiate observable heat.

But when scientists got around to actually measuring the amount of excess heat Jupiter radiates, it quickly became obvious that the "primordial heat model" was inadequate to account for Jupiter's infrared emissions. Even today, the current ratio of absorbed solar energy to emitted (five billion-year-old) internal Jovian energy is still almost *two to one*. This is far in excess of what would be expected after such an enormous amount of time.

After the Voyager fly-bys of the 1980s, the second "internal heat" proposal—the "helium drip model"—was favored for the heat excess observed from Saturn. But, because of their relatively light masses (less than thirty times the Earth's), only the third possibility—massive internal radioactive decay—has been seriously attempted as an explanation for Uranus' and Neptune's more puzzling anomalous infrared emissions. There are, however, serious problems with all of these conventional explanations for all planets less massive than Jupiter.

During the Voyager encounters of Uranus and Neptune, spacecraft instruments detected a barely measurable (but significant) "infrared excess" for Uranus of about 1.14 to 1. Whereas for Neptune, essentially Uranus' planetary "twin," the ratio of internal heat to intercepted sunlight was a striking *three to one.* [31]

However, simultaneous Doppler tracking gravity measurements conducted during the fly-bys detected no anomalous central concentrations of heavy elements in either planet. This would be *required* if the excess observed IR radiation is, in fact, caused by excessive internal radioactive element concentrations.

Having failed to confirm the radioactive decay model, the conventional planetary physicists searched around for an alternative explanation for Uranus' excess energy output. They quickly seized on the one characteristic that separates Uranus from all other bodies in the solar system—its pronounced "axial tilt."

Compared to all the other planets of the solar system Uranus has an "obliquity" (the technical term) some 98° to the plane of its orbit of the sun; Neptune's is much more "normal": about 30°. (In comparison, Earth's obliquity is about 23.5°.) This led to a new proposal, the "recent collision model." This asserted that Uranus—somehow, long after its formation—suffered a massive impact with another major object, perhaps an errant moon. This, according to the theorists, in addition to accounting for the current "tipped over situation" of the planet, would have also added a significant amount of geologically "recent" internal energy to Uranus, driving up internal temperatures by equivalent amounts. This model argues that these resulting elevated temperatures in Uranus, derived from a massive cosmic collision, could thus account for Uranus' current infrared excess, as observed by Voyager in 1986.

Unfortunately, this explanation also quickly ran into difficulties. For one thing, Uranus is barely radiating "over unity" (more energy coming out than is going in) at its distance from the sun, while Neptune is radiating almost three times more energy "out" than it's getting from the sun. When these two planets are "normalized" (i.e., when their differing distances from the sun are taken into account), their *absolute* internal "over unity" energy emissions are, in fact, just about the same. If the recent collision model was correct, Uranus should be radiating far more than Neptune. In fact, there is virtually no difference. If a moon or other large object had recently struck Uranus, it certainly was not responsible for the planet's excess heat emissions.

What Hoagland then began to consider was that there might be an *external* source for this anomalous heat output, perhaps the same source that was fueling the 19.5º upwellings—but what could be the source of this mysterious excess energy that seemed to defy conventional explanation and conform to occult geometric rules?

Hidden History

At this point, Hoagland and Torun had a conundrum. They had certainly developed a series of observations and connections that demanded closer scrutiny—but in what context? It was not enough to argue that the Cydonia ruins imparted knowledge of tetrahedral geometry and that this geometry seemed to have some reflected physical effects in the rotating bodies of the solar system. They had to have a coherent model of the *mechanism* that was driving all of these observed planetary upwellings and the anomalous heat output. The pattern itself implied that there was an underlying physics involving the energy upwellings.

There is a perfectly natural explanation for such "anomalous energy" appearing in celestial bodies—unfortunately, it hasn't been seriously considered by science for over a hundred years. Hoagland found that the idea that "forces"—like gravity or magnetism—could be modeled geometrically was becoming quite popular within modern mathematics. With that in mind, he began to look back at physics theories from the 1800s, and found that the very father of modern day physics, James Clerk Maxwell, had indeed dabbled in some equations that seemed to fit what he and Torun had observed on the outer planets. Maxwell had routinely argued that the only way to solve certain problems in physics was to account for some phenomena as 3D "reflections" of objects existing in higher spatial dimensions. When Maxwell died, this higher dimensional or "scalar" component was stripped out of his equations by Oliver Heaviside, and the resulting "classic Maxwell equations" became the basis for our modern day models of electromagnetic force. But if Maxwell's original work was correct—even though it had been discarded—then that meant his original concepts could potentially explain the various planetary phenomena Hoagland and Torun had observed. Tentatively, Hoagland began to examine more closely this early model of "hyperdimensional physics."

Hoagland found that a number of modern day mathematicians had actually begun to model these possible higher dimensions geometrically. "Topologists" like H. S. M. Coxeter had done a significant amount of work in mapping the higher dimensional properties of a rotating "hypersphere," a sphere that exists in more than just three spatial dimensions. The arcane mathematics describing this "hypersphere," and the multiple dimensions above it, are so complex that they are virtually unintelligible to all but the most educated math buffs. Yet the *signatures* of this higher dimensional physics—as reflected in our own 3D universe—are much easier to spot and predict. His equations predicted that such a figure—if it was *rotated*—would create roiling vorticular motions (exactly like the observed dynamics of Jupiter's "Great Red Spot") in the sphere's 3D geometry, and at a *specific* latitude—

19.5°.

This was exactly what Hoagland and Torun were seeing in their own observations of the rotating planets and moons of the solar system. If these observations were indeed linked to the higher dimensional properties of the "rotating hypersphere," then it meant not only that rotating *planets* existed simultaneously in multiple, higher spatial dimensions, but also that this new physics could provide potentially limitless amounts of energy to drive the observed dynamics of their atmospheres, their internal fluid motions, their surface geological "upwellings"—everything! Even, ultimately, "life" itself....

The cornerstone of the hyperdimensional model is the notion that "higher" spatial dimensions not only exist, but are also the underlying foundation upon which our entire 3D reality exists. Beyond that, everything in our observable 3D world is, in fact, driven by mathematically modeled "information transfer" from these higher dimensions. This "information transfer" might simply be the result of changes in the geometry of a connected system, say a change in the orbital parameters of a planet, like Jupiter or the Earth. Since we are limited in our perceptions to the 3D universe we live in, we cannot "see" these higher dimensions. However, we can see (and measure) *changes* in these higher dimensions that have *simultaneous* effect on *our* reality. By definition, this change in higher dimensional geometry is perceived in our 3D universe as an "energy output"—like the various 19.5° planetary energy upwellings we have noted previously.

The hyperdimensional model therefore implicitly argues that, contrary to Einstein's dictates, instantaneous "action at a distance" in our universe is indeed possible—in fact, it is a given—because of this higher dimensional information transfer. The model predicts that the effects of a "cause" in our three dimensions, whatever it may be, can be felt in measurable and predictable ways in our perceived universe at speeds incomparably greater than that of light. The universe accomplishes this seemingly impossible feat by the transformation and transfer of information (as a different "energy") through "hyperspace," i.e. through these higher spatial dimensions. In our familiar three dimensions, this information/energy is then transformed back into familiar energy forms, like light, heat and even gravity itself.

Changes in one gravitationally-connected system on a large scale, like the planetary scale of a solar system, can therefore have an instantaneous, measurable effect on other bodies in the same system—providing there is a "resonance condition" (a "matched" connection) between those two objects via hyperspace. Thus, the hyperdimensional model argues that everything, even widely separated three-space objects like remote planets, are ultimately connected through this four-space interaction; meaning that a "cause" in one place (like Jupiter) can have an "effect" in another (like the sun)—without any measurable 3D force (such as an electromagnetic wave) having measurably traversed the three-space distance *in between*.

Mainstream physics teaches us that this phenomenon—called "non-locality," which has been observed for decades in laboratory experiments[32] is simply a baffling "quantum reality" limited to ultra-short distances at the sub-atomic level, but that it does not, and *cannot*, affect larger objects at great distances (like planets, stars or galaxies themselves). Because the speed of light is theorized to be the absolute speed limit of our 3D macro-universe, no cause is supposed to have a measurable effect on any object anywhere sooner than it takes light to travel the distance in between. Yet these "magical," faster-than-light signals between atomic particles, and even communication between photons, has now been overwhelmingly confirmed at macro distances as well. Our current understanding of the limiting speed of light, based on Maxwell's equations of electromagnetism, asserts that only certain kinds of energy, like electromagnetic radiation, can even travel long distances through the vacuum of space.

In this classical "Einsteinian" view of physics, there is no "aether"—as it was called in Maxwell's day—to carry electromagnetic radiation's transverse waves across the vacuum. In the hyperdimensional model, the aether is *back*—as the actual transformation medium between the higher spatial realities and *our* dimension—through something called the "torsion field" (The word "torsion" is formed from the same root as "torque"—and means "to spin").

So a torsion field is a "*spin* field"— a crucial point we will come back to.

The torsional-aetheric field is, therefore, *not* the 19th-century's "electro-magnetic aether" at all—but rather a spin-sensitive, geometric aetheric *state*—whereby hyperdimensional information/energy *is* detectable in our dimension via spinning vorticular (rotating and/or precessing) physical systems.

Contrary to mainstream physics dogma, a rich series of experiments—for over a hundred years—have overwhelmingly confirmed various aspects of this *non-electromagnetic* "spin-field aether." The still developing mathematics and graphs that model this theoretical cosmology now, unfortunately, are as Byzantine and arcane as anything in science today. However, this mathematics is reinforced by a veritable explosion of theoretical papers and fascinating laboratory experiments which have been secretly carried out in *Russia* for over 50 years—only becoming widely available here (via the Internet) with the collapse of the Soviet Empire.

So even though there is substantial and growing reason to suspect that the hyperdimensional/torsion model may ultimately be correct as the "Theory of Everything," most modern physicists (certainly in the West) still reject the notion, and stubbornly don't want to "go there."

Yet that was not always the prevailing sentiment in Western physics.

Hyperspace

The mathematical and physical parameters required for such "information/energy gating" into this spatial dimension from potential "n-dimensions" were primarily founded in the work of several 19th-century founders of modern mathematics and physics. Among these were German mathematician Georg Riemann; Scottish physicist Sir William Thompson (who would eventually be knighted by the British crown as "Baron Kelvin of Largs" for his scientific and technological contributions); Scottish physicist James Clerk Maxwell; and British mathematician Sir William Rowan Hamilton.

Riemann mathematically initiated the 19th-century scientific community (if not the rest of Victorian society) into the unsettling idea of "hyperspace" on June 10, 1854. In a seminal presentation made at the University of Göttingen in Germany, Riemann put forth the first mathematical description of the possibility of "higher, unseen dimensions" under the deceptively simple title "On the Hypotheses Which Lie at the Foundation of Geometry."

Reimann's paper was a fundamental assault on the two-thousand-year-old assumptions of "Euclidian geometry"—the ordered, rectilinear laws of "ordinary" 3D reality. In its place, Reimann proposed a 4D reality (of which our 3D reality was merely a "subset"), in which the geometric rules were radically different, but also internally self-consistent. Even more radically, Reimann proposed that the basic laws of nature in three-space, the three mysterious forces then known to physics—electrostatics, magnetism and gravity—were all fundamentally *united* in four-space, and merely "looked different" because of the resulting "crumpled geometry" of our 3D reality. In essence, he argued that gravity, magnetism and electricity were all the same thing—energies coming from higher dimensions.

Reimann was suggesting a major break with Newton's "force creates action-at-a-distance" theories. These had been proposed to explain the "magical" properties of magnetic and electrical attraction and repulsion, gravitationally curved motions of planets and falling apples for over 200 years. In place of Newton, Reimann was proposing that such "apparent forces" are a direct result of objects moving through three-space geometry, *distorted* by the intruding geometry of four-space.

It is clear that Maxwell and other "giants" of 19th-century physics (Kelvin, for one), as well as an entire contemporary generation of mathematicians (like Cayle, Tait, etc.), took Reimann's ideas very much to heart; Maxwell's original selection of four-space "quaternions" as the mathematical operators for his force equations and descriptions of electrical and magnetic interaction, clearly demonstrate his belief in Reimann's approach—as does his surprising literary excursions into poetry, vividly extolling the implications of "higher-dimensional realities," including musings on their relationship to the ultimate origin of the human soul.[33]

In 1867, following decades of inquiry into the fundamental properties of both matter and the space between, Thompson proposed a radical new explanation for the most fundamental properties of solid objects: the existence of "the vortex atom." This was in direct contradiction to then-prevailing theories of matter, in which atoms were still viewed as infinitesimally "small, hard bodies [as] imagined by [the Roman poet] Lucretius, and endorsed by Newton..." Thompson's "vortex atoms" were envisioned, instead, as tiny, self-sustaining "whirlpools" in the so-called "aether"—which Thompson and his contemporaries increasingly believed extended throughout the universe as an all-pervasive, incompressible fluid.

Even as Thompson published his revolutionary model for the atom, Maxwell, building on Thompson's earlier explorations of the underlying properties of this "aetheric fluid," was well on the way to devising a highly successful "mechanical" vortex model of the "incompressible aether" itself, in which Thompson's vortex atom could live—a model derived in part from the laboratory-observed elastic and dynamical properties of solids. Ultimately, in 1873, he would succeed in uniting a couple hundred years' worth of electrical and magnetic scientific observations into a comprehensive, overarching electromagnetic theory of light vibrations, carried across space by this "incompressible and highly stressed universal aetheric fluid."

Maxwell's mathematical basis for his triumphant unification of these two great mystery forces of nineteenth century physics were "quaternions," a term invented (adopted would be a more precise description) in the 1840s by mathematician Sir William Rowan Hamilton, for "an ordered *pair* of complex numbers." Complex numbers themselves, according to Hamilton, were nothing more than "pairs of real numbers which are added or multiplied according to certain formal rules." In 1897, A.S. Hathaway formally extended

Hamilton's ideas regarding quaternions as "sets of four real numbers" to the idea of *four spatial dimensions*, in a paper entitled "Quaternions as Numbers of Four-Dimensional Space."[34]

According to Maxwell, action at a distance *was* possible through the "aether," which he defined as higher spatial dimensions —or what we now call "hyperspace." In other words, the father of modern terrestrial electromagnetic physics had come to the same conclusions as Hoagland's theorized "Martian architect" at Cydonia.

This may seem to be a tenuous connection at first, but when you read certain lines from his poem presented to the Committee of the Cayley Portrait Fund in 1887, it becomes quite clear, that he *knew*:

> *"Ye cubic surfaces! By threes and nines, draw round his camp your seven-and-twenty lines—the Seal of Solomon in three dimensions..."*

This clear description of the "Seal of Solomon in three dimensions," is an overt reference to the geometrical and mathematical underpinnings of the infamous "circumscribed tetrahedral geometry" memorialized all over Cydonia. If you take the base figure of a tetrahedron—the equilateral triangle—and add a second equilateral triangle to the figure exactly opposite the first, and then circumscribe that figure with a circle, you get the familiar "Star of David," the "Seal of Solomon" that Maxwell describes. And in this figure, the tips of the double triangle touch the circle at the poles and at 19.5° north and south, directly linking the identical, *hyperdimensional quaternion geometry* whose physical effects we have now rediscovered all across the solar system. And, of course, taking this "Seal of Solomon" and drawing it in three dimensions would give you a double-star tetrahedron circumscribed by a sphere [Fig. 2-4].

The reference to "seven and twenty lines" is also a not-so-subtle reference to a 2D sketch of a double tetrahedron encompassed by a "hypercube," shown as a base 2D form of a hexagon [Fig. 2-5].

The Heavy Hand of Heaviside

In a tragedy for science, two other 19th-century "mathematical physicists," Oliver Heaviside and William Gibbs, "streamlined" Maxwell's original equations down to four simple (if woefully incomplete) expressions after Maxwell's death. Because Heaviside openly felt the quaternions were "an abomination,"—never fully understanding the linkage between the critical scalar (a directionless measurement, like speed) and vector (a directionally defined value, like displacement) components in Maxwell's use of them to describe the energy potentials of empty space ("apples and oranges," he termed them)—he eliminated over *twenty quaternions* from Maxwell's original theory in his attempted "simplification."

Oliver Heaviside was once described (by *Scientific American*) as "self-taught and... never connected with any university... [he] had [however] a remarkable and inexplicable ability to arrive at mathematical results of considerable complexity without going through any conscious process of proof." According to other observers, Heaviside actually felt that Maxwell's use of quaternions and their description of the "potentials" of space was "mystical, and should be murdered from the theory." By drastically editing Maxwell's original work after the latter's death, excising the scalar component of the quaternions and eliminating the hyperspatial characteristics of the directional (vector) component, he effectively accomplished this goal single-handedly.[35]

This means, of course, that the four surviving classic "Maxwell equations"—which appear in every electrical and physics text the world over, as *the* underpinnings of *all* 20th-century electrical and electromagnetic engineering—never appeared in *any* original Maxwell paper or treatise. And every invention, from radio to radar, from television to computer science to every "hard" science from physics to chemistry to astrophysics that deals with these electromagnetic radiative processes, are based on these supposed "Maxwell equations."

They are, in fact, *Heaviside's* equations, *not* Maxwell's. The end result was that physics lost its promising theoretical beginnings as a truly "hyperdimensional" science over a century ago, and instead was saddled with a very limited subset of that potentially unifying theory, thanks to Heaviside.

The advocates of an aether-based model of force were dealt a greater blow in 1887, when the Michelson-Morley experiments effectively proved there was "no material aether." What was lost, however, because of Heaviside, was that Maxwell had never believed in a material aether himself—he was assuming a *hyperspatial* aether simultaneously connecting everything in the universe.

The major source of confusion surrounding Maxwell's actual theory, versus what Heaviside reduced it to, is its math—a notation system perhaps best described by H. J. Josephs:

"Hamilton's algebra of quaternions, unlike Heaviside's algebra of vectors, is not a mere abbreviated mode of expressing Cartesian analysis, but is an independent branch of mathematics with its own rules of operation and its own special theorems. A quaternion is, in fact, a generalized or hypercomplex number."

In 1897, Hathaway published a paper specifically identifying these hypercomplex numbers as "numbers in four-dimensional space." Thus, modern physics' apparent ignorance of Maxwell's nineteenth century success—a mathematically based, 4D "field-theory"—would seem to originate from a basic lack of knowledge of the true nature of Hamilton's quaternion algebra itself.

And, unless you track down an original 1873 copy of Maxwell's "Treatise," there is no easy way to verify the existence of Maxwell's "hyperdimensional" quaternion notation; for, by 1892, the third edition incorporated a "correction" to Maxwell's original use of "scalar potentials," thus removing a crucial distinction between four-space "geometric potential," and a three-space "vector field" from all subsequent Maxwellian theory—which is why modern physicists, like Michiu Kaku, apparently don't realize that Maxwell's original equations *were*, in fact, the first geometric four-space field theory, expressed in specific four-space terms—the language of quaternions.

Rediscovery

One of the difficulties of proposing a "higher dimension" is that, inevitably, people (and scientists are people), will ask "well, where *is* it?!"

The most persistent objection to the four-space geometries of Reimann, Cayley, Tait and Maxwell was that no experimental proof of a "fourth dimension" was readily apparent. One of the more easily understandable aspects of "higher dimensionality" was that a being from a "lower dimension" (a 2D "Flatlander," for instance) entering our "higher" 3D reality, would appear to vanish instantly from the lower-dimensional world (and, consequently, appear just as suddenly in the higher dimension, albeit geometrically distorted.) When she returned to her own dimension, she would just as "magically" reappear.

To the scientific mind, however, people in our dimension don't just turn a corner one day and promptly vanish into Reimann's fourth dimension. While mathematically derivable and beautifully consistent, to "experimentalists" (and all real science ultimately *has* to be based on verifiable, independently repeatable experiments) there seemed no testable, physical proof of hyperdimensional physics. Thus hyperspace—as a potential solution to unifying the major laws of physics—quietly disappeared, not to resurface until April 1919.

At that time, a remarkable letter was delivered to one Albert Einstein. Written by Theodr Kaluza, an obscure mathematician at the University of Konigsberg in Germany, the letter's first few lines offered a startling solution (at least to Einstein, who was unknowledgeable of Maxwell's original quaternion equations) to one of physics' still most intractable problems, the mathematical unification of his own theory of gravity with Maxwell's theory of electromagnetic radiation via introduction of a fifth dimension. (Because Einstein, in formulating the General and Special Theory of Relativity in the intervening years since Reimann, had already appropriated time as the fourth dimension, Kaluza was forced to specify his additional spatial dimension as a fifth. In fact, this was the same spatial dimension as the four-space designations used by Maxwell and his colleagues in their models over 50 years before.)

Despite its stunning mathematical success in apparently finally uniting gravity and light, the same question was asked of Kaluza as had been asked of Reimann over sixty years before. Because there was no overt experimental proof of the physical existence of another spatial dimension, Kaluza got the same "OK, where is it?" challenge to his assertions. Kaluza had a very clever answer: he proposed that the fourth dimension had somehow collapsed down to a tiny circle, "smaller than the smallest atom."

In 1926, another essentially unknown mathematician, Oskar Klein, was investigating the peculiar implications of Kaluza's ideas in the context of the newly invented atomic theory of quantum mechanics. Klein was a specialist in the truly arcane field of mathematical topology—the higher dimensional surfaces of objects. Quantum mechanics had just been proposed a year or so before Klein's further topological investigation of Kaluza's ideas, by Max Planck and many others rebelling against perceived limitations of Maxwell's (now sanitized) electromagnetic theory. The "quantum mechanics" theory would eventually become a highly successful (if bizarre, by common-sense standards) non-geometric effort to describe interactions between "fundamental particles," exchanging "forces" through discrete "quantitized" particles and energy in the sub-atomic world. Eventually, combining the two inquiries, Klein theorized that, if it truly existed, Kaluza's new dimension had likely collapsed down to the Planck length itself—supposedly the smallest possible size allowed by these fundamental interactions. However, that size was only about "ten to the *minus* thirty-three" centimeters across. Thus, the main obstacle to experimental verification of the Kaluza-Klein Theory (and the reason why people simply didn't "walk into the fourth dimension") was that quantum mechanics calculations affirmed that the *only way* to physically probe such an infinitesimally tiny dimension was with an atom smasher. There was only one small technical problem: The energy required would exceed the output of all the power plants on Earth, and then some.

Thus, the brief blip of new interest in hyperdimensional physics—the discussions of Kaluza-Klein among physicists and topologists—dropped through the floor by the 1930s. This occurred both because of Klein's proof of the apparent impossibility of any direct experimental verification of additional dimensions, and because of the dramatic revolution then sweeping the increasingly technological world of big science.

There was at that time a flood of verifications gushing forth from atom smashers all around the world, feverishly engaged in probing quantum mechanics. The rapidly multiplying populations of "fundamental particles" spawned by this bizarre mathematical world led Einstein to refer to the theory as "sorcery." Later, even though he accepted some of the experimental results, he remained skeptical that it was a complete answer to the questions posed by the physical universe.

Thirty more years would pass before scientific interest in hyperspace would be reborn as superstring theory—in which fundamental particles and "fields" are viewed as hyperspace vibrations of infinitesimal, multi-dimensional strings. For most physicists currently interested in the problem, the superstring hyperdimensional model has overwhelming advantages over its predecessors. Besides effectively unifying all the known forces of the universe, from electromagnetism to the nuclear force, in a literally beautiful "ultimate" picture of reality, it also makes a specific prediction about the total number of n-dimensions that can form: ten or twenty-six, depending on the rotation of the "strings." The bad news is that they can't be tested either, because all ten dimensions are curled up (in the model) inside the experimentally unreachable Planck length.

The hottest mainstream scientific theory to come along in more than half a century, the closest thing we have to a "Theory of Everything," is not only a hyperdimensional model of reality, it is yet another theory which, by its fundamental nature, *can't be scientifically tested*—while a hyperdimensional model which *can* be tested (and was, apparently, for decades behind the Iron Curtain) has been systematically ignored in the West for over a hundred years.

Tesla, Bearden and DePalma

As Hoagland continued to make new connections of the geometry of Cydonia with the historical treatment of hyperspatial realities, he encountered a number of independent, rouge experimentalists who had been working along these same lines. Foremost among these were Dr. Bruce DePalma, an M.I.T. physicist and researcher, and Lt. Col. Thomas Bearden, a nuclear engineer and physicist who had been working on Maxwell's original model since his days on the U.S. Army's scalar weapons programs.

Bearden had tirelessly researched Maxwell's original writings, and concluded that Maxwell's original theory is, in fact, the Holy Grail of physics—

the first successful unified field theory in the history of science. Bearden had done dogged detective work to uncover Maxwell's papers, and from them had concluded that Heaviside had literally hijacked Maxwell's theory and set modern science back almost a hundred years. According to Bearden, not only would modern physics *never* find the single unifying element for gravity, electricity and magnetism (because it was all based on Heaviside's broken version of Maxwell's model), but that if the original model were restored, it had the potential to unleash nearly limitless amounts of energy, and to allow humanity the means to actually "engineer" forces like gravity at the quantum level.

This radical view was supported by Bearden's own research, which was based on papers and experiments carried out by Sir Edmund Whittaker and Nicola Tesla in the early 20th-century, and later confirmed in the so-called "Aharonov-Bohm" experiments.[36]

Tesla, the inventor of modern civilization (through his discovery of alternating current), had conducted a number of relevant experiments in his lab in Colorado Springs in 1899. During one experiment, he observed and recorded "interfering scalar waves." Via massive experimental radio transmitters he had built on a mountaintop in Colorado, he was broadcasting and receiving (by his own assertion) longitudinal stresses (as opposed to conventional EM "transverse waves") through the vacuum. This he was accomplishing with his own, hand-engineered equipment (produced according to Maxwell's original, quaternion equations), when he detected an interference "return" from a passing line of thunderstorms. Tesla termed the phenomenon a "standing columnar wave," and tracked it electromagnetically for hours as the cold front moved across the west. Tesla's experiments were suddenly stopped when his benefactor, J. P. Morgan, discovered the true purpose of his experiment—to generate unlimited amounts of electricity "too inexpensive to charge for."

Bearden was also interested in generating energy by creating "longitudinal stress" in the vacuum using Maxwell's quaternion/hyperdimensional potentials. Bearden wrote several papers on the theory, eventually published by the Department of Energy on their official website[37]

Bearden then set about the task of actually constructing a device that could draw "energy from the vacuum," eventually patenting a device (The "Motionless Electromagnetic Generator") that seems to generate energy from literally *nothing*.[38]

Of course, you can't really get something from nothing, and Hoagland quickly realized that the effect Bearden was describing was the same "hyperdimensional" effect he was seeing in the heat generation of the outer planets.

There is now much fevered discussion among Western physicists on the Quantum Electrodynamics Zero Point Energy of space—"the energy of the vacuum." To many familiar with the original works of Maxwell, Kelvin, et. al., this sounds an awful lot like the once-familiar "aether," merely updated and now passing under an assumed name. Described as some sort of exotic quantum effect to make it seem acceptable, this "zero-point energy" is nothing more than Maxwell's hyperdimensional physics in another guise.

Thus, creating and then relieving a "stress" in Maxwell's vorticular aether is precisely equivalent to tapping the energy of the vacuum that, according to current quantum mechanics models, possesses a staggering amount of such energy per cubic inch of empty space. Even inefficiently releasing a tiny percentage of this "strain energy" into our three dimensions, or into a body existing in 3D space, could make it appear as if the energy was coming from nowhere—something from nothing. In other words, to an entire generation of students and astrophysicists woefully ignorant of Maxwell's original equations, such energy would appear as the dreaded "P" word ... "perpetual motion."

As we shall show, it is this "new" source of energy—in a far more controlled context—that seems to also be responsible for not only the anomalous infrared excesses Hoagland has noted in the so-called giant outer planets of this solar system, but for the radiated energies of stars themselves.

Yet how does one create a "stress in the aether" to generate energy, or test this hyperdimensional physics theory? The theoretical notions of Maxwell and Bearden had already been tested, albeit inadvertently, by the aforementioned Dr. DePalma.

DePalma, brother of the famed film director Brian DePalma, long before he and Hoagland met, had been running (since the '70s) a series of ground-breaking "rotational experiments" which remarkably confirmed much of what Hoagland would be theoretically rediscovering twenty years later. One practical invention was DePalms's "N-Machine"—a high-speed, "homo-polar generator" which is able to pull measurable electrical power literally out of "thin air" (the vacuum) with no expenditure of fuel ....

Among DePalma's many other radical results was an experiment in which he simultaneously ejected two metal balls—one spinning at over 27,000 rpm and one not spinning at all—from the same test rig (below). He then measured the rate at which they both rose and fell. In shocking contrast to the expected result if standard "Newtonian" mechanics are at work, the spinning ball *rose farther and faster, and fell to the ground faster,* than the non-spinning ball—even though exactly the same upward force (momentum) had been applied to both [Fig. 2-5].

The implication was that the *spinning ball* had somehow managed to gain energy from somewhere—which simultaneously altered both the effects of gravity *and* inertia on it... exactly as Bearden's model had independently proposed.

DePalma conducted countless additional rotational experiments, including with massive gyroscopes, throughout the 1970s. In the course of these mechanics, he discovered that gyros, when spun up—and simultaneously—induced to mechanically precess (wobble on their axis of rotation), could also be used to substantially negate the effects of gravity. In one experiment, a 276-pound "force machine" was reduced in weight by *six pounds*—about a 2% loss—when the gyros were switched on.

Depalma also discovered that these massive rotating systems, even when carefully isolated, could induce "anomalous rotational motions" in other gyroscopic set ups, even in other rooms ... but only if they were *also* rotating.

As a result of these years of painstaking laboratory experimentation, with a great variety of different spinning systems, DePalma ultimately argued that all rotating objects—including planets and stars—*must intrinsically* precess. "Precession" is the tendency of spinning objects, like a child's top or a planet like the Earth, to wobble on their rotational axes. In conventional mechanics, precessional motion is explained as coming from an outside force (like the Moon's gravity tidally tugging on the Earth's slightly bulging equator) unbalancing an object's spin.

Based on DePalma's empirically-derived measurements of rotation, he predicted that even apparently isolated spinning objects would precess—by virtue of their interaction with other rotating objects. They drew energy from some kind of non-magnetic, non-gravitational field (termed "the OD field"...), which he proposed had to exist to explain his perplexing "energy accreting" spinning ball experiment. Ironically, totally unknown to DePalma—because of the Cold War and the KGB's strict secrecy controls—what he was observing was also being simultaneously observed by his Russian counterparts, and termed "a torsion field"—based on the same *rotational* interactions.

The idea of "isolated precession"—the logical conclusion of DePalma's decades of observations of all kinds of anomalous spinning systems—has never been experimentally examined under controlled laboratory conditions (at least, in any published literature in the West ...), as DePalma needed a "zero gravity" (force-free) environment to conduct the proper test. Unfortunately, before Hoagland (with his NASA associations and connections) could conclude arrangements for just such test of Bruce's remarkable hypothesis, to be conducted at NASA-Lewis's vacuum zero-gravity drop test facility, DePalma suddenly died in 1998.

Ultimately, Depalma's fascinating prediction concerning *intrinsic* rotational precession—which can now be elegantly explained by invoking multiple rotational motions and *torsion* interactions occurring *simultaneously* in *higher dimensions*—will come to have great significance in our story later on.

With Bearden and DePalma's experimental input in hand, Hoagland began to consider serious methods of doing the one thing that all the other hyperdimensional theories had been unable to accomplish: carry out actual *testing* of their core assumptions.

A Testable Theory

The true scientific method is something that is woefully misunderstood in our modern world, even by many scientists today. The history of science is replete with debates that have raged back and forth in colossal wars of ego and self-interest—but the method itself is supposed to protect us from scientists becoming a new priesthood, by making sure that when models don't fit the new data, they are discarded, no matter how appealing they may be to some interests. Unfortunately, it rarely works that way.

Hoagland wanted to immediately separate his concept of hyperdimensional physics from the earlier models in one distinct way—prediction. Only if his new ideas could be either confirmed or falsified would his modern version of

Maxwell's revolutionary ideas ever gain traction—and in order to accomplish that, the first order of business for any valid scientific model is to produce *testable predictions* based on that model. Fortunately, there were several implicit tests of the hyperdimensional model almost immediately suggested by the original observations themselves. Eventually, Hoagland settled on four additional key predictions that would determine if Cydonia's "embedded tetrahedral physics" and the resulting "hyperdimensional model" would be falsifiable. All of these tests would invariably spring from the same, somewhat surprising source.

Angular Momentum

Hoagland's first focus was on the anomalous heat emissions he and Torun had noted in the outer planets. Since in three dimensions *all* energy eventually "degrades" to random motions, via Kelvin and Gibb's laws of thermodynamics, then "stress energy" of the aether (vacuum) released inside a material object, even if it initially appears in a coherent form, will eventually degrade to simple, random heat, ultimately radiated away as excess infrared emissions into space. In the end, all energy, whatever its source, ends up looking the same.

Because of this, Hoagland focused on the initial, astrophysical conditions under which such "Maxwellian space potentials" can be released inside a planet or star. The idea was to predict a specific set of signatures that would uniquely identify the source of these energy emissions as hyperdimensional, versus a "normal" 3D effect.

In attempting to understand the anomalous IR radiation, one thing quickly became clear; to a first order, the infrared excesses of the giant planets all seemed to correlate very nicely with *one parameter* each has in common—their total system "angular momentum."

The mass of a body and the rate at which it spins, in classical physics, determines an object's angular momentum. In the hyperdimensional model, it's a bit more complicated, because objects apparently separated by distance in this dimension are in fact connected in a "higher" (four-space) dimension.

So, in the hyperdimensional model, one also adds in the orbital momentum of an object's gravitationally tethered satellites—moons in the case of planets, planets in the case of the sun or companion stars in the case of binary star systems.

In this view, then, the one advocated by Hoagland and derived from his "meaningless" observations of the mathematics of Cydonia, this total system angular momentum was the key to understanding how things really worked in our 3D universe. This is in stark contrast to the currently accepted view of field theory and electromagnetic force, which views a planet or star's mass as the most important characteristic dictating astrophysical behavior. Because mainstream physicists are working with Heaviside's version of Maxwell's original concepts, the most significant "force" they can observe is gravity. Since gravity is driven by mass, modern physics has always assumed that mass is the single most influential aspect of astrophysical interactions.

However, when we measure total solar system *angular momentum*, we get something of a surprise... [Fig. 2-6]

Jupiter as it turns out, which has less than 1% of the mass in the solar system, somehow possesses 60% of the angular momentum, whereas the sun, which possesses 99% of the mass, has only 1% of the angular momentum. If the conventional view of the solar system is correct, then in fact the distribution of angular momentum to mass should be roughly equivalent. Instead, it is completely flip-flopped. Various notions have been floated to explain this theoretical discrepancy, including the idea that the sun somehow magically "transferred" its angular momentum to the planets—but there are big problems with all of these ideas, and formation theoreticians are not close to providing an answer.

Hoagland had first begun to examine the role angular momentum might play in his emerging theory when he made one significant association—the common link connecting all the objects for which the Cydonia "embedded tetrahedral model" works, from the planets to the sun, seemed to be based on a relationship between angular momentum and magnetic fields. Before the adoption of the present, complex "self-excited dynamo theory" (with internal, circulating, conducting "fluids" as the mechanism for general planetary and stellar magnetism), another, strictly empirical, hypothesis was proposed—a strikingly simple relationship between the observed total angular momentum of the object, and a resulting magnetic dipole.

Termed "Schuster's Hypothesis" (after Sir Arthur Schuster [1851–1934], who made the original correlation and whose very name and empirical discovery is, inexplicably, completely missing from all NASA literature on planetary magnetism), it has been successful in predicting magnetic field strengths (Blackett 1947, Warwick 1971) ranging from the Earth's, to the sun's, to Jupiter's vast field (20,000 times the terrestrial dipole moment). Schuster's prediction, made over sixty years prior to the 1973–74 Pioneer 10 and 11 close-up confirmations (Warwick 1976), led Warwick to comment thusly on the remarkably predictive power of "Schuster's hypothesis" in 1971:

"Dynamo theory has not yet successfully predicted any cosmical fields. Its use today rests on the *assumption* that no alternative theory corresponds more closely to observations."

Indeed, after Mariner 10 detected a magnetic field around Mercury that not only fit Schuster's hypothesis, but also directly contradicted the dynamo theory, even Carl Sagan admitted that there was a need for serious revision of science's view of planetary magnetism.[39]

Taking Schuster's 1912 proposal, Hoagland and Torun plotted contemporary parameters for angular momentum versus the observed magnetic dipole moment (for all planetary objects now visited by spacecraft with magnetometers), and found Schuster's hypothesis confirmed—with the exception of Mars (which had its magnetic field stripped away in a recent catastrophe, which we will deal with later) and Uranus (discussed below). The truth is, the dynamo theory has never produced a single correct prediction of any planetary magnetic field, and Schuster's Theorem has been correct in almost every case.

Uranus, as the sole exception to the accuracy of Schuster's Theorem, may in fact be the exception that proves the rule. Uranus has a rotational period almost exactly the same as Neptune's, and should by definition have a similar magnetic field strength. However, Neptune has a magnetic field strength about half of Earth's while Uranus' is almost two thirds that of Earth's. If Schuster's Theorem is correct, then the two planets' magnetospheres should be about the same intensity. Instead, there is an almost two-to-one ratio. Uranus, however, is exceptional in many other ways—its obliquity is almost 90º from solar vertical, implying that it may have suffered a significant pole shift in the recent past that could account for this discrepancy. If this happened in the geologically recent past, then there would logically follow a period of non-compliance with the theorem. Given that there was also a significant hyperdimensional factor (related to DePalma's tests with spherical precession) that may have accounted for Uranus' present condition, and since Schuster's observation works quite well for the other planets, it seems likely that there was an external factor not yet fully understood for Uranus' non-compliance—but clearly, if Schuster's Theorem is valid in seven out of nine cases and the Dynamo theory is zero for nine, Schuster's observation would appear to be a far closer fit.

The fascinating, observed correlation between angular momentum and magnetic dipole moment led Hoagland to make a similarly simple empirical connection to his own work. When he looked at the relationship between the anomalous infrared emissions and angular momentum, he discovered that it too correlated precisely with each planet's total system angular momentum.

When one graphs the total angular momentum of a set of objects, such as the radiating outer planets of this solar system (plus Earth and the sun) against the total amount of internal energy each object radiates to space, the results are striking [Fig. 2-7].

The more *total* system angular momentum a planet (or any celestial body) possesses, the more anomalous energy it is capable of generating (actually, the energy—like the "anomalous energy" DePalma observed in his spinning ball experiment—is being "ducted" into the interior of the rotating mass from a higher dimension, via the 3D "aetheric/torsion field").

For the "hyperdimensional physics" model, this simple but powerful relationship seems to be the equivalent of $E=MC^2$: a celestial object's total internal luminosity seems dependent upon only *one* physical parameter: luminosity equals total system angular momentum (object, plus all satellites).

What this implies is that the amount of energy a given object radiates is dictated by the force exerted on it through hyperspace, and that this hyperspatial energy is measurable in our 3D world as angular momentum. In graphing this fairly obvious correlation [Fig. 2-7], everything seemed to line up quite nicely. All the planets fell right where they should along the graph—the only exception was the sun. Somehow, the sun seemed to be missing a substantial quantity of its angular momentum.

The conventional view of the sun, and all stars like it, is that they are massive nuclear furnaces, fueled by the collapse of matter into a raging fireball of energy. This process causes a literal fusion of atoms in the sun's interior, and should by implication produce fusion byproducts. One such expected fusion byproduct is neutrinos, sub-atomic particles that lack an electrical charge. However, experiments to measure the sun's neutrino output have discovered that the sun is not emitting *anything* like the number of neutrinos required by the standard solar model for its observed energy emission. If its energy is due to "thermo-nuclear reactions" (as the standard model *demands*), then the observed "neutrino deficit" is upwards of 60%. Even more remarkable, certain kinds of primary neutrinos (calculated as required to explain the bulk of the solar interior's fusion reactions, based on laboratory measurements) turn out to be simply missing altogether.

Recently, a combination of theoretical readjustments to existing quantum theory, coupled with the results of new neutrino detectors brought on-line, seem to have reconciled the observed "neutrino deficit" for the sun—bringing the observed number (and "flavor") back in line with the *adjusted* theory. But, there is the nagging suspicion on our part that this tinkering with the original neutrino/standard solar model—remember, created *before* the anomalous solar neutrino deficit had been observationally discovered—is a kind of "academic cheating" ....

The answer to the sun's apparent violation of the standard solar model, ironically, is contained in its striking "violation" of our key angular momentum/ luminosity diagram. In the hyperdimensional model, the sun's primary energy source—like the planets'—must be driven by its total angular momentum, its own "spin momentum" plus the total angular momentum of the planetary masses orbiting around it. As we pointed out above, though the sun contains more than 98% of the mass of the solar system, it contains less than 2% of its total angular momentum. The rest is in the planets. Thus, in adding up their total contribution to the sun's angular momentum budget, if the hyperdimensional model is correct, we should see the sun following the same line on the graph that the planets, from Earth to Neptune, do. It does not.

The obvious answer to this dilemma is that the hyperdimensional model is simply wrong. The less obvious conclusion is that we're missing something— like additional planets.

In trying to account for the missing angular momentum, Hoagland had found his first testable prediction of the hyperdimensional model. By adding another big planet (or a couple of smaller ones) *beyond* Pluto (several *hundred* times the Earth's distance from the sun), the sun's total angular momentum could move to the right on the graph, until it almost intersects the line (allowing for a percentage, about 30%, of internal energy expected from genuine thermonuclear reactions). This creates the specific prediction that the current textbook tally of the sun's angular momentum is deficient for one seemingly obvious reason: we haven't found *all* of the major planets in the solar system yet.

So the first implicit prediction of the hyperdimensional model was that eventually, observatories would find one massive planet in a prograde orbit, or two smaller solar system members in retrograde orbits. Either observation, within certain limits, would allow the sun to move to its predicted location on the graph and confirm the correlation between energy output and angular momentum.

Yet there was one even bigger implication of the relationship between angular momentum and energy output. If the connection was real, then it means that our view of the hierarchy of the solar system is completely backwards. In the hyperdimensional model, the tail (the planets and moons) wags the dog (the sun)... an assumption that would begin to have far-ranging implications.

Confirmation?

The next step in testing this aspect of the model was to begin a search for any evidence that another member of our solar system might conceivably exist. For years, there has been an ongoing search for "Planet X" in astronomy. This search has grown out of the observation that something, presumably a large undiscovered planet, seems to exhibit an influence on the orbits of Neptune and Uranus. The search for this "perturber" eventually led to the discovery of Pluto, but no such large planet has yet been found, at least not officially.

There have, however, been some very interesting "unofficial" discoveries that may have a bearing on this prediction. In 1982, a front-page article appeared in the *Washington Post*,[40] including an interview with JPL's Dr. Gerry Neugebauer about an object spotted in Orion by the IRAS infrared satellite, at an estimated 50 billion miles from Earth. This object fit Hoagland's prediction within very tight parameters. To date, no follow-up observations or papers have been published on this object, and in inquiries to Dr. Neugebauer, he stated that the quotes in the story were "taken out of context. I do not know about it nor any follow-up."

Neugebauer strained credulity with this reply. How many of us would claim they didn't know anything about a subject for which we had been featured in a *Washington Post* article? When you read the actual interview and story, which is clearly based on Neugebauer and Dr. James Houck's information, it seems clear that they are not coming clean. The article describes the object as small and dark, about the size of Jupiter, and at a range of 50 billion miles from Earth. At this distance (about 537 A.U. Astronomical Units), the object would probably have to be a brown dwarf, a body roughly the size of Jupiter but 50 times as massive. The *Post* article goes on to say that "two different telescopes" were enlisted to do a visual search for the object—yet Neugebauer insists it was *never* followed up. Clearly, there seems to a blanket of denial that has now fallen around this early IRAS observation.

In 1999, evidence for yet another solar system member, this time in Sagittarius (exactly opposite Orion in the celestial sphere) was cited in several news reports.[41] In this case, the object was assumed to be about the same size as the IRAS object, but many times more distant—between 25,000 and 32,000 A.U. Its presence was implied by the patterns of long-period comets.

Both of these observations show that Hoagland's prediction of additional planet(s) found at great distances from the Earth is at least somewhat supported by various observations. What separates Hoagland's model from all the other Planet X theories is the specific prediction that the object in question—when it is finally officially acknowledged—will possess enough angular momentum to move the sun to its proper place on the graph.

But what has to be questioned is if NASA *does* find our missing tenth major planet –Pluto's recent demotion to mere "solar system object" notwithstanding—will they tell us? There are numerous rogue researchers who have long predicted the existence of Planet X, and to admit that it had been discovered would give credence to their models and theories. As we dug in deeper on this mystery, it became obvious that NASA's reasons for quieting Neugebauer had far more to do with politics (as usual ...) than science.

Infrared Variability

The next implicit test of the hyperdimensional model also grew out of the infrared emissions observation. If Hoagland and Schuster's observations were right, and there was a direct connection between luminosity, magnetic field intensity and angular momentum, then this had some fairly stunning implications. Since the infrared emissions in Hoagland's model were implicitly presumed to be hyperdimensional in nature—i.e. connected to higher spatial geometries—then orbital changes in the configuration of the "system" (the constantly moving planets and moons of the solar system) should by implication lead to *varying* outputs of energy—like turning a rheostat up and down to control the output of a light. This is a crucial point, since conventional physics, by rote, demands that energy outputs for these planets be "constant," perceptibly decaying only in the *very* long term.

If the ultimate source of planetary (or stellar) energy is this vorticular (rotating) spatial stress between dimensions (*a la* Maxwell), then the constantly changing pattern (both gravitationally and dimensionally) of interacting satellites in orbit around a major planet/star, coupled with its equally changing geometric configuration vis-à-vis the other major planets of the system, *must* modulate that stress pattern as a constantly changing, geometrically twisted "aether." In Hoagland's hyperdimensional model, it is this constantly changing hyperspatial geometry that is capable of extracting energy from this underlying rotating, vorticular aether and then releasing it *inside* material, rotating objects.

Initially, this excess energy can appear in many different forms—high-speed winds, unusual electrical activity, even enhanced *nuclear* reactions—but, ultimately, it must all degrade to simple excess heat. Because of the basic physical requirement for *resonance* in effectively coupling a planet (or a star's) rotating 3D mass to the underlying 4D aether rotation, this excess energy generation must vary with time, as the changing orbital geometry of the "satellites" and other major members of the solar system interacts with the spinning primary (and the underlying vorticular aether) in and out of phase.

For these reasons, time variability of this continuing energy exchange must be a central hallmark of this entire hyperdimensional process. It should also be straightforward to determine. All that is required is to take measurements of Jupiter's IR output at various points in time along its orbit and at different positions relative to the other planets. If the hyperdimensional model is correct, then differing orbital configurations should make Jupiter's (and all other "gas giant") IR output(s) highly variable. At certain times, it should be more than its canonic two to one output. At others, it should be less [Fig. 2-8].

Again, the historical scientific record provided considerable support for this. Dr. Frank J. Low made the first observations of Jupiter's anomalous heat output from highflying aircraft in 1966 and 1969. Low, considered the father of modern infrared astronomy, published the early results that showed Jupiter's output was in the 3-1 range.[42] He later went on to create the proposal that led to the creation of IRAS, the first infrared space telescope that made the IR observations of the possible "Planet X" in Orion that we discussed earlier.

Three years later, Low made further observations that reduced the figure from 3-1 to 2-1, a better than 30% difference that was far beyond the margin of error of the instruments in use on both occasions. Later ground based telescopic observations during the 1970s revised the figure even further downward, to about a 1.67–1.00 ratio, a further 30% reduction![43] The Voyager missions largely confirmed the 1.67 ratio in the early 1980s. The assumption, despite the wide discrepancy, was simply that the earlier instruments were crude and their measurements were just way off. Since the energy output had seemed to stabilize around the 1.67 figure in the late 1970s and early '80s, it was generally assumed this was the case and that the earlier results were written off.

Fortunately, there have been a couple of probes to the outer solar system since the Voyager days, Galileo and Cassini, and both carried instruments that could measure the infrared emissions of the outer gas giant planets. So the only thing keeping Hoagland from testing this aspect of the Model is in getting someone to take the measurements, or to release them if they did take them.

That has proven to be a harder task than one might imagine. Calls to the universities where the infrared data for both spacecraft are gathered and archived resulted in a distinct lack of co-operation. Hoagland was told that in order to get the figures for the measurements, he would have to "prove" his affiliation with an "approved" research center or university. However, a search of NASA's online astrophysics abstract database yielded some interesting information.

In a recent paper, observations made with Cassini's Composite Infrared Spectrometer (CIRS) seemed to offer confirmation of Hoagland's prediction.[44]

The team of researchers found that Jupiter's infrared emissions did not fit the expected "1.67 to 1.00" ratio that had been the canonic number since Voyager. While it does not give exact numbers, the abstract states that "the equatorial temperature minimum is much more pronounced than observed at the time of Voyager," and that "it is likely related to the temporal variation in the equatorial stratospheric temperatures that have been reported from ground-based observations." Not only does Jupiter appear to be continuing to display variable heat output consistent with Hoagland's model, but the latter statement also implies that more recent ground based observations have been seeing the same thing.

While we won't go so far as to claim a cover-up of the more recent thermal imaging data for Jupiter and the other outer planets, clearly the results are anomalous for the conventional models and consistent with Hoagland's key HD prediction. Unfortunately, we will have to wait for actual published numbers before we can rack this prediction firmly in the "confirmed" category.

Short-Term Amplitude Variations

This same aspect of the model could also be applied on a smaller scale to produce yet another prediction. In our own planetary system, all the "giant" planets possess a retinue of at least a dozen satellites: one or two major ones (approximating the size of the planet Mercury), with several others ranging down below the diameter and mass of our own moon, in addition to a host of smaller objects. Because of the "lever effect" in the angular momentum calculations, even a small satellite orbiting far away (or at a steep angle to the planet's plane of rotation) can exert a disproportional effect on the "total angular momentum" equation—just look at Pluto and the sun [Fig. 2-6].

Even now, Jupiter's four major satellites (which have collective masses approximately 1/10,000th of Jupiter itself), during the course of their complex orbital interactions, are historically known to cause *time-altered* behavior in a variety of well-known Jovian phenomena—including "anomalous" latitude and longitude motions of the Great Red Spot itself.

As Hoagland presented at the U.N. in 1992, the Great Red Spot (GRS)—a mysterious vortex located for over 300 years at that infamous 19.5º S. Latitude, via the circumscribed tetrahedral geometry of the equally infamous twenty-seven line problem is *the* classic "signature" of hyperdimensional physics operating within Jupiter.

The existence of decades of recorded anomalous motions of this Spot, neatly synchronized with the highly predictable motions of Jupiter's own "Galilean" moons are clearly *not* the result of conventional gravitational or tidal interactions, given the relatively insignificant masses of the moons compared to Jupiter itself. Rather, they seem to be following the models of Maxwell, Schuster and Whittaker; the result of powerful hyperdimensional modulation from the changing geometric configurations of these same satellites. It is the long lever of angular momentum and harmonic torsion resonance on the constantly changing, vorticular scalar stress potentials (torsion field states) *inside* Jupiter that are causing the changes in the GRS.

So, hyperdimensional test number three: look for small, short-term amplitude variations in the infrared emission levels of *all* the giant planets, synchronized (as are the still-mysterious, but *cyclic* atmospheric motions of the GRS on Jupiter) with the orbital motions and conjunctions of their moons, and/or, the motions of these outer planets relative to the *other* major members of the solar system.

Confirmed short-term variations in the current planetary IR outputs of a few hours (or even a few days') duration—and synchronized with the orbital periods of the planets' satellites themselves—would thus be stunning evidence that all the mainstream explanations are in trouble, and that the hyperdimensional model deserves much closer scrutiny. Rising or falling output over years and decades (as strongly implied by the historical IR observations of Jupiter, from Frank Low to Cassini) would support a longer-term, planetary modulation of these internal HD energy releases.

In fact, of course, both sets of modulations should be occurring simultaneously—easily separated via a suitably written computer observation program ... providing someone even looks.

These changing interactive stresses in the boundary between hyperspace and "real" space (in the hyperdimensional model) now also seem to be the answer to the mysterious "storms" that, from time to time, have suddenly appeared and disappeared in the atmospheres of several of the outer planets. The shrinkage and virtual disappearance in the late 1980s of Jupiter's Great

Red Spot is one remarkable example; as was Saturn's abrupt production of a major planetary "event," photographed by the Hubble Space Telescope in 1994, as a brilliant cloud erupting at *19.5º N* (where else?); Neptune's "now you see it now you don't" major "19.5 storm"—the Great Dark Spot—is yet another.

And, in the latest solar system mystery to confound NASA theorists, there is the sudden formation of a *second* red spot, nicknamed "Junior" which occurred on Jupiter in 2006. This massive (Earth-sized), atmospheric vortex coalesced over a few weeks from three smaller vortices (each about the size of Mars) and then immediately began also turning "GRS red."

And NASA is admittedly "clueless" (again) to what is actually going on ....

Since the prevailing NASA view is that these planets' excess IR output *must* be constant over time, no one has bothered to look for any further correlations between a rising or falling internal energy emission and the (now historically well-documented) semi-periodic eruptions of such "storms"—and they should.

This whole notion, that the changing configuration of a planet's (or star's) system members relative to the "primary," can have an effect on its total energy output, is revolutionary to current thinking, but hardly without precedent.

There is a very well known, long period and still mysterious variability associated with the largest "hyperdimensional gate" in our own neighborhood—the sun.

Its complex changes, which include a host of related surface phenomena—solar flares, coronal disturbances, mass ejections, etc.—is termed "the sunspot cycle," because the number of simultaneous "spots" (lower-temperature vortices appearing dark against the hotter solar surface, as this activity occurs) waxes and wanes over about 11 years. The full magnetic reversal of the sun's polarity takes two complete sunspot cycles to return to "zero"—thus the complete "solar cycle" is somewhat over twenty years.

In the 1940s, the Radio Corporation of America (RCA) hired John Nelson, a young electrical engineer, in an effort to improve the reliability of short-wave radio communications around Earth. Such radio transmissions had been observed to be more reliable in the "lulls" in between solar activity associated with "peak" sunspot years.

To his surprise, Nelson soon specifically correlated this rising and falling radio interference with not only the sunspot cycle, but with *the motions of the major planets* of the solar system. He found, to his increasing astonishment, a very repeatable—in essence, *astrological*—correlation between the inexorable orbits of all the planets (but especially, Jupiter, Saturn, Uranus and Neptune, which hold essentially all the solar system's known angular momentum) and major radio-disturbing eruptions on the sun.[45]

The hyperdimensional model finally provides a comprehensive theoretical explanation—a "linking mechanism"—for these (to a lot of astronomers) still embarrassing decades-old RCA observations. For, in essence, what John Nelson had *rediscovered* was nothing short of a "hyperdimensional astrology"— the ultimate, very ancient, now highly demonstrable angular momentum foundations behind the *real* influences of the sun and planets on our lives. Nelson also "rediscovered" something else:

"It is worthy of note that in 1948, when Jupiter and Saturn were spaced by 120º, and solar activity was at a maximum, radio signals averaged of far higher quality for the year than in 1951 with Jupiter and Saturn at 180º and a considerable decline in solar activity. In other words, the average quality curve of radio signals *followed the cycle curve between Jupiter and Saturn rather than the sunspot curve...*" [Emphasis added.]

These decades-old observations are very telling ... not only *confirming* Jupiter and Saturn as the primary "drivers" behind the sun's known cycle of activity (in the hyperdimensional model), but strongly implying an additional direct effect of their changing angular relationship on the electrical properties of Earth's ionosphere. This, of course, is totally consistent with these changing planetary geometries affecting not just the sun, but the other planets simultaneously as well, just as "conventional" astrologers have claimed, via Maxwell's "changing scalar potentials."

At this point, then, *only* the hyperdimensional theory:

1. Points to the deepest implications of the simple astronomical fact that the "tail wags the dog"—that the planets in this physics are fully capable of exerting a determinant influence on the sun, and each other, through their disproportionate ratio of total solar system angular momentum: over 100 to 1, in the (known) planets' favor.

2. Possesses the precise physical mechanism—via Maxwell's "changing quaternion scalar potentials"—accounting for this anomalous planetary angular momentum influence.

3. Has already publicly identified, at the United Nations in 1992, a blatant geometric clue to this entire hyperdimensional solar process: the maximum sunspot numbers (those large, relatively "cool," rotating vortices appearing on the solar surface), rising, falling and methodically changing latitude, during the course of the familiar twenty-two-year solar cycle—and peaking every half-cycle (around eleven years), at the solar latitude of about 19.5°.

Pulsars

Pulsars are yet another area where the hyperdimensional physics theory can be put to the test. Hoagland and Torun predicted that pulsars, because of their incredible angular momentum and magnetic properties, should be excellent hyperdimensional physics test beds. In fact, in the case of one specific pulsar, they may be a major key in validating the overall Hoagland/Torun model [Fig. 2-9].

B1757-24, a pulsar first observed in July 2000, was found to possess far more angular momentum than it should have. In fact, the object defies all known and accepted "laws of physics" and seems to be tapping additional angular momentum from an unseen source. That unseen source, according to Hoagland and Torun's prediction, is higher dimensional energy released by the rapid rotation of the pulsar.

Under the conventional physics model, stars are assumed to be "born" from spinning gas and dust nebulae. As they contract (under gravity), like an ice skater tucking in her arms, they must spin faster. This is the central tenet of a fundamental law of (current) physics, called "the conservation of angular momentum." The only way a star is supposed to be able to get rid of this fixed quantity of angular momentum transferred to it at birth, is to "re-transfer" the momentum to space through one of basically two means: direct mass loss and/or magnetic interactions (accelerations) between the star and any surrounding nebulae or bodies (such as a companion set of planets or another orbiting star).

For most of a star's "main sequence" life, the period when it is assumed to be relatively stable in its spin and energy output (although the hyperdimensional model states that this output also is not "constant" or completely "stable"—but that's another argument), these mechanisms are supposed to be able to transfer at best a few percent of the star's original angular momentum. So, a star at the end of its billions-of-years-old life is supposed to have pretty much the same quantity of angular momentum as it was originally born with.

When a massive star (between five and twenty times the mass of the sun) reaches the end of its life (defined in the conventional models as "the exhaustion of its nuclear fuel"), it goes supernova. In these models, roughly ninety percent of the outer parts of the star leaves by this means (ultra-rapid mass transfer into space—in excess of 5,000 miles per second!), leaving the remaining, collapsed, ultra-dense core behind, as a now rapidly-spinning "neutron star." Such spinning, incredibly dense objects (essentially, the mass of the sun and the density of an atomic nucleus, smashed into a volume about the width of a small city) are supposed to be at the heart of the "pulsar phenomenon."

Thus, when "born" in this violent end-process of stellar evolution, such a rapidly spinning object is supposed to have been given (through the previous mechanisms) a *finite* quantity of angular momentum—not as much as the original star (because of the large fraction of mass lost in the explosion, taking *that* angular momentum with it), but just as finite.

In the ensuing "pulsar phenomenon," such a spinning, highly magnetized object is far more likely to interact with other nearby gas clouds, etc., than the original star. This is because the original magnetic field of the whole star is also supposed to be conserved, and is now collapsed down to the new volume of a "city-sized" object, from an original volume perhaps several *trillion* times as large. Such incredibly high-strength magnetic fields are then supposed to be able to accelerate matter still in the vicinity of this newly-born, rapidly spinning object (the outwardly exploding shell of the original star) and fling some of it away from the star via "magnetic acceleration" at an appreciable fraction of the speed of light. This phenomenon is what's supposed to create the accelerating beams of matter that spin with the rotation of the star (up to a hundred times per second), producing the rapidly rotating, ultra-stable "light house effect" of radio, gamma ray and optical emissions that characterize the "pulsar phenomena" seen even thousands of light years away.

If a planet, like the Earth, is in line with these beams of matter, then we can "see" the lighthouse effect. If it is not, we will never spot the pulsar. In this model, because such an exotic, rapidly spinning, comparatively tiny object (but with the mass of the sun) is heavily interacting (through its now incredibly strong surface magnetic fields) with the still slowly (comparatively speaking) expanding shell of its own outer layers (from the original supernova explosion), it should also be transferring—at a measurable rate—its own *finite* amount of angular momentum to the larger cloud. This has to inevitably result in a slow, steady and observable "spin down" of the neutron star.

Radio, optical and x-ray/gamma ray observations of the almost 1,000 known pulsars discovered since 1968 have measured this "spin down" effect in a wide variety of situations. The incredibly regular radio, optical and x-ray/gamma ray pulses emitted by such stars have been observed time after time to slowly lengthen by a tiny but measured amount over several years—an indication of an ultra-slow "despinning" of the tiny stars. Using the (assumptive) law of the conservation of angular momentum, this steady spin down is viewed as confirmation not only of the known laws of angular momentum, but also as a means to "date" the ages of these stars, a kind of "pulsar clock" with a presumed constant half-life.

Because about half of all known stars are binaries, when one of these stars explodes as a Supernova it releases itself and its companion in opposite directions, at whatever the original orbital velocity was between them. In the Sagittarius pulsar, the fleeing pulsar eventually flew right out of the slowly expanding shell of gas from the original explosion (the expanding blast wave ran into an interstellar cloud and slowed way down, the neutron star core didn't). Using the known distance, space velocity and geometry of the pulsar/cloud relationship, the new VLA measurements of the actual space velocity of this pulsar was discovered to be only about 300 miles per second—way below the estimated 1,000 mps previously assumed.

From the observed "spin down rate," the previous estimate of the age of the neutron star/pulsar (when the original supernova exploded) was about 16,000 years—but from the "kinemetic" age of the star (measured by its known velocity beyond its own expanding envelope), the age of the original explosion is now estimated as happening about 170,000 years ago—a factor of *ten* disparity.

Since the now-measured space velocity of the pulsar is not open to any alternative interpretation (it's a very simple measurement, as compared to the model for a despinning pulsar), the age since the pulsar's formation (and separation from its binary companion) must be about the same: 170,000 years. So, for 170,000 years this pulsar has been on its own, yet the rate at which its rotation is slowing indicates a *much* younger age. Obviously, something is radically wrong with the pulsar model of a finite amount of angular momentum slowly being expended.

The simplest explanation for this "impossibility" is that the star has been able to tap into a previously unknown source of angular momentum, which has been "trickle charging" the spin of the neutron star even as the acceleration of charged particles in its beams has been draining it at a rate which has extended the pulsar's active life approximately *ten times* the observed "rate" of deceleration. Such an "unknown source" of energy is precisely predicted by the hyperdimensional model, which says that the more angular momentum an object initially possesses, the more it can "tap" into this invisible source of energy to maintain that momentum against known 3D transfer mechanisms. The actual mechanism for maintaining the pulsar's spin is probably the conversion of the star's precessional energy (which, in DePalma's experiments, is *not* contingent on a nearby gravitating companion) into rotational energy. An apt analogy would be a bathtub with a hole in it. Water is flowing out of the hole at an observed rate—but, unknown to the "observers," there is a hidden plumbing network, refilling the tub at a rate which almost, but not quite, replaces the water lost through the hole. The result is a significantly extended "lifetime" for the bathtub reservoir, but with no obvious sources of the refilling. Result: the water in the tub drains out a lot slower than it should, even though the rate of water through the hole is well known.

There is, flatly, no other explanation for this "extra" angular momentum in pulsar B1757-24. Unlike anything that may be dreamed up by the conventional theorists in an after-the-fact attempt to patch up their broken theory, the hyperdimensional model not only implicitly—but Hoagland and Torun specifically—predicted *exactly* this sort of finding. This now makes five specific predictions of Hoagland's hyperdimensional physics model, a model based on the supposedly meaningless tetrahedral alignments of the monuments of Mars that have been confirmed by empirical observations.

There is another "test bed pulsar"—PSR B1828-11—which also seems to be on the verge of proving the hyperdimensional model right, this one via a very different set of measurements: Bruce DePalma's still-to-be-laboratory-tested hypothesis regarding "free precession."

PSR B1828-11 is an "isolated" pulsar (meaning, it is not a member of a binary star system), and is also located in the direction of Sagittarius. In late 2000, a set of radio telescope measurements made by three Jodrell Bank astronomers revealed a remarkable property about this rapidly spinning neutron star: it had *three* simultaneous radio pulsar "periods," as opposed to the usual one—a "fundamental period" of about 1000 days, and three "sub-harmonics" of 500, 250 and 167 days each.

The first interpretation of this data by the discoverers was that the pulsar—despite being *totally isolated*—was somehow exhibiting "free space precession"; its radio beams sweeping by us on Earth at increasingly different geometries and times ... in these repeating cycles ... revealing the physical *precession* of the spinning neutron star itself!

Several alternative theoretical explanations for this remarkable behavior were immediately proposed by other astrophysicists:

"The B1828-11 pulsar, even though composed of a superdense sea of free neutrons compressed inside a sphere only about 20 kilometers across, with a surface gravity *a hundred billion times* the surface field of Earth, is not *perfectly* round; it is this slight intrinsic deformation (by less than a *tenth of a millimeter*) which causes PSR B1828-11 to spin slightly off-axis ... to "precess.""

Or:

"The pulsar's dense 'neutron sea,' flowing superconductively under its brittle surface 'crust,' hasn't quite kept pace with the solid surface's decreasing rotation (caused by the pulsar's intense magnetic braking forces); this, in turn, is causing 'procession' to occur—essentially from 'sloshing' of the lagging, internal neutron sea."

From another paper:

"The pulsar is potentially surrounded by a close-in 'accretion disc' of gas and dust, orbiting at a significant angle to the pulsar's equator. This, then, causes a hidden 'forced precessional torque,' through simple gravitational tidal effects from the orbiting material ...."

And:

"PSR B1828-11 is potentially orbited by a strange matter 'quark planet' (denser than even a collapsed neutron star), which is causing the precession by its significant tidal interactions," etc., etc., etc.

Each of these theoretical attempts to explain PSR B1828-11's bizarre behavior has serious objections—beginning with the "neutron sloshing model"; according to other astrophysical theoreticians, any internal fluid motions ("slosh" in the neutron sea) should "damp out" (dissipate energy) after only a few hundred rotations of the pulsar. Since PSR B1828-11 is rotating *two and a half times per second*, and the pulsar has an estimated age of a hundred thousand years, those astrophysicists have a real problem explaining how such lagging fluid behavior could still persist ... after over eight *trillion* neutron star rotations.

Hoagland has a very different (and simpler) explanation for this baffling behavior—drawn directly from hyperdimensional theory and DePalma's empirical experience with "rotating systems":

That PSR B1828-11 could simply be the most unambiguous Galactic evidence, so far, of actual HD precession.

In other words, a stunning astronomical example of exactly the type of laboratory "HD test" Hoagland was trying to arrange with NASA for DePalma ... when DePalma died.

When asked the most fundamental rules of Hyperdimensional Physics, Hoagland will often wryly respond: "rotation ... rotation ... *rotation*."

Courtesy of long-time associate, David Wilcock, a few years ago Hoagland was sent several papers on an almost unknown field of Russian science. As he began reading the translations—from decades of past and current researchers, and their intensely controversial experiments in the former Soviet Union—he suddenly realized that *here was a completely separate database*, with literally thousands of published scientific papers, all *totally consistent* with DePalma's equally baffling 1970s observations of his "OD field" around rotating masses.[46]

As one of these Russian reviews (written by Yu.V.Nachalov and A.N.Sokolov) noted:

> *"...Over the course of the XX century, various investigations in different countries, representing a variety of professional interests, repeatedly reported the discovery of unusual phenomena that could not be explained in the framework of existing theories. Since these authors could not understand the physics of the observed phenomena, they were forced to give their own names to the fields, emanations and energies responsible for the creation of these phenomena. For instance, N.A.Kozyrev's 'time emanation,' W.Reich's*

'O-emanation' or 'orgone,' M.R.Blondlot's 'N-emanation,' I.M.Shakhparonov's 'Mon-emanation, A.G.Gurvich's 'mitogenetic emanation,' A.L.Chizhevsky's 'Z-emanation,' A.I.Veinik's 'chronal field,' 'M-field, A.A.Deev's 'D-field,' Yu.V.Tszyan Kanchzhen's 'biofield,' H.Moriyama's 'X-agent,' V.V.Lensky's 'multipolar energy,' 'radiesthesietic emanation,' 'shape power,' 'empty waves,' 'pseudomagnetism,' H.A.Nieper's 'gravity field energy,' T.T.Brown's 'electrogravitation,' 'fifth force,' 'antigravitation,' 'free energy.' This list can be easily continued ...."

The Russians, in this review, realized that all these apparently disparate anomalous phenomena were actually only different manifestations of *the same phenomena*—ultimately termed "Torsion Field Physics."

Torsion, as noted earlier, is still essentially unknown to Western science ... and not by accident; until the collapse of the Soviet Union in 1991 and the sudden flood of torsion scientific literature appearing on the World Wide Web, "torsion" was literally a forbidden subject to export to the West; now, more than 20,000 research papers on torsion physics have been published in the open scientific literature—over half of them by Russian (and former Soviet Bloc) scientists.

Here's what one Western engineer, Paul Murad,[47] currently employed by an American institution that is researching space propulsion applications of torsion field theory has to say about the current "state-of the art":

"... The only field that [can] support faster than light phenomenon according to some Russian physicists [is] the spin or torsion field. Torsion is different from these other three fields [electrostatics, magnetics and gravitics] that [have] spherical symmetry. Torsion [can] be right-handed or left-handed and is based upon a cylindrical field and can be created by large accumulations of electricity and rotation of a body that if, above a certain speed, [will] enhance the torsion field. Torsion can lead to other phenomenon to include frame dragging. Here in a vacuum, frame dragging occurs when a rod is inserted concentrically inside of a cylinder and has no physical contact with that body. If the rod is suddenly removed, the cylinder will also move or is dragged along with the rod. Other examples exist regarding rotational bodies that would also influence adjacent rotating bodies due to the interaction of one spin field ...with another...

"Obviously one would like to find a theory that relates all of these effects

with the result of better understanding gravity. The closest thing I could find [in reading the existing Russian literature] is a comment made by Matveeko that the torsion field is identical to the transverse spin polarization of the physical vacuum and a gravitational field is identical to the longitudinal spin polarization of the physical vacuum. Thus, these two fields, gravity and torsion, appear to be related and may be the key [relationship we must understand] before we learn how to harvest [limitless] energy from the physical vacuum or the zero point field. These issues are all interesting theories and definitely should be further explored if mankind wishes to get serious about space travel to planets and the far horizons."

The theoretical "father" of torsion physics is generally considered to be French mathematician, Dr Elie-Joseph Cartan, who in 1913 published a refinement of Einstein's General Theory of Relativity, whereby curved "spacetime" could flow in spiral patterns around *rotating* objects, a phenomenon not originally dealt with in Relativity, termed "torsion."

Ultimately known as "Einstein-Cartan Torsion (ECT)," the initial physical predictions were extremely limiting and disappointing; the resulting forces from ECT were calculated to be some "27 orders of magnitude (27 powers of ten!) *smaller* than the gravitational effects of Relativity." Further, these effects were calculated to be restricted to *static* (non-moving) field geometries around rotating objects, fields that could *not* propagate through space as "waves."

Because of these severe limitations, ECT is thought by most physicists (those who even know about Cartan's contributions to Relativity theory) to be, at best, a minor curiosity—and to play an almost infinitesimal role in the universe at large, even at the sub-atomic level.

However, later Russian theoreticians (like Dr. Gannady Shipov),[48] applying separate torsion ideas originally propounded by the 17th-century philosopher, Rene Descartes—that, ultimately *all motion* (even apparently linear motion) is "rotation" (in a "curved" universe)—would go on to prove that torsion fields are ultimately *not* static (as Cartan had calculated from his erroneous assumptions regarding what constitutes "rotation"), but actually *dynamic.*

Dynamic torsion (also called "Ricci torsion"—after the 19th-century Italian mathematician who refined Decartes ideas and combined them with the pre-Relativity spatial geometry of Bernhard Riemann) is generated by any *moving and simultaneously rotating* object [from spinning atoms, to entire planets (especially those precessing); from orbiting stars, to entire galaxies stars ...]. The calculated strength of dynamic torsion fields is something like "*21 to 22 orders of magnitude stronger*" than Cartan's "static fields." Not only that, but these fields can *travel* ... as a "torsion wave" through spacetime—capable (in the equations of some Russian theorists cited by Murad—above) *of exceeding the vacuum speed of light by at least a billion times.* (That's the lower bound, the actual speed could be a *lot* higher; the theoretical maximum velocity at which a dynamic torsion wave can travel is actually still quite unknown [Fig. 2-11]....)

For those having trouble visualizing how "torsion" works, how it compares to more familiar forms of information and energy transmission—such as electromagnetic radiation—perhaps a couple of analogies will help; if spacetime (Maxwell's "aether") is pictured as "a 2D porous, geometric structure"—like a very thin sponge, or maybe a paper towel—then electromagnetic energy can be pictured as water seeping *through* the sponge or towel at a finite rate of speed (as a substitute, in our analogy, for "C"—the "velocity of light in a vacuum").

Now, in this thought experiment, allow a drop of water to fall on the towel/sponge, entering its 2D surface (and bringing in additional energy ...) from "a higher dimension."

Two things will simultaneously occur:

On impact, the droplet will create ripples in the water in the towel or sponge (remember, our fluid analog to electromagnetic radiation), similar to raindrops on a pond [Fig. 2-10]; simultaneously, the impact will also create *invisible* sound waves within the *material structure* of the towel/sponge (our analogy with the geometric *structure* of our 3D aether).

Since the speed of sound in this material structure is much faster than the speed of pressure waves (ripples) in the water ... the information regarding the input of new energy into the sponge/towel structure, arriving from a "higher dimension," will be communicated almost instantly *throughout* that structure via the *sound* waves its appearance triggers ... while the tiny water ripples, set up by the same impact, will take much longer to physically flow to every portion of the towel/sponge ....

In our analogy, this difference in relative speeds represents the vast gap in velocity between electromagnetic radiation—limited to "the speed of light" in our 3D reality—and dynamic torsion, which (according to Kozyrev's own astronomical measurements) can travel—as a *spiraling* wave—through the aether incomparably faster ....

The reality of "torsion physics"—information communicated through the aether from a higher dimension, analogous to invisible and much faster sound waves, compared to "physical ripples on a pond"[Fig. 2-10]—changes *everything*.

Suddenly, the bizarre "OD field effects" observed by Depalma surrounding his rotating gyroscopes, able to somehow influence the spin of other rotating objects even in distant rooms, and the equally mysterious "non-Newtonian pendulum anomalies," discovered by Nobel Laureate Dr. Maurice Allais during a total solar eclipse over Paris, in 1954, all are suddenly revealed to stem from an *identical* foundation—

Modifications to Einstein's fundamental "Theory of General Relativity."

If Einstein and Cartan are the "god fathers" of current torsion theory, the late Russian astronomer—Dr Nikolai A. Kozyrev—is definitely the "building architect" of this new science.

Kozyrev, a Soviet astrophysicist, became world-famous in 1958 for his controversial spectroscopic detection of the first apparent gas emissions from the Moon (thus indicating it was, at some level, still active geologically).

In parallel to his major astronomical career, Kozyrev also quietly conducted 33 years of empirical laboratory investigations of "rotation on rotation" behind the Iron Curtain.[49] This work was completely independent of DePalma's eerily similar, equally painstaking efforts in the West.

Revealing the astronomical entry point for his "new physics," Kozyrev wrote in 1963:

> "... It is of interest that even such a concrete question—namely, why do the Sun and the stars shine, i.e., why are they out of thermal equilibrium with the ambient space—cannot be answered within the known physical laws ...."

In the end, all these scientists—DePalma , Kozyrev and Hoagland, separated by half a world and two completely different ideologies—independently confirmed the same inexplicable phenomena surrounding "rotation," and the simultaneous appearance of anomalous energy into all rotating objects as a result ... an energy somehow, in Kozyrev's prophetic words, coming from beyond "the known physical laws."

The difference is Kozyrev's 33 years of countless laboratory demonstrations of this physics (and consistently anomalous results) eventually inspired a new generation of Russian mathematical physicists—like Shipov, decades later—to search out the *theoretical* foundations of these multiple "torsion phenomena."

It is safe to say that, without the major work of Nikolai Kozyrev, the currently exploding field of "torsion physics"—based on decades of his *repeatable* experiments—simply would not have occurred.

And, without Hoagland's serendipitous discovery of Kozyrev's work in 2005, "hyperdimensional physics" would still be lacking the sweeping *experimental and mathematical foundation* of "torsion physics" it has now suddenly discovered is its direct heritage.

For, surprise, surprise—

The energy and information existing in higher physical dimensions, accessible in three dimensions *only* through the physical "rotation" of mass, is the ultimate *source* of all "torsional phenomena" that Kozyrev observed ....

In 1993, the Angstrom Foundation, in Stockholm, Sweden, awarded the "International Angstrom Medal for Excellence in Science" to Hoagland for his role in rediscovering the hyperdimensional physics upon which Maxwell's original treatises were built.

You might think, based on all the information just outlined, that the theory of "hyperdimensional physics"—replete with correlated observations, testable predictions and significant experimental successes—should be making quite a bit of noise in the world of advanced theoretical physics. Regardless of the reductionist arguments, Hoagland and Torun formed a coherent, productive and eminently *testable* model of the reality of the Cydonia artifacts. This model contains no less than eight specific, testable predictions, five of which have already been confirmed or have been supported by initial observations. By any reasonable standard, that would be more than sufficient (one would think) for conventional science to at least consider taking the ideas and their source (Cydonia) seriously.

Instead, with the exception of the Angstrom Foundation, the political reaction has been stone silence.

Hoagland, who at one time was warmly embraced by various NASA facilities and programs, suddenly found himself on the outs with those same institutions when he pressed the issue of the tetrahedral mathematics of Cydonia. His ideas were welcomed, it seemed, as long as there was no real means of proving his hypothesis. It was only when he ventured into the realm of hyperdimensional physics and sought for it the same status as any other testable theory that NASA suddenly decided it would no longer lend an ear to his ideas.

It was at this point—as we entered the 1990s—that we began to suspect that there was something seriously wrong with this picture.

Chapter Two Images

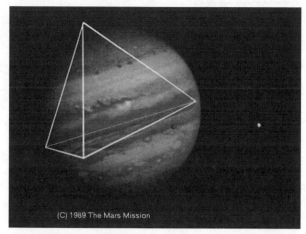

(C) 1989 The Mars Mission

Fig. 2-1 – Jupiter's "Great Red Spot" at 19.5° S – positioned precisely according to Hoagland and Torun's proposed "Hyperdimensional Physics" Model.

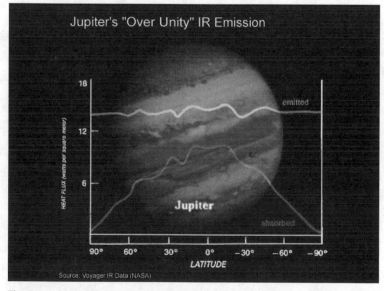

Fig. 2-2 – Jupiter's anomalous "over unity" energy emission. Bottom curve references energy Jupiter absorbs from sunlight; top curve, Jupiter's internal energy emission. The observation that Jupiter is emitting more energy into space than it receives from the sun, is one of the major, continuing planetary mysteries of the solar system: "Where is the excess energy coming from?" (NASA data; graphic, Enterprise Mission).

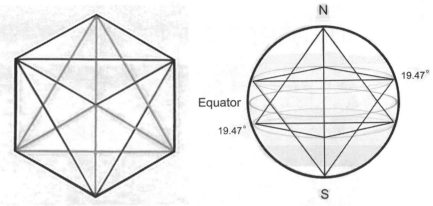

Fig. 2-3 - The figure of "seven and twenty lines" as defined by Maxwell is a 2D representation of a 3D double tetrahedron, encompassed by a hypercube.

Fig. 2-4 - Maxwell's "Seal of Solomon in three dimensions"— a double tetrahedron circumscribed by a sphere.

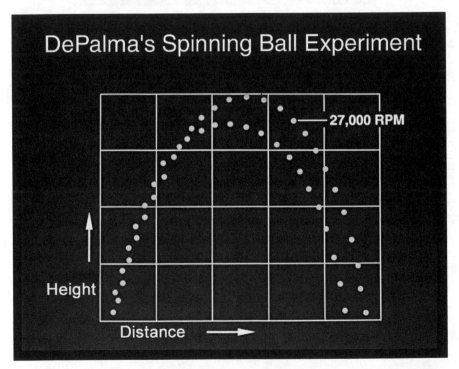

Fig. 2-5 – Dr. Bruce De Palma's "Spinning Ball Experiment." Metal ball spun up to 27,000 RPM, rose higher and faster and fell faster and further, than non-rotating ball. This result violates basic "Newtonian laws of motion."

Distribution Standard Model
(Relative Percentages of Total Angular Momentum)

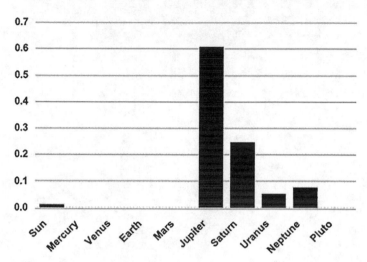

Fig. 2-6 – Total Solar System angular momentum. Even though the sun holds 99% of the mass of the solar system, it possesses less than 1% of its total "angular momentum" (orbital and rotational energy); the rest (the other 99%) resides in the planets – primarily in Jupiter.

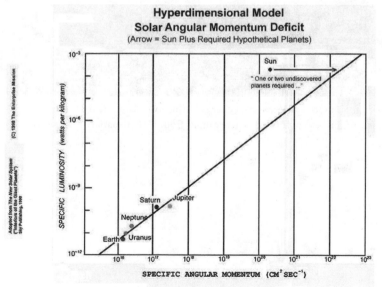

Fig. 2-7 - Solar System Luminosity vs. Angular Momentum.

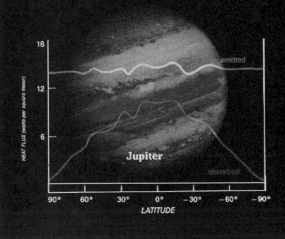

" The [Jovian] energy balance derived in this paper, ~1.67, is at the **low end of the range of previously accepted values,** which range from 1.6 to approximately 2.5. Our [1980 Voyager] measurement of the infrared emission ... is in agreement with the [1974] Pioneer results and the [1978 ground-based] measurement of Ericson et al. **but lower than all other previous [1960's] ground-based and airborne measurements.**"

"Albedo, Internal Heat, and Energy Balance of Jupiter: Preliminary Results of the Voyager Infrared Investigation"

R. A. Hanel et al. [1980]
Journal of Geophysical
Research

Fig. 2-8 – Jupiter's measured, variable excess energy output – explainable (in the Hoagland/Torun HD Model) as the direct result of a modulated energy input from higher spatial dimensions.

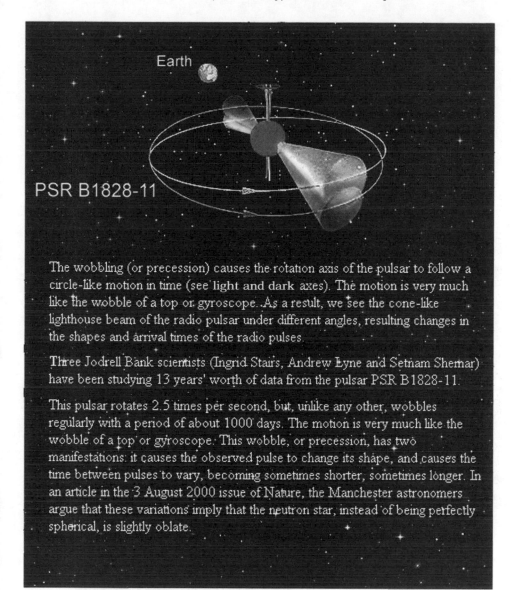

The wobbling (or precession) causes the rotation axis of the pulsar to follow a circle-like motion in time (see light and dark axes). The motion is very much like the wobble of a top or gyroscope. As a result, we see the cone-like lighthouse beam of the radio pulsar under different angles, resulting changes in the shapes and arrival times of the radio pulses.

Three Jodrell Bank scientists (Ingrid Stairs, Andrew Lyne and Setnam Shemar) have been studying 13 years' worth of data from the pulsar PSR B1828-11.

This pulsar rotates 2.5 times per second, but, unlike any other, wobbles regularly with a period of about 1000 days. The motion is very much like the wobble of a top or gyroscope. This wobble, or precession, has two manifestations: it causes the observed pulse to change its shape, and causes the time between pulses to vary, becoming sometimes shorter, sometimes longer. In an article in the 3 August 2000 issue of Nature, the Manchester astronomers argue that these variations imply that the neutron star, instead of being perfectly spherical, is slightly oblate.

Fig. 2-9 – Diagram of precessing pulsar PSR B1828-11 "beaming" geometry, relative to Earth radio telescopes.

(C) 2003 Joanna Phillips

Fig. 2-10 – Raindrops falling through a 3D "higher dimension," transfer energy in expanding ripples when they "appear" on the 2-dimensional surface of a pond. A process similar to "4D, higher dimensional energies" creating "torsion ripples" in a 3-D "aether" – when they "appear" (via rotation of atoms, planets, stars, etc.) in our 3-dimensional reality.

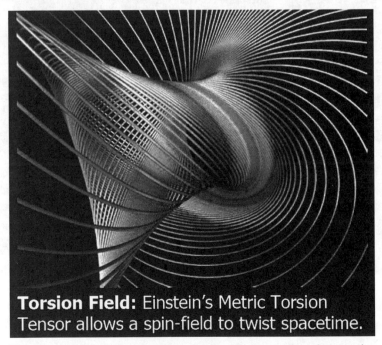

Torsion Field: Einstein's Metric Torsion Tensor allows a spin-field to twist spacetime.

Fig. 2-11 – Modified Einstein-Cartan Torsion Theory predicts that rotating mass uniquely distorts space time ("the aether") – sending measurable torsion waves of energy spiraling outward from the center of rotation; the Hoagland-Torun Hyperdimensional modification proposes that the ultimate source of this "spin-field energy" is a rotating hyperspatial "gate" – allowing the energy to enter 3-space from higher dimensions.

Chapter Three

Political Developments

"The question is not whether you are right or wrong, sir. You are not even in the conversation."—Dr. Carl Sagan to Dr. John Brandenburg, regarding Brandenburg's work on Cydonia.

Throughout the 1980s, though many NASA officials and prominent scientists expressed interest and curiosity about the results of Hoagland's independent research behind the scenes, the Agency's quarter-century-long position on Cydonia—as well as the position of its planetary scientists—was uniformly caustic. While the official Agency position on the reality of Cydonia was far from encouraging, various NASA labs and facilities were often far more open and accommodating, at times even complimentary.

Yet as interest in the Cydonia issue began to reach unprecedented levels, NASA began to march out troops to try and quiet the unrest. One of these was Dr. Steven Squyres, Carl Sagan's protégé at Cornell. In 1988, Hoagland and Squyres faced off in a nationally televised debate on the extraterrestrial artifacts issue on CBS's *Nightwatch* news program, hosted by Charlie Rose [Fig 3-1].

During the debate, Squyres made various erroneous claims, including the tired "disconfirming" photos charge, the notorious assertion that the measurements done by Hoagland and Torun had not been made on ortho-rectified versions of the data (they had) and that Cydonia did not meet established NASA standards for possible artificiality (no such standards have ever been published). Hoagland, by now well-versed in the standard NASA diversionary tactics and misrepresentations, out-argued Squyres on each and every issue, until Rose finally pinned Squyres on one crucial point—he had never actually *looked* at a single Viking image from Cydonia!

That effectively ended the televised debate—but it did not quell the resistance from NASA to honestly examine the issues raised by the independent investigation. In fact, the results of this very public debate only seemed to harden that resistance in some quarters of the Agency.

As recounted by Hoagland in *Monuments,* he was invited on no less than five occasions by various NASA facilities to make presentations to the Agency employees on the subject of Cydonia. One of these appearances, at Cleveland's NASA/Lewis (now NASA/Glenn) facility was videotaped and eventually released as *Hoagland's Mars, Vol. 1.*

Critics, like *Skeptical Inquirer's* Gary Posner (in a vicious personal attack published in that magazine in 2001) have, in the intervening years, downplayed the significance of Hoagland's appearances, claiming various individuals involved with them have now recanted their interest in Hoagland's work. Foremost among them is Dr. John Kleinberg, who now claims (at least according to Posner) that Hoagland's appearances were nothing out of the ordinary.

The reality is that Hoagland's initial NASA/Lewis presentation, on March 20, 1990, was quite significant.

Not only was the Center Auditorium filled with NASA engineers and scientists (even to the point of overflowing to the aisles), but special viewing rooms were set up around the complex to allow other NASA/Lewis personnel still on the job to view the presentation via closed-circuit NASA television. There was even an official "charge number" for Lewis employees who came to the Auditorium to use so they could be compensated for the time they spent watching Hoagland's Cydonia presentation live.

Three video cameras (and cameramen) were in place in the main Lewis Auditorium to both broadcast the event live to all the other Center buildings as well as to officially record it for NASA's archive. Joyce Bergstrom, of NASA-Lewis's Public Affairs Office, had promised to provide subsequent broadcast quality copies of the presentation to ABC News, among others, due to requests from the media. The night before the presentation, Bergstrom also set up a special NASA television interview for Hoagland—by Dr. Lynn Bondurant, director of NASA/Lewis' Educational Programs Office.

Not only did Bondurant personally conduct the interview, he also arranged to professionally record it for a later PBS broadcast. He requested that Hoagland come in after hours the night before his scheduled presentation, and proceeded to set him up in a teleconference room—with a huge backdrop of the official NASA/Lewis logo framed behind him, so that during the entire interview it would appear in virtually every shot.

Now, if Hoagland was "just another normal guest," with no more status than any other outside party that might get invited to speak at Lewis, why would he get such red carpet treatment (and we haven't even gotten around to mentioning the limousine service from the airport, the executive lunch with all the senior staff, and the full tour of NASA-Lewis before his Presentation in the afternoon ...)? Did *all* of NASA/Lewis's guest speakers get brought in the night before, to be interviewed for a PBS special with the *official* NASA logo prominently featured over their shoulders? And, if the presence of the NASA seal behind Hoagland during that extensive interview on his Cydonia research was *not* meant as a tacitly implied *endorsement*, why not conduct the interview in the visitor parking lot, or some other equally "unidentifiable" location?

Yet not only did Bondurant conduct the interview himself, from the actual interview tape it is obvious that he had read *Monuments* cover to cover. The Director of the NASA/Lewis Education Office spent over two and a half hours asking a series of serious, sober and highly detailed questions, based on an obviously extensive knowledge of the work of not only Hoagland, but of the other Mars anomalies investigators as well. He knew the details—some of them quite obscure—of almost a decade of research on Cydonia carried out by DiPietro, Molenaar, Carlotto and Torun. This hardly seems the behavior of someone just being a genial host and having no real interest in Hoagland's ideas or published works.

A few months after his appearance at NASA/Lewis, Hoagland was invited again to the facility by the same Dr. Bondurant who had so thoroughly interviewed him back in March. The intent this time was to hold a full briefing and educational workshop for representatives from various high schools and universities from around the country—and even NASA Headquarters itself—on *The Monuments of Mars*. In Posner's article, he again tried to downplay the significance of this invitation, claiming that it was no big deal and was "only" attended by fifty people.

In reality, it was certainly a major event, as all the attendees were leaders in their fields, and the Workshop came complete with pre-printed workbooks and background references (prepared by NASA/Lewis). Since this was a special session for educators, rather than a general presentation for the whole facility, it was held in a room with a capacity of about fifty because that's how many educators from around the country were invited.

Posner actually didn't argue with any of this. He simply used a statement by NASA/Lewis' Director of Internal Affairs, Americo F. Forestieri (who wasn't even employed there when these events took place) to imply that Hoagland is "stretching it a bit" by claiming that his second appearance at NASA was a "major national NASA education conference" at "a packed auditorium full of teachers and scientists and engineers and educators." He apparently bases this solely on the fact that "only" fifty educators attended the conference. What's the implication? That a conference cannot be "major" unless more than fifty people attend it? And if those fifty people are top-flight educators, including from NASA Headquarters itself, then is it too much to assume that this is a fairly major event? Is Hoagland wrong or self-serving to have described it that way?

Yet if we use Posner's standard, which is apparently that an event sponsored by a major NASA division is not "major" unless it is attended by more than fifty people, then isn't Hoagland's *previous* NASA/Lewis appearance, viewed by over a thousand NASA scientists and engineers in the NASA/Lewis Main Auditorium live and shown to literally thousands more via closed circuit television, to be considered "major?"

It can be argued that it was Forestieri, not Posner, who made the claim that Hoagland was "stretching it a bit." Yet if Posner didn't agree with Forestieri's standard, why did he use it in his article? He clearly wanted to give the impression that Hoagland (at the least) exaggerates the importance of his frequent invitations to NASA/Lewis. The evidence would seem to argue to the contrary, that it is Posner that is "stretching it a bit" to try to make something disingenuous (on Hoagland's part) out of these events.

And, not only did Bondurant's NASA-Lewis Office of Education officially put on this Conference, but he also used it to announce to the assembled scientists, engineers and educators that this session and Hoagland's previously taped appearances (from March) were going to be part of an upcoming PBS miniseries to be called "Hoagland's Mars."

Bondurant had evidently been planning (obviously at the behest of his boss, Dr. Klineberg), since that initial "night before" interview in front of the NASA/Lewis logo, to create this program. Hoagland, along with everyone at the Conference, was surprised at this announcement, since they hadn't been in the loop on the plans at all.

Subsequent to Bondurant's announcement, the process of creating the series went forward with no input from Hoagland (other than providing some of his own Cydonia images and graphics); it was 100 percent a NASA/Lewis production and was being prepared for broadcast on January 6, 1991. Then, less than three weeks before the scheduled air date, on December 13, 1990, Bondurant called Hoagland with bad news. Sounding (according to Hoagland) "like death warmed over," he somberly informed Hoagland that the plug had been had pulled on the planned "Hoagland's Mars" series, and he was to report to NASA Headquarters in Washington, D.C. immediately with all the tapes, scripts and graphics for the programs. When Hoagland asked what had happened, Bondurant told him that JPL had somehow "got wind of the series," and had absolutely "raised hell" back at Headquarters about it. Later, Hoagland confirmed as much from another long-term source within NASA Headquarters itself.

So what had happened?

The problems evidently began with something known as the original "Enterprise Mission." In early 1990, Hoagland had begun an educational project of his own in Washington, D.C., at Dunbar Senior High. Drawing unabashedly upon the *Star Trek* motif of his friend Gene Roddenberry, the *U.S.S. Dunbar* was designed by Hoagland and colleagues to stimulate interest in science among the students at this 99% black inner-city school by focusing their research on various real space science issues and arguments within NASA—such as the Hubble Telescope; the Magellan Venus Mission; and Mars. The prototype Dunbar experiment for a National program (located just off Capitol Hill) was to end by tackling the thorny issues swirling around the Face and Cydonia itself.

This educational project, with the able contributions of both national corporations and local community volunteers (including Keith Morgan, of ABC News, and his entire family), eventually received a nomination for a "Point of Light" award from President Bush's (41) own Point of Light Foundation. With the program catching the attention of the White House itself, and after several months of negotiating, the "U.S.S. Dunbar" got a chance to welcome its most important visitor in October, 1990—Barbara Bush, First Lady of United States.

Hoagland promptly sent a tape of her appearance (shot by the students themselves) to Bondurant, and suggested it be included at the end of his production of "Hoagland's Mars," because of the specific references to the Washington, D.C. "Enterprise" experiment that Bondurant had included in the "night before" interview months before.

However, that was when it literally "hit the fan"—It seems that the notion of the First Lady of the United States, the wife of the President of the United States, tacitly endorsing the notion of artificial ruins at Cydonia by her sheer presence at a Hoagland project, was just a bit too much for the folks at JPL.

Perhaps this is also why Klineberg's formal introduction of Hoagland, back on March 20, was somehow mysteriously excluded from the official NASA/ Lewis versions of the presentation video tapes (including those that went—very late—to ABC News). The cause, according to the Lewis technical department, was due to "simultaneous failure of all three cameras." Miraculously, they all came back online just in time for Hoagland to begin speaking.

In the end, Bondurant's proposed NASA-Lewis Mars series was reduced to a single half-hour program of "talking heads," featuring a "balanced response" from such unbiased figures as Michael Carr and his JPL cohorts (the same ones who, according to two "inside" NASA sources, had killed the much more extensive "Hoagland's Mars" series). It had nothing to do with a "lack of technical quality," as the NASA Headquarters' Public Affairs Office would later claim.

The point of all this is that the actual events, as described in *Monuments,* and when viewed with any sort of objectivity, clearly support Hoagland's version of the events, rather than Posner's disingenuous characterizations. Hoagland certainly did not exaggerate the importance of his appearances at NASA/Lewis, and indeed it seems he was on track to a significant, official NASA endorsement of his work until "JPL happened."

It was shortly after this series of events; Hoagland's appearances at NASA facilities, the publication of "The Message of Cydonia," the curious endorsement of Hoagland's work by various entities connected with Bush Sr. that things began to turn ugly.

NASA and its various sub-agencies and facilities began to respond to increasing public and Congressional inquires on the Cydonia question with vitriolic rhetoric and even outright falsehoods. When Hoagland and Erol Torun began making inroads in the United States Congress to make Cydonia an imaging priority for the upcoming Mars Observer program, the response began to get harried. NASA seemed intent on avoiding the simple testing of the Cydonia hypothesis at all costs. Various documents were issued, seemingly using Carl Sagan's *Parade* hit piece as a model, in response to letters and requests from (among others), Representative Robert Roe, Chairman of the House Committee on Science, Space and Technology. Besides the simple continuing misstatements concerning the existence of the supposed "disconfirming" photographs of the Face, several more extensive, formal responses were issued.

Foremost among them was an anonymous document from inside NASA titled "Technical Review of the Monuments of Mars." This paper began to make the rounds of various NASA facilities and public affairs offices and was widely circulated by the Agency to individuals and government officials as a justification for not making re-imaging of Cydonia a priority. Dr. Stanley McDaniel, Epistemologist and Professor Emeritus at Sonoma State University, had this to say about the memorandum in his book *The McDaniel Report*:

"This memorandum cannot be taken seriously as a responsible scientific evaluation. It refers only to a limited selection of claims in a single work on the subject (a popular book not intended as a strict scientific report). The claims that are dealt with are taken in isolation, generally misrepresented, and the evaluations are cursory and significantly flawed. Although the paper is characterized as a technical review, it does not deserve the title by any reasonable standard. The use of it in an official communication sent out by NASA in response to an inquiry by a United States Congressman raises a very serious concern about the integrity of NASA's treatment of the subject."

Eventually, McDaniel learned that Paul Lowman, a Goddard Spaceflight Center geologist, had authored the memo. As to why Lowman didn't have the guts to openly acknowledge his authorship, one can only speculate.

The cancellation of the "Hoagland's Mars" PBS series was the beginning of a new and much more acrimonious relationship between Hoagland, NASA and the independent researchers. Presumably, this had something to with a major new development on the horizon: NASA's follow-up to Viking, the Mars Observer Program.

Mars Observer

Mars Observer was announced in the late 1980s as the next generation follow-up to Viking. The Mission would be the first new unmanned reconnaissance of Mars in over 20 years, with a host of vastly improved scientific instruments. However, initial specs for the spacecraft were highly disappointing to anyone seeking a resolution to the Cydonia issue, since the Mission was not originally designed to include a camera. Eventually, the Mission Planners came to their senses, and it was decided fairly late in the game to include a one-meter-per-pixel resolution gray-scale camera. That, however, was where the problems actually began.

The man who would build, point and control the camera was a former JPL employee named Dr. Michael Malin. Malin, among other interesting affiliations, had once been part of a project to analyze the purported UFO photographs of infamous "contactee" Billy Meier. In that capacity, Malin, then an associate professor at Arizona State University, had concluded that Meier's controversial photographs were not fakes.

"I find the photographs themselves credible, they're good photographs," he commented at the time.[50]

The Meier photo investigation had been organized by well-known UFO investigator Wendelle Stevens (Lt. Col., USAF, ret.) and also focused on similar prime cases. From 1978-83, the person orchestrating photo testing for Stevens was Jim Dilettoso, a long time UFO buff himself and also the Director of Special Projects for APRO (the Aerial Phenomena Research Organization), the leading UFO research group of that era.

Dilettoso was something of a Renaissance man, dividing his time between developing high-end computer processing equipment and designing tours for the rock group the Moody Blues as a day job. There were two primary missions for the photo testing activities: first, develop a methodology for analysis of the photos (size, distance, hoax, mistakes, et. al.), and second, test photos using this process in association with recognized experts.

Dilettoso went to dozens of manufacturers of image processing equipment, including government contractors like EG&G and TRW, government labs like the United States Geological Survey and JPL, and certain universities that were known for their image processing and analysis capabilities; USC and Arizona State University among them. When he found someone of merit and interest that he felt might be able to contribute to the rather atypical projects he was directing for Stevens, he followed a very exacting protocol that included security considerations for both sides.

All selected scientists signed non-disclosure agreements (which later came back to haunt Stevens and Dilettoso when cynical "skeptics" contacted some of these people to verify their participation, and the scientists predictably denied they'd taken part). Lee and Brit Elders of Intercep Security, a leading firm of the time, handled the security. The scientists that Dilettoso recommended for the testing were fully scrutinized by Intercep before any contact, briefing or testing could begin.

During the Voyager fly-by of Saturn, Dilettoso was at JPL as a contractor and it was there he met Richard C. Hoagland—who was also at the Lab working as a reporter covering the Voyager mission for American Airlines' *American Way* Magazine. There he also met Dr. Michael Malin, the future "Principal Investigator" on the Mars Observer camera.

Dilettoso had already visited Bob Nathan (who had helped develop JPL's early VICAR imaging processing software)[51] at JPL a few times to have him analyze four "legitimate" Meier photos and two "control pictures," and it was Nathan that steered him to Malin. Malin was working at JPL at the time, but was planning a move to Arizona soon thereafter to take a position as a professor and scientist in the Geology Department. After making initial contact at JPL, Dilettoso made arrangements with Malin to meet up with him at ASU a few months later.

Malin was working on the study of volcanoes, earthquakes and other landmass spectacles and his imaging work included satellite images tiled on 3D topographic maps, and computer simulations of seismic events.

Stevens and Dilettoso went back to Malin's lab at ASU around 1980, with the same four Meier photos and two controls that they had taken to everyone else. Malin digitized them and did some preliminary analysis while they were there, and did further study in the weeks following. He told Dilettoso in a subsequent visit that he had spent quite some time with the photographs. Simply put, Malin's findings were that he found no evidence of superimposition, or "dubbing," as he called it. He gave them no written report, as they had not asked for one. They never went back to retrieve or erase the images, which were now digitized and in Malin's system, since they thought it would be good for him to have them and pass them around to his colleagues, which they assumed that he did.

In 1985, Gary Kinder was writing the book *Light Years*, about the Meier case investigation. Kinder interviewed Malin and included Malin's comments that were in the (skeptical) affirmative. (Also included were Malin's comments about Dilettoso, which were mildly supportive.) Malin reiterated that he had found no evidence of a hoax, and he said so to Kinder, but he was not yet convinced that the objects in the images were extraterrestrial spacecraft.

It is interesting that Malin even admitted to Kinder that he had tested the Meier photos, let alone that he did not make a negative comment about his findings, given his later hostility to the notion of the Cydonia artifacts. If anything, the Meier case was significantly more "far out" than anything Hoagland ever proposed, and unlike the Cydonia investigation, was not inherently falsifiable. Malin later got a MacArthur grant and dropped from sight until he re-appeared on the Mars Observer project as the man behind the camera.

There is an interesting sidebar here: over the years, Dilettoso kept regular contact with Malin's secretary/lab assistant from ASU, who would visit his Village Labs processing facility in Tempe, Arizona every few months in the 1990s. She had actually become a major investigator in the arena of crypto-archeology, the study of possible ancient ruin sites on Earth. She frequently dropped photos of "artifacts" she had spotted in satellite images off to Dilettoso, asking for his comments.

It is easy to conclude that "Barbara" likely had no real scientific training of her own, and was being guided by Malin in these endeavors. Since Malin himself was a geologist and had no experience in engineering or archaeology, using his secretary as a public surrogate would enable him to become expert in the techniques necessary to spot artifacts in the images set to be received by Mars Observer (and later Mars Global Surveyor) without arousing suspicion as to his true intentions regarding Cydonia. In fact, it was a perfect cover.

Malin chose to reserve judgment on the more spectacular aspects of Meier's story, but this early foray into such arcane territory showed that he was at least willing to consider unusual or even bizarre claims like Meier's. But what the entire independent Mars investigation community wanted to know, circa 1992, was just what his position was on Cydonia and the Face.

Malin quickly asserted (Barbara's new hobby notwithstanding) that he had no interest in even testing the Cydonia hypotheses by merely targeting the formations with his new camera. In fact, he stated his outright opposition to making even a minimal effort to re-photograph Cydonia on numerous occasions. Because the camera was a "nadir pointing" device, meaning that it could not swivel or aim at specific targets without the entire spacecraft being repositioned (and hence using valuable fuel), Malin argued that at best he might get "one or two" random opportunities to target a specific object like the Face or D&M during the regular science mission. However, as the specs evolved, Mars Observer soon became a much more capable mission, with additional fuel added to the Mission Plan to enable an extension of the original two-year science acquisition phase of the project.

Hoagland and Dr. Stanley McDaniel began to dig into Malin's contentions, and quickly discovered that Malin's claim of at best "one or two" opportunities to target the Face was greatly understated. After consulting with Mission Planners at JPL and reviewing the technical specs, they found that there would be more on the order of *forty plus* chances to target the Face during the regular two-year science phase. So why would Dr. Malin—if he was honest—underestimate the imaging opportunities by a factor of *twenty*? Hoagland and McDaniel smelled a rat, and they decided to try an end run.

Hoagland and the other researchers then began to lobby NASA and Congress to target the formations, only to make an extremely unpleasant discovery. Neither NASA nor Congress had anything to say about where Michael Malin had pointed his Mars Orbiter camera.

In an unprecedented move, NASA had decided to sell the rights to all of the data collected by the Observer to Malin himself, in an exclusive arrangement that gave Malin godlike powers over when, or even *if*, he decided to release any data the camera collected. This private contractor arrangement not only neatly absolved NASA from any responsibility as to what was photographed with an instrument and mission paid for by the taxpayers of the United States, but it gave Malin the right to embargo data for up to half a year, if he so chose.

This marked the first time in NASA history that data returning from an unmanned space probe would not be seen "live," as it had all throughout the preceding 30 plus years of Mariner, Lunar Orbiter, Surveyor, Apollo, Viking and Voyager Missions. The logic of the arrangement was tenuous, at best. NASA claimed that in order to assure that private contractors would bid on future space projects like Mars Observer, they had to guarantee an "exclusive rights period" to the private contractors/scientists, so that they could write the first scientific papers from the data collected "without unfair competition from other, non-project scientists."

Of course, it was not required in any way to grant Malin the right to withhold some or all of the data completely, which he could, under a clause that gave him the right to delete "artifacts" from any or all of the images. In essence, Malin could release a blacked-out image, and then simply claim the image had been filled with "artifacts." It also meant that for a period of up to six months, he could do literally anything at all to the images, and no one—not even NASA—would be the wiser.

Malin even moved his entire private company, "Malin Space Science Systems" (which held the actual Mars Observer camera contract) away from ASU in Arizona and JPL in California to San Diego (over 300 miles south of JPL in Pasadena). This effectively insulated Malin from the Mars planetary science community. Visitors—other scientists within the community, or even co-investigators with Malin on Mars Observer—were quite unlikely to "drop in" unannounced if everyone had to drive four or five hours from JPL just to get to Malin's offices. And, when they did get there, if they didn't get directions beforehand, they'd be out of luck. For some reason, Malin's company was never publicly listed on the shopping mall marquee where his offices were (and still are) actually located.

Curiously, however, the move did put him *right across the street* from one of the worlds largest "supercomputer" facilities ... where he could literally hand-carry digital imaging tapes back and forth ....

To Hoagland and the other independent researchers, this was an untenable situation. It was anathema to Hoagland that a publicly funded program could be subject to such an obvious sell-out of the public's right to know, and their faith in the integrity of the data. Instead, the total control was in the hands of a man who had expressed outright hostility to the idea of even *testing* the Cydonia hypothesis. So the whole issue was subject to Malin— without oversight of *any* kind—having the scientific integrity *not* to alter or withhold data that might make him look like a fool.

By 1992, with the September launch of Mars Observer approaching, McDaniel entered the fray. Using various political and academic contacts, he put pressure on NASA and JPL from several directions, forcing them to address, *on the record*, just why they were not able to target Cydonia or the Face specifically. NASA responded with various contradictory, if not disingenuous (McDaniel's words) arguments, including those by Dr. Malin. At every turn, McDaniel and Hoagland shot down the arguments, finally getting NASA Headquarters Public Affairs' spokesman, Don Savage, to officially admit (in a Headquarters letter) that the infamous "disconfirming photos" of the Face never existed.[52]

Mars Observer was a troubled mission almost from the very beginning. Besides the various political controversies swirling around the Cydonia question, it had a series of technical mishaps that made even casual observers wonder if the mission was cursed, or if somebody just didn't want it to succeed. Even the Mission's Project Office described Mars Observer's journey to the Red Planet as "traumatic."[53]

In late August 1992, during a routine inspection of the spacecraft on the launch pad, NASA technicians discovered severe contamination, inexplicably *inside* the protective shroud, consisting of "metal filings, paint chips and assorted trash." NASA publicly speculated that the damage was done when the spacecraft had been hastily unplugged from an outside air-conditioner and the payload shroud hermetically sealed, a measure actually designed to protect it from the imminent effects of Hurricane Andrew. But the Agency never actually cited a specific cause for the contamination from its (brief) investigation. With an immovable launch window looming just weeks away, the Orbiter payload was hurriedly removed from the pad and taken back to the payload integration clean room—for disassembly, inspection and possible "aggressive cleaning."

It was there that Program technicians made a second, even more disturbing discovery.

According to Mars Observer Project Manager David Evans, during the inspection process NASA discovered the presence of an unspecified "foreign substance" inside the spacecraft's (Malin's) camera assembly that would have made the resultant images blurred and virtually worthless for resolving the Cydonia issue.[54] According to Evans, since the camera was a sealed assembly, the mysterious contamination could only have been introduced into the camera in disassembly and check-out after it came assembled from Malin's facility, in the *JPL clean room itself.*

How such a basic "mistake" could have been made, given the nearly $1 billion price tag of the Mission, is hard to fathom. Checking the cleanliness of the camera optics is invariably the top priority for a mission that has a visible light camera as its primary scientific instrument. Had this bizarre "Vaseline smearing" of the lens not been discovered at the Cape (serendipitously, because of a hurricane), Mars Observer would have been an embarrassment on the scale of the original Hubble Telescope debacle.[55] Fortunately, NASA engineers at Cape Canaveral (the "honest" ones) were able to clean the spacecraft and get it back to the pad in record time for its September 25 launch.

Meanwhile, NASA management was no longer simply insisting that the terms of Malin's private contract with the Agency gave him the "right" to target or ignore Cydonia at his whim (as well as embargo the images and legally remove "artifacts" from the data); Program Scientist Bevan French was additionally defending the notion that the Face and other objects were "too small" to be effectively targeted by the Malin camera in the first place. This was despite the fact that there was a defined Mission objective to target the sites of the two previous Viking Landers which, as opposed to the mile-wide Face, were each less than 15 feet wide.

They continued to insist, in correspondence and in open debate in various public forums, that Malin had the final say, and that they were powerless to influence him. Beyond that, they defended the practice of exclusivity as the only means of achieving scientific results; despite the fact that no prior mission—manned or unmanned—in the agency's history had utilized this private contractor status. In the past, the taxpayers who paid for the mission had always owned the data.

Now it seemed they were lucky to even see it.

As the launch date arrived, the political pressure was reaching a fevered level; Hoagland was live on CNN, reminding viewers of all this strange history even as the spacecraft lifted off. Fortunately, the actual launch itself seemed to go off without a hitch. Then, something truly bizarre happened: all contact was lost with Mars Observer and its still-attached second stage rocket, for almost 90 minutes.

Just twenty-four minutes into the Mission, with the spacecraft set to fire a second-stage rocket after separating from its first stage Titan booster and set out on its mission to Mars, all radio and telemetry went dead. Aircraft over the Indian Ocean reported seeing a brilliant red-orange flash—possibly the second stage firing, possibly the spacecraft exploding—coinciding with the timing of the critical rocket firing. Given that the spacecraft had gone inexplicably silent, flight controllers assumed the worst. Imagine their relief when a little over an hour later, Mars Observer just as suddenly and inexplicably reappeared, apparently none the worse for wear.

So what exactly had happened during those lost eighty-five minutes?

It's impossible to know for sure, but on two subsequent attempts to retrieve the onboard telemetry, recorded during the "missing time" event, there was absolutely nothing to be heard. Then, on a third attempt –several *days* later— suddenly, a completely normal data stream appeared. There was only one problem: the first two attempts had received a carrier signal and "timing code," indicating that a recording was made, but the tape simply contained *no* data.

How did the data from the missing-time episode suddenly find its way onto a tape that had been *blank* only days before? It was as if someone had erased the actual recording, then subsequently uploaded a manufactured "nominal" data stream a few days later. The Deep Space Network (DSN) engineers were *insistent* that they hadn't simply missed something the first two times around. "There was *no data* on that tape the first two times!" JPL's Deep Space Network manager angrily declared.

The news media, of course, had little knowledge or understanding of just how impossible the whole situation was, and quickly dropped the issue. It did, however, become considerably more relevant eleven months later.

By that time, after a relatively quiet trip to the Red Planet, Mars Observer was nearing its goal and the debate over Cydonia was once again gaining steam. News stories mentioned Cydonia as a matter of course. Buoyed by the imminent publication of Dr. McDaniel's three-year long investigation into the Cydonia controversy, Hoagland and the other independent researchers had been very successful in pressuring the Agency through newly found political and media contacts. Then, just weeks before Mars Observer's scheduled orbital insertion burn and the delivery of McDaniel's report to both Congress and NASA, the Agency suddenly decided to change plans. NASA indicated a willingness to reconsider not only its position on the data embargo and the lack of live televised images from the Orbiter, but also announced that they were considering a radical new science plan.

Because the first few weeks of the planned mapping orbit period would occur during a solar conjunction and just before the beginning of dust storm season on Mars, there was a chance it could be months before any pictures of Mars were returned at all, much less targeted images of Cydonia. NASA's solution was to try a "power in" maneuver that would place the spacecraft in a mapping orbit some twenty-one days early. However, in other documents and letters to Congress, NASA inexplicably added almost as many days to the "check out" and calibration phase upon reaching this science mapping orbit, meaning that no useful images of the planet could be expected until after the conjunction, at the least.

To Hoagland and McDaniel, the sudden lengthening of the unnecessary "calibration" phase was an obvious ruse. If JPL was going to take extra time to "calibrate" the instruments, effectively negating the advantage of the power-in maneuver, why bother "powering in" at all? The answer seemed simple: by powering in, NASA could buy themselves twenty-one priceless days to secretly examine whatever Martian real estate they wished (obviously Cydonia) without any public or media pressure to release the data they were gathering.

Any and all of the images acquired in this time period could be "officially" denied—since the spacecraft was simply being "calibrated" and not really gathering science quality data at all.

Predictably, Hoagland and McDaniel raised a stink, and NASA suddenly found itself under additional pressure from various sources to provide live images of Cydonia. Hoagland upped the ante by scheduling a press conference for the day that Mars Observer was scheduled to achieve orbit around the Red Planet. The briefing would be held at the National Press Club in Washington, D.C., and would be attended by many of the principals involved in the independent investigation, including Dr. Mark Carlotto, Dr. Tom Van Flandern, Dr. David Webb and architect Robert Fiertek.

And then, four days before Mars Observer was scheduled to make its orbital burn and commence operations, McDaniel delivered his final report simultaneously to NASA, Congress, the White House and the media. Mars Observer mission director Bevan French got a personal, hand-delivered copy. The following Sunday, August 22, 1993, French was scheduled to debate Hoagland on national TV, on ABC's *Good Morning America* [Fig. 3-2]

Just as he had with Cornell's Dr. Steven Squyres, Hoagland destroyed French in the open forum. Having been given an astonishing six minutes, more than twice the usual time allotted for such segments, Hoagland used the opportunity to bludgeon French's weak and sometimes contradictory arguments. Forced to defend an indefensible position—that NASA should willfully allow one man to have godlike powers over data paid for by the American taxpayers— French wilted under the pressure. The final insult came at the end, when the exasperated host, Bill Ritter, finally just confronted French point blank. "Dr. French, why don't you just take the pictures, immediately release them and *then prove these guys wrong?*" French, unsurprisingly, had no real answer.

Then, at exactly 11 a.m. Eastern Time, just moments after Hoagland had creamed French on national television, AP science reporter Lee Siegel got a call from a JPL spokesman. The NASA rep informed him that Mars Observer had simply *disappeared*, some fourteen hours earlier!

The timing of this announcement, just moments after the Mars Observer Program Scientist had badly lost a very public nationally televised debate with the leader of a highly critical Agency opponent, seemed a bit too coincidental. Why hadn't French simply admitted that the Mars Observer was lost at the top of the segment? It is inconceivable that he, the Program Scientist, didn't know for over fourteen hours that "his" spacecraft had been lost.

French could have saved himself a lot of heat and needless embarrassment by simply announcing on *Good Morning America* that the Mars Observer was in trouble. This would have neatly changed the subject of the segment, and shifted any discussion of Cydonia and artifacts to the back burner.

In hindsight, it isn't difficult to figure out what actually happened—after other high NASA officials (and their bosses) watched French's lame Cydonia spin control fail miserably—and on *live* television—NASA went to Plan B. They either pulled the plug on the Mission outright—out of fear of what *uncensored* images of Cydonia would reveal—or NASA (remember, an official *"defense agency of the United States ..."*) simply took the entire Mission "black."

The extraordinary scrutiny the agency was under at the time would have made it nearly impossible to conduct a survey of Cydonia in secret. Under this pressure, the most viable solution was to either scrap the program, or come up with a way to conduct the preliminary reconnaissance in secret—not only from general public and the press, but from its own "honest" employees at JPL as well.

As it happened, NASA pulled off exactly that scenario, under rather suspect circumstances. With Mars Observer officially "lost," they could conduct a highly detailed survey that could tell them either how to take future "public" images to ensure minimum political impact, or even how to whitewash the images believably.

A commission was formed to determine what had caused the spacecraft to cease operations. Unfortunately, the investigation was doomed from day one for one simple reason: there was *no engineering telemetry to analyze.*

NASA, in another unprecedented move, had inexplicably ordered Mars Observer to shut off its primary data stream prior to executing a key pre-orbital burn. Resultantly, there was no data at all from the spacecraft's final few nanoseconds of existence (if indeed it had been lost). This is crucial, since even if a chemical fuel explosion had taken place, it would obviously travel much slower than a speed of light radio signal, and the spacecraft's destruction sequence could have been recorded. Such a recording could have been used to reconstruct those final moments in detail and make an educated determination as to exactly what (if anything) had gone wrong. Instead, because NASA had violated the first rule of space travel—you *never* turn off the radio—no cause for the probe's loss was ever satisfactorily determined.

Regardless, Hoagland and the others decided to go on with their press briefing the following Tuesday, as there was still a remote chance that communications could be re-established.

He was also able, on short notice, to put together a placard-waving, public demonstration against NASA's potential censorship of Cydonia. Through the overtime efforts of long-time friend, independent Mars investigation supporter and colleague on the West Coast, David Laverty, they managed to pull together a reasonable gathering right outside NASA's Mars Observer Control Center—three thousand miles away from Washington, at JPL.

The local and national TV shots of "the people"—vocally demonstrating against NASA's planned Cydonia secrecy, and for the *first time* in the Agency's decades-long *history*—dominated CNN (and other network) coverage of the "missing Mars Observer story" throughout the remainder of the day.

Meanwhile, on the East Coast, the Press Club briefing was also being extremely well received (for a room full of skeptical reporters ...), with Hoagland later landing various network follow-ups—including in-studio conversations with Robin McNeil, on PBS' prestigious *McNeil/Lehrer News Hour*, and a couple days later with Larry King, on CNN's "Larry King Live."

Ultimately however, none of this changed anything, as Mars Observer stayed permanently "disappeared."

There is a curious postscript to this mystery.

A few days after returning from Washington and the Press Club briefing, Hoagland discovered several messages on his answering machine. There were four especially intriguing ones—from four separate individuals—each independently claiming to be "JPL employees." Each had a similar story to tell: Mars Observer was "still alive," but had been taken "black" by a cabal that was operating inside JPL. The anonymous voices told Hoagland that he and McDaniel had placed "too much heat on JPL," and they (NASA) could not risk showing the real ground truth at Cydonia "live," without having a chance to preview what it might reveal first. The plan, they said, was to miraculously "find" Mars Observer some months later and bring it back into public operation. There was, however, one condition: if what Hoagland and the independent researchers suspected were down there (i.e. genuine ET *artifacts*) could be reasonably confirmed, "You'll never hear from Mars Observer again," one of them promised.

Hoagland was never able to verify their identities, but each of the men seemed to be unaware of the others and beyond that, each had the technical expertise and inside knowledge of the JPL system and facilities to be who they claimed to be. One piece of information later turned out to be quite interesting: one of the men claimed that since the Deep Space Network was being used to look for Mars Observer, JPL could not risk sending the "missing" spacecraft data back to Earth over the conventional DSN antenna network This source claimed that "they" would use "alternative methods" to get the data back to Earth, without elaborating.

A few months later, *another* anonymous source told Hoagland that the Hubble Space Telescope was being used to "photograph UFOs" using a light gathering device called the "high speed photometer," and that the (then) imminent "Hubble Repair Mission" was going to secretly bring back a load of videotapes of the event. Then another caller, a couple days later, called with an even more extraordinary tale ... that Hubble was to be used in a future "New World Order '*laser* light display in the clouds ... to fake the *Second Coming!*'"

Hoagland had little faith that these reports were true, but it did get him thinking. If his supposed JPL sources were right, then how could Mars Observer send surreptitiously obtained images of Cydonia back to Earth without being detected? If the DSN was too "hot," then could a different data transmission really be used? After a little digging, he realized that the spacecraft had carried a second instrument, a laser altimeter that was the precursor to Mars Global Surveyor's MOLA instrument. This powerful laser could, indeed, be used to send a data stream over a very narrow infrared beam, literally millions of miles back to Earth, where one very special instrument *could* secretly detect and relay the signal to the appropriate "audience" on Earth: the Hubble's high-speed photometer.

Hoagland never got any proof that this was done, but there was one more curious side bar. Months later, when STS-61 was sent up to rendezvous with Hubble and repair the telescope's crippled optics, the crew only brought one piece of equipment back with them when they returned to Houston—Hubble's suddenly, curiously "obsolete," high-speed photometer.

All of this may seem cloak-and-daggerish, but the facts are there to support the idea that something very fishy was going on with the Mars Observer from the beginning. From the inexplicable pre-launch "sabotage" to the mysterious loss of signal for over an hour (when an alternative set of instructions could have been uploaded to the spacecraft unbeknownst to the regular spacecraft launch crew or flight controllers), to the ill-conceived "power-in" deception, followed by the bizarre behavior over the loss of the spacecraft (withheld by the project head until minutes after he had lost a crucial debate with Hoagland), nothing seemed normal about this mission.

And the reality is that the question at hand—are there the remains of an ancient, extraterrestrial civilization now visible on the surface of Mars—is only *the most crucial question in the two-million-year history of the entire human race.*

The idea that NASA, or its Pentagon handlers, might go to the trouble and expense of fiddling with two highly visible missions just to have a surreptitious "first crack" at the ground truth of Cydonia only seems preposterous when taken in isolation. In the context we are about to put it in, it becomes not only plausible, but perhaps even imperative.

In the end, the whole Mars Observer debacle had been enough to convince Hoagland that the "honest but stupid" model of NASA's behavior was simply no longer tenable. He gave up any notion that there was a logical, non-conspiratorial explanation for the Agency's erratic and unethical behavior, and gave himself fully to the concept of an out-and-out cover-up of the whole Cydonia question.

But he paid a price. For simply publicly admitting what any logical person would conclude, given the same evidence, Hoagland was forever ostracized from the independent Mars research community he played such a major role in creating. His decision to go it alone, in the face of opposition from all his previous colleagues, left him with one now-overwhelming question to confront—why had they done it? What was so crucial, so destabilizing about Cydonia, that NASA would take such enormous political risk?

It would take the better part of the next decade to find that answer.

The Brookings Report

"I'm sure you're aware of the extremely grave potential for cultural shock and social disorientation contained in the present situation, if the facts were prematurely and suddenly made public without adequate preparation and conditioning. Anyway, this is the view of the Council... there must be adequate

time for a full study to be made of the situation before any thought can be given to making a public announcement. Oh yes... as some of you know, the Council has requested that formal security oaths be obtained in writing from everyone who has any knowledge of this event..." —Dr. Heywood Floyd, *2001: A Space Odyssey*

In mid-1993, Professor Stanley V. McDaniel was seeking additional documentation for his then-ongoing study into NASA's new imaging and data policy surrounding the Mars Observer program. As we have shown, *The McDaniel Report* played a key role in pressuring NASA to abandon its position that the principal investigator holds all data rights from future space probes.

In the final stages of his study, McDaniel asked Richard C. Hoagland for some assistance in locating difficult-to-find historical NASA documents and research papers relating to its SETI (Search for Extraterrestrial Intelligence) project. Hoagland advised McDaniel of the long-rumored existence of an official NASA report—supposedly commissioned by the space agency in its early years and relating to prospective NASA censorship of SETI evidence if it was ever discovered.

At McDaniel's urging, Hoagland began actively searching for the document, polling various contacts and eventually having a conversation with former police detective Don Ecker. Ecker, a consultant to *UFO Magazine*, called in a couple of favors and not only confirmed the existence of this highly controversial study—but came up with the actual title: "Proposed Studies on the Implications of Peaceful Space Activities for Human Affairs."

Hoagland then called upon another friend, Lee Clinton, who after considerable effort tracked down an actual copy of the several-hundred-page NASA Report in a Federal Archive in Little Rock, Arkansas. Clinton made several copies of the 300-page study, and duly forwarded sets of the complete document to Hoagland, as well as McDaniel, who featured it in his final Report as strongly indicating a long-standing potential NASA policy of "cover-up" on this specific issue.

The Brookings Institution was probably the world's foremost "think-tank" of its day, and the contributors to the NASA study were a veritable "who's who" of the leading academics of the time. MIT's Curtis H. Barker, NASA's own Jack C. Oppenheimer, and famed anthropologist Margaret Mead were all consulted for contributions to the final Report.

After scouring the document, Hoagland and McDaniel found several passages that they felt were particularly relevant—and potentially explosive—to their recent experiences with NASA over Mars Observer. The most stunning remarks came on page 215, where the Report mentions the possibility that artifacts may be found by NASA in their coming explorations:

"While face-to-face meetings with it [extraterrestrial intelligence] will not occur within the next twenty years (unless its technology is more advanced than ours, qualifying it to visit Earth), artifacts left at some point in time by these life forms might possibly be discovered through our space activities on the Moon, Mars or Venus."

Later on the same page, the document considers the implications of such a discovery:

"Anthropological files contain many examples of societies, sure of their place in the universe, which have disintegrated when they had to associate with previously unfamiliar societies espousing different ideas and different life ways: others that survived such an experience usually did so by paying the price of changes in values and attitudes and behavior... the consequences of such a discovery are presently unpredictable..."

It then suggested, obviously, that further studies were needed, and that NASA must consider the following questions:

"How might such information, under what circumstances, be presented or withheld *from the public? ...the fundamentalist (and anti-science) sects are growing apace around the world... For them, the discovery of other life—rather than any other space product—would be electrifying... If super-intelligence is discovered, the [social] results become quite unpredictable... of all groups, scientists and engineers might be the most devastated by the discovery of relatively superior creatures, since these professions are most clearly associated with mastery of nature."* (p. 225) [Emphasis added.]

The Report then references an obscure work by psychologist Hadley Cantrell, titled *The Invasion From Mars: A Study in the Psychology of Panic* (Princeton University Press, 1940). The Rockefeller Foundation under a grant to Princeton University commissioned this little known book. Its subject was the 1938 Orson Welles *War of the Worlds* broadcast (which it is estimated that more than a million people in the northeast United States panicked over). The implication is that the broadcast was a warfare psychology experiment, and that America dramatically failed the test.

It isn't difficult to interpret the Brookings Report. Among its wide-ranging analysis and conclusions are the following:

1. Artifacts are likely to be found by NASA on the Moon and\or Mars.
2. If the artifacts point to the existence of a superior civilization, the social impact is "unpredictable."
3. Various negative social consequences, from "devastation" of the scientists and engineers, to an "electrifying" rise in religious fundamentalism, to the complete "disintegration" of society are distinct possibilities. The *War of the Worlds* broadcast provides an excellent example.
4. Serious consideration should be given to "withholding" such information from the public if, in fact, artifacts are ever discovered.

So here we had the proverbial smoking gun.

Not only was NASA advised—almost from its inception—to withhold any data that supported the reality of Cydonia or any other discovery like it, they were told to do so for the good of human society as a whole. Most especially, they should withhold the data from their own rank and file engineers and scientists, since they were the most vulnerable members of all of human society.

It didn't take a proverbial rocket scientist to conclude that NASA took these recommendations and transformed them into policy at the highest levels. Nor would it be surprising if the whole question of "artifacts" were considered a national security issue—given (again) NASA's founding charter position as "a *defense* agency of the United States [emphasis added]."

Although the document itself is fairly obscure, it has had a major social impact. The Brookings Report was the basis for Arthur C. Clarke and Stanley Kubrick's seminal film *2001: A Space Odyssey.* In fact, according to a 1968 *Playboy* interview, Kubrick could quote from the Report chapter and verse. In the interview, he quoted the exact passages shown above, and declared that the whole question of covering up the discovery of artifacts to be the central theme of his groundbreaking film.

Critics felt that the discovery of the Brookings document didn't really change anything. They argued that the document was too old to be relevant, the passages dealing with extraterrestrial artifacts were too small a part of the overall Study, and that there was no "proof"—despite NASA's well-documented and duplicitous behavior (via McDaniel's meticulously referenced study)—that the recommendations had ever been implemented.

However, the notion that a forty-year-old document is "too old" to still be relevant would come as a great surprise to constitutional lawyers and scholars, who regularly debate and actively argue the merits of our Founding Document—which is now over 230 years old. As to so little of the document actually dealing with the question of artifacts, it is true that the report is a vast, far-sweeping overview of the future of space exploration, but that hardly makes any one part of it irrelevant. The First Amendment to the Constitution is only a small portion (just 45 out of 11,713 words) of the overall document, yet no one would rationally argue against it being the most important section of the entire manuscript.

The Brookings Report itself recommends that the key questions we cite should be "further studied," but as yet no one has uncovered such a formal study, even though one presumably took place. As to the question of the actual impact these recommendations may have had, read on...

John F. Kennedy's "Grand NASA Plan"

"The very word 'secrecy' is repugnant in a free and open society; and we are as a people inherently and historically opposed to secret societies, to secret oaths and to secret proceedings. We decided long ago that the dangers of excessive and unwarranted concealment of pertinent facts far outweighed the dangers which are cited to justify it."

– President John F. Kennedy, April 27, 1961

One of the criticisms we have endured in referencing the Brookings Report is that we can't "prove" that the document was ever implemented, other than to continually point out NASA's conduct which is consistent with the passages we cite. The argument is that there is no other evidence that it had any impact on the *realpolitik* of the day. We will now argue that this is not the case, and that Brookings may have had a great deal of influence on one of the seminal events of the twentieth century.

As we cited in the introduction, President John F. Kennedy had made a proposal shortly before his death that the United States and the Soviet Union should consider merging their respective space programs. Not only was this idea a radical one for its day given the deep suspicions both countries held of each other, but it may have been the last straw that ultimately got him killed.

On April 12, 1961, Yuri Gagarin had become the first human in space aboard a Soviet spacecraft. Six days later, NASA finally delivered a report they had commissioned on the proposed plan for space exploration—the aforementioned Brookings Report—to Congress. The delivery of the Report, which had been languishing on the desk of the NASA Administrator since November 30, 1960, suddenly had a new urgency.

Just about two weeks later, as if he was responding directly to the calls in the Report for NASA to consider suppression of the discovery of ET artifacts, Kennedy made a speech in which he signaled that he intended his administration to be an open one. He took the opportunity of a speech before the American Newspaper Publishers Association at the Waldorf-Astoria hotel in New York City to make the comments cited above.

His speech, titled "The President and the Press,"[56] was clearly an attempt to reach out to the assembled publishers and editors in order to not only protect official secrets whose revelation might harm the national security of the United States, but to also help him in *revealing* secrets that were unnecessarily being kept. His opening comments, speaking of "secret societies" and the dangers of "excessive and unwarranted concealment" of things he felt the American people had a right to know, was an unmistakable shot across the bow of these secret societies, and we take it as a direct reference to the recommendations contained in the Brookings Report. It is also very obvious from his statement that he considered these dark forces of "concealment" to be very powerful. Why else would he ask for the press's help in fighting this battle?

Within a little over a month of drawing this important "line in the sand" Kennedy addressed a Joint Session of Congress and issued his ringing call for "landing an American on the Moon" before 1970:

"First, I believe that this nation should commit itself to achieving the goal, before this decade is out, of landing a man on the Moon and returning him safely to the earth. No single space project in this period will be more impressive to mankind, or more important for the long-range exploration of space; and none will be so difficult or expensive to accomplish," he said on May 25, 1961 [Fig. 3-3].

This sequence of events implies that his "President and the Press" speech may have been influenced by the Brookings Report. Gagarin's flight obviously sent shockwaves through the U.S. space and security agencies. They'd known that the Soviets were ahead in space technology, but the U.S. wasn't even remotely close to being able to put a man in orbit. The immediate reaction was to finally send the report to Congress for review, as the game plan for the U.S. response.

The inclusion of the key phrases, about withholding any discoveries which may point to a previous and superior presence in the solar system, might have easily prompted Kennedy's speech just a few days later. It was by then a foregone conclusion that the U.S. would enter into a manned space race with the Soviets, but Kennedy was practically begging the press to help him make public the discoveries NASA might make.

Soviet premier Nikita Khrushchev's son, Sergei Khrushchev (now a senior fellow at the Watson Institute at Brown University) has stated that after the May 25th public call to "go to the Moon," Kennedy then did an extraordinary thing: less than ten days later, he *secretly* proposed to Khrushchev at their Vienna summit that the United States and the Soviet Union merge their space programs to get to the Moon together.[57] Khrushchev turned Kennedy down, in part because he didn't trust the young President after the Bay of Pigs fiasco, and also because he feared that America might learn too many useful technological secrets from the Russians (who were, clearly, still ahead in "heavy lift" launch vehicles—useful in launching nuclear weapons).

Although the offer was not made public, it's easy to imagine the consternation it might have caused at the Congressional level if it had leaked. Powerful congressmen, like Albert Thomas of Texas (a close political ally of Vice President Lyndon Johnson and a staunch anti-communist) who was Chairman of the Appropriations Committee in the House of Representatives, might have blown their tops if they had known about it. Thomas quite literally controlled all of the purse strings for the NASA budget and, along with LBJ, later got the Manned Spacecraft Center located in his home district in Houston. It is hard to imagine him, just a few weeks after receiving the Brookings study which called for keeping certain discoveries *from* the American people, agreeing to share these same discoveries with our Cold War enemy.

For that matter, it's hard to imagine *Kennedy* supporting such an idea. He had always spoken of the space race in stirring, idealistic, *nationalistic* terms:

> "... *Those who came before us made certain that this country rode the first waves of the industrial revolution, the first waves of modern invention, and the first wave of nuclear power, and this generation does not intend to founder in the backwash of the coming age of space. We mean to be a part of it—we mean to lead it. For the eyes of the world now look into space, to the Moon and to the planets beyond, and we have vowed that we shall not see it governed by a hostile flag of conquest, but by a banner of freedom and peace. We have vowed that we shall not see space filled with weapons of mass destruction, but with instruments of knowledge and understanding.*
>
> *"Yet the vows of this Nation can only be fulfilled if we in this Nation are first, and, therefore, we intend to be first. In short, our leadership in science and in industry, our hopes for peace and security, our obligations to ourselves as well as others, all require us to make this effort, to solve these mysteries, to solve them for the good of all men, and to become the world's leading spacefaring nation [emphasis added]."*

The situation was surely made worse in 1962 by the Cuban Missile Crisis, in which both nations stared down the barrel of nuclear annihilation and carefully stepped back from the brink. Far from discouraging him, these events may have emboldened Kennedy to try again. In August 1963, he met with Soviet Ambassador Dobrinyin in the Oval Office and once again (secretly) extended the offer. This time, Khrushchev considered it more seriously, but ultimately rejected it. On September 18, 1963, Kennedy then met with NASA Director James Webb. This is how NASA's official history describes that meeting:

"Later on the morning of September 18, the president met briefly with James Webb. Kennedy told him that he was thinking of pursuing the topic of cooperation with the Soviets as part of a broader effort to bring the two countries closer together. [Webb would have been unaware of Kennedy's previous two offers to Khrushchev, as they were made in private talks with the Soviet premier.] He asked Webb, 'Are you sufficiently in control to prevent my being undercut in NASA if I do that?' As Webb remembered that meeting, 'So in a sense he didn't ask me *if* he should do it; he told me he thought he should do it and wanted to do it...' What he sought from Webb was the assurance that there would be no further unsolicited comments from within the space agency. Webb told the president that he could keep things under control."[58]

Kennedy obviously wanted to avoid criticism from inside NASA on his new proposal. Selling the idea to the Soviets would be hard enough, but selling it to the American people and the Congress if there was "dissension in the ranks" might make it near impossible. If Webb couldn't hold discipline from inside NASA, the whole effort would collapse.

Kennedy then surprised the entire world when only two days later he went before the United Nations General Assembly and startlingly repeated his offer of cooperation, this time in public:

"Finally, in a field where the United States and the Soviet Union have a special capacity—in the field of space—there is room for new cooperation, for further joint efforts in the regulation and exploration of space. I include among these possibilities a joint expedition to the Moon. Space offers no problems of sovereignty; by resolution of this assembly, the members of the United Nations have foresworn any claim to territorial rights in outer space or on celestial bodies, and declared that international law and the United Nations Charter will apply. Why, therefore, should man's first flight to the Moon be a matter of national competition? Why should the United States and the Soviet Union, in preparing for such expeditions, become involved in immense duplications of research, construction and expenditure? Surely we should explore whether the scientists and astronauts of our two countries—indeed of all the world—cannot work together in the conquest of space, sending someday in this decade to the Moon not the representatives of a single nation, but the representatives of all of our countries."[59]

It is unclear what NASA Director Webb thought of the President's idea, but NASA insiders—as the President had feared—immediately expressed public doubts that the technical integration problems could be overcome.[60] The Western press was also very cautious. Many articles appeared resisting the idea of cooperating with a Cold War enemy that barely a year before had pointed first strike nuclear missiles at most of our major cities and sent our Nation to the brink of war. The Soviet government themselves did not make any official comment on the speech or the offer, and the Soviet press was equally silent.

But by far, the strongest objections came from within the U.S. Congress.

One of these objections came from a predictable source—Republican Senator

Barry Goldwater of Arizona. But, as foreshadowed earlier, another, even stronger protest came from a close political *ally* of the President and Vice President—Democratic Congressman Albert Thomas of Texas. Thomas made such a strong objection to the President that Kennedy personally wrote him on September 23, 1963 (just three days after his UN speech) to reassure him that a separate, American space program would continue, *regardless* of the outcome of negotiations with the Soviets: "In my judgment, therefore, our renewed and extended purpose of cooperation, so far from offering any excuse for slackening or weakness in our space effort, is one reason the more for moving ahead with the great program to which we have been committed as a country for more than two years."[61]

Within a couple of weeks, the lack of public support, even within the U.S., seemed to have scuttled the idea permanently, and Kennedy began to publicly back away from his own proposal.[62] Then, strangely, the idea abruptly resurfaced.

On November 12, 1963, Kennedy was suddenly reinvigorated about it and issued National Security Action Memorandum #271. The memo, titled "Cooperation With the USSR on Outer Space Matters," directed NASA Director Webb to personally (and immediately) take the initiative to develop a program of "substantive cooperation" with his Soviet counterparts in accordance with Kennedy's September 20th UN proposal. It also called for an interim report on the progress being made by December 15, 1963, giving Webb a little over a month to get "substantive" cooperation with the Soviets going.[63]

There is a second, even stranger memo which has surfaced, dated the same day. Found by UFO document researchers Dr. Robert M. Wood and his son Ryan Wood (author's of "*Majic Eyes Only: Earth's Encounters With Extraterrestrial Technology*") the document is titled "Classification Review of All UFO Intelligence Files Affecting National Security"[64] and is considered by them to have a "medium-high" (about 80%) probability of being authentic. The memo directs the director of the CIA to provide CIA files on "the high threat cases" with an eye toward identifying the differences between "bona fide" UFOs and any classified United States craft. He informs the CIA director that he has instructed Webb to begin the cooperative program with the Soviets (confirming the other, authenticated memo) and that he would then like NASA to be fully briefed on the "unknowns" so that they can presumably help with sharing this information with the Russians. The last line of the memo instructs an interim progress report to be completed no later than February 1, 1964.

Whether this second memo is genuine or not—and it certainly is consistent with Kennedy's stated plans—what is quite clear is that something dramatic happened between late September 1963, when Kennedy's proposal seemed all but dead, and mid-November, when it suddenly sprang back to life. What could have possibly occurred to motivate Kennedy to begin an unprecedented era of cooperation with America's Cold War enemy?

To put it simply, "Khrushchev happened."

Sergei Khrushchev, in an interview given in 1997 after his presentation at a NASA conference in Washington, D.C. commemorating the fortieth anniversary of Sputnik, confirmed that while initially ignoring Kennedy's UN offer, his father Nikita changed his mind and decided in early November 1963 to accept it. "My father decided that maybe he should accept (Kennedy's) offer, given the state of the space programs of the two countries (in 1963)," Khrushchev said.[65] He recalled walking with his father as they discussed the matter, and went on to place the timing of his father's decision as about "a week" before Kennedy's assassination in Dallas, which would date it right around November 12-15. Later, in a 1999 PBS interview, he repeated the claim: "I walked with him, sometime in late October or November, and he told me about all these things."[66]

We feel it is important to emphasize that Sergei Khrushchev has a unique perspective, if not bullet proof credibility as a first-hand witness to this virtually unknown—but absolutely documented—twist in space history. He is a well-respected and acknowledged scholar, serving at one of the most prestigious Ivy League universities in the United States. He has no motive to "make up" such history, as doing so would destroy all the credibility as a scholar he has spent a lifetime building.

So what logically happened is that sometime in early- to mid-November, Nikita Khrushchev communicated in some way that he was willing to consider Kennedy's proposal. Kennedy responded by ramping up the bureaucracies at his end, as reflected in the two November 12 memoranda. Unfortunately, there are no declassified documents to this point which confirm that the two men had any communication during this period. Still it seems quite unlikely that Kennedy would suddenly resurrect a seemingly dead policy without some hint from Khrushchev that it would be positively received.

One event we do know that actually happened, which may have finally tipped the balance in Khrushchev's mind: another very disappointing Soviet space failure had recently occurred. A Mars-bound unmanned spacecraft code-named "Cosmos 21" failed in low Earth orbit exactly *one day* (November 11) before Kennedy's sudden "Soviet Cooperation Directive" to James Webb.

All we can say for certain is that as of November 12, 1963, John Kennedy's "Grand Plan" to use NASA and the space program to melt the ice of the Cold War—and to *share* whatever Apollo discovered on the lunar surface with the *Russians*—was alive, vibrant and finally on its way to actual inception—

And, ten days later, Kennedy was dead.

The Third Rail of Conspiracy Theories

Whenever anyone brings up that fateful day in Dallas, November 22, 1963, and includes it in any dialog on any other subject, then that subject immediately becomes subject to scorn and ridicule. If you bring the Kennedy assassination into the conversation, you'd better be ready to have half the audience throw the rest of your ideas on to the trash heap of history. The Kennedy assassination is—to use a common political axiom—the "third rail" of conspiracy theories.

It is for this reason that we reluctantly began to look at the events of that morning in Dealey Plaza. We felt compelled to review the events surrounding John F. Kennedy's murder because so much of what we had uncovered *pointed* to a conspiracy to remove him from office.

By late 1963, Kennedy's personal popularity with the American people had grown stronger, and his chances of re-election in 1964 looked increasingly good. While he was generally unpopular in the South, he was actually more popular in Texas because of Lyndon Johnson, his showdown with Khrushchev over Cuba, and the dollars the space program was bringing to Texas. So there is the specter of a young, vigorous leader with rising popularity, who had openly declared his intention to reveal secrets he felt the American people had a right to know (thereby ignoring the cautions embedded in the Brookings Report), and who just happened to be threatening to bring this Nation's greatest enemy into the fold as an ally in our most technologically sensitive arena—and add to that the possibility that he was going to share "UFO secrets" with them as well.

Probably the hidden powers behind the scenes, the "secret societies" that

Kennedy spoke of in "The President and the Press," were quite willing to abide his radical ideas as long they could count on the Russians rejecting them. But, when Khrushchev abruptly changed his mind, and there was a possibility that the merged space programs might *actually happen*, Kennedy became far too much of a liability to tolerate. If indeed these forces of "unwarranted concealment" actually existed, they'd have had little choice but to eliminate him once he started issuing orders to begin the actual transfer of information and technology to the Soviets.

It makes little difference really whether it was a military-intelligence cabal that decided Kennedy had to go, simply because he was going to share our highly sensitive space secrets with the Russians (as the NSAM #271 makes clear) or if it was another, shadowy "secret society" that had other reasons for keeping any space discoveries from leaking out (as we shall document later). What matters is whether or not there is any credible evidence that Kennedy was killed by anything other than a single lone-nut gunman. By definition, if there was a second gunman in Dealey Plaza that sunny fall morning, then there was a conspiracy. Period.

Let us start by saying that we have little doubt that Lee Harvey Oswald was in Dallas that morning, that he was in the Texas School Book Depository sixth floor window, that he certainly fired at the President and that he may have even fired the fatal shot. That established, what evidence exists to support the idea of a second gunman, and therefore a true conspiracy?

In 1979, the House Select Committee on Assassinations conducted an exhaustive analysis of tape recordings made around the time of the shots fired in Dealey Plaza, and concluded that they contained evidence of two overlapping shots. They determined that four shots were fired, the first, second and fourth shots by Oswald, and a third near simultaneous shot from another location. Experiments conducted by the Committee in Dealey Plaza concluded that the third shot came from the direction of the infamous "grassy knoll."[67] This acoustic evidence has been called into question over the years, but rebuttals and counter arguments have left the question open, despite the official findings of the Committee.

The whole issue of a second gunman on the grassy knoll could be settled if there was just one photograph or segment of film footage that showed him there. Over the years, most of us have been led to believe that no such evidence exists. As we found out, that's not necessarily true.

In the early 1990s, the A&E cable network showed a nine-part series called *The Men Who Killed Kennedy*. It focused on a wide range of conspiracy theories and theorists, eventually concluding that Kennedy had been taken down by a French hit squad hired by Fidel Castro and endorsed by Nikita Khrushchev. Later episodes placed the focus on Vice President Johnson.

None of this was too impressive to the authors, except for the story of one (then) new witness, Gordon Arnold. Arnold gave the A&E show his first on-camera interview since first coming forward in the late 1980s. He claimed to have just arrived in Dallas from basic training in the army, and while on leave in Dallas (on his way to his station in Alaska) had decided to go down to Dealey Plaza to film what he thought was a parade. He had no idea until he arrived that President Kennedy was in town. When he tried to get a vantage point on a freeway overpass, a man in a business suit flashed a CIA ID and ordered him out of the area. He then made his way down to the picket fence area of the so-called grassy knoll, where he stood and waited for the President's limo to come by.

According to Arnold's story, he was in full uniform, including his pointed overseas army cap, and was filming using his mother's camera, which he had borrowed for the day. As the Presidential motorcade drove by, he suddenly felt a bullet zip past his ear very close, and heard a shot ring out. He hit the ground as quickly as he could. The next scene he described is completely bizarre.

According to Arnold, as he rolled back over amid the chaos, a man in a Dallas police officer's uniform confronted him, kicked him and ordered him to surrender his film. Since the officer was carrying a rifle and pointing it at him, Arnold complied. Arnold also noticed three other strange things about the man: even though he was wearing a uniform, he wore no policeman's hat, which would have been standard issue for a Dallas police officer. Arnold also testified that the man's hands were dirty, and that he was crying. According to Arnold, he walked away with the film behind the fence and off in the direction of the railroad yard behind Dealey Plaza. He evidently shortly met up with another man Arnold described as a "railroad worker." Arnold was so shaken by this experience that he never discussed it until the late 1980s. He figured no one would believe him anyway, since he had no proof of any of it.

But the A&E program was interested in testing Arnold's story against known photographs of the grassy knoll area. They decided to interview two researchers (Jack White and Gary Mack) who had done some work on one of the few known photographs taken of the grassy knoll area at the time of the assassination. The photograph they studied is known as the Mary Moorman photograph because it was taken by a witness named Mary Moorman, who was standing on the lawn just opposite the grassy knoll [Fig. 3-5].

Earlier in the same episode, A&E interviewed a witness who claimed to be the "babushka lady," so named because she wore a distinctive headscarf on that fateful day. In 1970, Beverly Oliver came forward to say she was the babushka lady, and that she had been filming the President when he was shot. She went on to claim that she gave her film to FBI agent Regis Kennedy, and that it was never returned. On the A&E show, she gave an interview in which she claimed to have heard a shot come from the grassy knoll, and when she looked up from her camera she saw a puff of smoke in the area of the fence. There are other films which show what may be a puff of smoke coming from the picket fence area of the grassy knoll.

In one film of the assassination, known as the "Marie Muchmore" film, you can certainly see both the babushka lady and Mary Moorman using their cameras at the instant the President is struck with the fatal shot. In frame by frame analysis, you can even see the first spray of blood from the president's fatal head wound. This would seem to be inconsistent with the medical evidence that dictates the head shot came from behind [Fig. 3-4].

The Mary Moorman Photograph

In any event, when White and Mack began to enlarge and enhance sections of the Mary Moorman photo, looking for any sign of Gordon Arnold, they got quite a surprise. An odd figure quickly stood out, right near the area Arnold said he was standing.

The figure appears to be a man in uniform, with a policeman's badge and shoulder emblem visible. His arms appear to be in a sniper's position, elbows out, as he would be if standing behind the fence and holding a rifle. Where the rifle should be is a bright flash of light, reminiscent of a muzzle flash, caught in an instant on film. In the enhancements, you can also clearly make out a receding hairline, prominent eyebrows and the fact that while the "badgeman" appears to be wearing a Dallas policeman's uniform; he is not wearing a hat [Fig. 3-6].

Just like the man Gordon Arnold had described, four years before this A&E

program aired. Later enhancements revealed another figure in the photo, just to badgeman's right. The figure is wearing an army summertime uniform, complete with the pointed overseas cap that Arnold said he was wearing. There is a bright spot where the unit pin on the hat would have been placed, and the figure seems to be holding something in front of his face—perhaps the movie camera Arnold had said he was using?

Oddly, the figure is also leaning to his right, as if he is just beginning to react to the muzzle blast behind and to his left. This is also consistent with what Arnold said he did that day. Later, yet more work revealed a third figure in the image, behind and to the right of the badgeman, wearing a hard hat and looking off to the frame right, as if scanning for anyone who may have been looking in their direction.

So here, finally, was visual evidence confirming not only the presence of a second gunman on the infamous grassy knoll—as so many witnesses testified to—but also of a witness who gave very specific details about both the gunman, his accomplice and his own disposition that day. There is flatly nothing in the Moorman enhancements which contradict Arnold's story, and assuming the techniques are valid, every reason to conclude it is a credible eyewitness account of a true event. To this day, while many have nitpicked Arnold's story (one debunker claimed it lacked credibility because on one occasion he mentioned the policeman's "dirty fingernails" as opposed to "dirty hands"), no one has yet repeated and challenged the photographic enhancements.

There are other details, too numerous to mention here, which support Arnold's story. But most compellingly, when he was shown the "badgeman" photo for the first time (on camera) he became very upset, teared up and said he wished he'd never brought the whole thing up. Not exactly the reaction of a publicity seeker, in the authors' opinion.

There are some that have pointed out a striking resemblance between the visible facial features of the badgeman and slain Dallas police officer J. D. Tippet. While we find these resemblances intriguing, we cannot say here that we endorse them. Tippet, according to the official cannon, was killed in the line of duty by Lee Harvey Oswald a short time after the assassination, and it is for that crime that Oswald was originally arrested in a nearby movie theater. What we do find interesting is if somehow Tippet, by all accounts a loyal police office and an admirer of Kennedy, was convinced to participate in the assassination out of some sense of higher duty to his country, then very conveniently, both "shooters," Oswald and Tippet, were dead within twenty-four hours of the assassination.

It might also explain why the badgeman was crying when he confronted Gordon Arnold.

The Wink of an Eye

So, satisfied that we now had evidence of a conspiracy in Dallas, the next question became: who was behind it? The plans for Kennedy to go to Texas had been made the previous spring, when Vice President Lyndon Johnson stated that Kennedy might visit Dallas in the summertime. It wasn't until September that a letter from Johnson aide Jack Valenti announced the Texas campaign swing. The trip centered around a special testimonial dinner for none other than Congressman Albert Thomas, the man who held the NASA purse strings and who Kennedy, by all accounts, adored. Thomas was dying from terminal cancer, and Kennedy was greatly relieved that he had decided to run for re-election and had avoided having an open seat in Congress to contest. Originally proposed as a one-day trip for November 21, by October Lyndon Johnson had become involved in the planning and a second day was added, November 22.

Kennedy was in a festive mood the evening of the 21st, pointing out Thomas' many contributions to the space program (which he was now about to hand over to the Russians!) and declaring him to be a good friend [Fig. 3-7].

"Next month, when the U.S. fires the world's biggest booster, lifting the heaviest payroll into... that is, payload..." Here the President paused a second and grinned.

"It will be the heaviest payroll, too," he quipped. The crowd roared.

"The firing of that shot will give us the lead in space," the President resumed in a serious vein. "And our leadership in space could not have been achieved without Congressman Albert Thomas. Your old men shall dream dreams, your young men will see visions, the Bible tells us. Where there is no vision, the people perish. Albert Thomas is old enough to dream dreams and young enough to see visions..."[68]

Kennedy departed after his speech, followed soon by Thomas and Vice

President Johnson. They both accompanied him to Dallas the next morning on Air Force One.

After the shooting, Kennedy was rushed to Parkland Memorial Hospital, but was obviously already dead. Doctors tried in vain to revive him, and the *Houston Chronicle* noted that Congressman Thomas waited outside the emergency ward until word came that Kennedy was dead. Vice President Johnson was whisked away to an undisclosed location. Later that evening, once Kennedy's body was aboard Air Force One, Johnson took the oath of office.

We've all seen the iconic photo, with a somber Johnson, his hand on the Bible, standing next to a dazed Jacqueline Kennedy as various aides looked on [Fig. 3-8]. One of the most prominent men in the background is a distinguished, bow-tied gentleman who is watching the proceedings very closely. Of course, it is Congressman Albert Thomas. What most of us have never seen is the next photo [Fig. 3-9], taken immediately after the oath was completed. In it, LBJ has turned immediately to his right. His facial muscles appear to be contorted into a broad smile as he makes eye contact with Congressman Thomas. Thomas, also smiling, returns the gesture with—of all things—a *wink*. While everyone else remains somber, Thomas and Johnson are the only two people in the picture who are smiling. The unspoken message between the two men could not be more clear: "We got him!"

Over the next few weeks, Johnson made a show of arguing to continue Kennedy's plans for Soviet cooperation in space. But in December, Congress, led by Representative Thomas, passed a new NASA funding bill expressly forbidding the use of NASA funds for cooperation with Russia, or any other nation:

"No part of any appropriation made available to the National Aeronautics and Space Administration by this act shall be used for expenses of participating in a manned lunar landing to be carried out jointly by the United States and any other country without consent of the Congress."[69] The same provision was repeated in subsequent NASA appropriations, continuing until the death of Congressman Thomas in 1966.

Keep in mind that Johnson had enormous political capital to continue any initiative of the martyred Kennedy that he so chose in those days and weeks following the assassination. Obviously, continuing the space cooperation initiative wasn't much of a priority, or he could have easily had it passed.

There are a couple of curious postscripts to this story.

By most accounts, Johnson should have still been President by 1969 when Neil Armstrong and Buzz Aldrin first walked on the Moon. He was constitutionally able to stand for re-election in 1968, but his great unpopularity because of his mishandling of Vietnam convinced him to forsake a second elected term and retire from public life. You would have thought, after being the head of the space program for so many years as Vice President and then continuing Kennedy's vision after his death, that Johnson would have been keenly interested in the events of July 20, 1969. But, as reported by presidential historian Doris Kearns Goodwin, Johnson not only didn't watch the Lunar Landing himself, he refused to let anyone at his Texas ranch watch it either, and ordered all the TVs to be turned off.

Perhaps, in the twilight of his life, with ample time to reflect on his own actions, the space program was no longer a source of pride for him, but of shame.

Recently, Saint John Hunt, the surviving eldest son of E. Howard Hunt—an infamous CIA operative actively involved with Watergate and long-rumored to have also been a key player in the Kennedy assassination—released a "deathbed confession tape" from his father. In a story published in *Rolling Stone* magazine, Saint John Hunt stated his father admitted to being one of the famous "three tramps" in photos of Dealy Plaza taken after the assassination and detailed specific players involved in the Kennedy assassination. The tape contains a remarkable "confirmation" in light of the completely independent evidence presented here that, above the CIA operatives (and contractors) who actually planned and carried out the plot to kill Kennedy, including E. Howard Hunt himself, they were all directed by one "top man."

Lyndon Baines Johnson.

We are left to contemplate our own accusations here. If men like Johnson and Thomas were willing to go so far as to orchestrate the murder of the President in order to protect the United States' own, singular (and singularly expensive) space program, then they must have expected to find wonders beyond imagining over the course of their voyages.

The only question for us was what, in fact, *did* they find, and was it worth the price that that the country (and History) ultimately paid?

Chapter Three Images

Fig. 3-1 – The Hoagland-Squyres debate, 1988 (CBS News)

Fig. 3-2 – The Hoagland-French debate, August 23, 1993 ABC's "Good Morning America" (ABC News)

Fig. 3-3 – President John F. Kennedy addressing a Joint Session of Congress, May 25, 1961 – calling for the United States to land a man on the Moon by the end of the decade "and return him safely to the earth." Note Vice President Lyndon Johnson's curious reaction (top, left) (NASA).

Fig. 3-4 - Video capture from the so-called "Marie Muchmore film" of the JFK assassination, just a microsecond after the fatal head shot. Note the spray of blood emanating from the President's forehead. The "babushka lady" is at far right, with the headscarf. Mary Moorman is standing at the far left in the black coat. (A&E)

Fig. 3-5 – The Mary Moorman photograph, taken just a split second before the frame capture in Figure 3-4. "Badgeman" and Gordon Arnold would be at the top of the stairs, above Kennedy's slumped figure, behind the white concrete wall atop the "grassy knoll" (A&E).

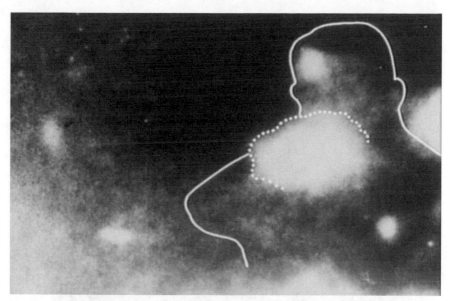

Fig. 3-6 - Close-up of "badgeman" – from enhancement of the Mary Moorman photo. Note position of arms, prominent eyebrows, badge reflection, lack of a policeman's hat and possible muzzle flash (A&E).

Fig. 3-7 – President Kennedy sharing a moment with Congressman Albert Thomas (D) of Texas, the night before the assassination. Thomas held the key purse strings in Congress for all NASA funding (Kennedy Library).

Fig 3-8 – Looking appropriately solemn for the cameras, Lyndon Johnson takes the oath of office aboard Air Force One, only hours after the President's assassination. Note Jacqueline Kennedy, right, Lady Bird Johnson (to Johnson's left) and Vice Presidential Aide Jack Valenti, kneeling, next to flowers. Between Valenti and Ladybird is Congressman Albert Thomas of Texas (wearing bow tie) (Kennedy Library).

Fig. 3-9 – Taken just moments later, the now infamous "wink" photograph. Johnson turns away from the cameras and smiles at Congressman Thomas, who winks back at him and smiles broadly in return. Is this the behavior of resolute leaders who have just experienced a national tragedy, or of two conspirators who just pulled off a coup (AP)?

Chapter Four

The Crystal Towers of the Moon

"It is very strange the way the ejecta... from Proclus crosses Crisium. It is almost like flying above a haze layer and looking down through the haze... It looks like it is suspended over it." – Al Worden, Apollo 15 Command Module Pilot

By the time Mars Observer had (apparently) taken its leave of the solar system, Hoagland was quietly into a new, parallel line of "artifacts" research. With no foreseeable new images of Mars for at least a half decade to come—Hoagland was able to focus full time on this promising new investigation, begun artound 1990.

Working on the assumption that an advanced, space-faring civilization had flourished and "cared enough to leave the very best" on Mars in the distant past, Hoagland concluded early-on that they would also likely have visited other planets or moons in the solar system. The most obvious place to find these telltale signs of prior visitation, it seemed to him, was Earth's own Moon. With much of the satellite already photographed at medium and high resolution by NASA for and by Apollo, and with Brookings's recently-acquired independent projections pointing to the same location, "Luna" seemed the best bet to continue the search for additional NASA confirmation of ancient archeological ruins in the solar system.

After considerable successes with speeches at the UN and also at NASA, which had attracted a loose, informal group of other talented researchers to the project from a variety of disciplines, Hoagland felt confident in his ability to gather additional contributors of the diverse backgrounds he considered essential for expanding the current "Mars" investigation to the Moon.

However, independently, Dr. Stan McDaniel had decided to form his own, separate working group, to be called "SPSR" (Society for Planetary SETI Research). Hoagland, along with all the academic folks who had gravitated to this unusual research *because* of Hoagland, had been asked to join. All immediately said "yes" … except Hoagland.

He was reluctant for several reasons.

First, he objected to the use of the term "SETI." This was a moniker developed by Carl Sagan for the use of radio astronomy to scan for stray radio signals from other stellar systems, and stood for "Search for Extraterrestrial Intelligence." Hoagland felt that the use of SETI gave unnecessary credibility to what he had come to believe was a waste of time at best or an outright scam at worst. No truly advanced ET culture would use communications as simple radio when they could use "scalar," or hyperdimensional technologies instead.

Hoagland also felt using this term gave Sagan the political playing field, and would permit him to hold an unearned sway over the philosophical development of what was essentially a brand new science—"*extraterrestrial archaeology.*"

He also could not abide by one of the dictums of McDaniel's new organization—that no "NASA conspiracies" now existed.

For some reason, McDaniel had decided to back down from the conclusions of his own voluminous Report, following Mars Observer's bizarre "disappearance," and retrench to the "honest but stupid" model to explain NASA's behavior vis-à-vis Cydonia. After twelve years of playing the game NASA's way, and after the blatant skullduggery and deception around the whole Mars Observer debacle, Hoagland simply couldn't stomach that. The two men had one final conversation in which McDaniel implored Hoagland to join the group, but when Hoagland turned him down, McDaniel parted with him bitterly. After working closely together on McDaniel's seminal, brutally honest expose' on NASA and Cydonia for over a year, the two now rarely communicated.

Hoagland realized that everything up to this point had been merely "foreplay"; from here on, he was going to have to pursue the wider truth about what NASA had been consistently, consciously, deliberately hiding for *decades*— not only from the American people, but from the people of the entire world—

Alone.

With his gaze now squarely fixed on the Moon, Hoagland faced only one considerable technical problem: where to start?

The Moon is a big place, with a land area larger than the African continent, and there were literally tens of thousands of photos from the various Lunar Orbiter, Surveyor, Ranger and manned Apollo missions. There was an extensive database, including an official NASA Study, of so-called "transient lunar phenomena"[70] (brilliant but brief flashes of light that have been recorded by observers for centuries), but the TLPs didn't seem to really follow any discernable pattern.

From one of his sources, Hoagland had been provided a lunar photo catalog from the 1960s, containing state-of-the-art ground-based telescopic images of various lunar features. It was on one of those Earth-based images that Hoagland spotted his first clue as to where to begin looking.

Hoagland acquired this historic lunar atlas, compiled by the Space Sciences Laboratory of the old "North American Aviation, Inc." in September 1992, a few days after the lift-off of the Mars Observer. At first glance, each photo resembled the next—distance and close-up shots, craters and maria. Then his eye fell to the southeast corner of page 241, where the crater Triesnecker first appeared in the collection. One Xerox enlargement later, and Hoagland knew he had taken his first step on what was to become a new research project bridging the gap between the Earth, Mars, and now the Moon. This extraordinary photo revealed an unmistakable image of an *equilateral triangle*—the 2D foundation of tetrahedral geometry and hallmark of hyperdimensional physics—immortalized in a lunar crater ... *in the center of the Moon* [Fig. 4-1]

Was it "a trick of light and shadow," as NASA likes to say ... or was it real?

Armed with a 486 personal computer, a 1600 dpi scanner and a variety of then state-of-the-art digital photo-processing software—more powerful than all the millions of dollars in equipment and manpower that NASA had twenty-five years before—Hoagland began anew to analyze, from his desktop, old NASA photos and negatives—beginning with Apollo 10 and the unmanned Lunar Orbiter and Surveyor Programs.

Hoagland had already discovered he was not the first to consider the anomalous aspects of various NASA lunar images. There was a sizable database of clearly disingenuous "UFO literature" out there on the subject, which in the years following Apollo had been taking increasingly misleading aim at NASA's missions to the Moon. Much of this "investigative literature" claimed, with the aid of the inevitably blurry "official NASA photographs," that NASA had discovered, and then promptly hidden, conclusive evidence of contemporary extraterrestrial bases on the Moon. Since these same authors demonstrably could not tell the difference between genuine artifacts and simple photographic defects, the idea of taking any of this evidence as indicative of any kind of authentic alien lunar presence seemed absurd. One writer even insisted that water-droplet "pull-apart marks" on an unnamed NASA Lunar Orbiter frame (from the film's development in zero-g) represented arrays of "lunar clouds" and "dammed-up ponds of water" on the airless Moon.

Unfortunately, while a source of much amusement, none of this helped him narrow down the best areas to search for genuine artifacts.

There was one interesting possibility: perhaps potential lunar artifacts, like the ruins at Cydonia, might be locatable through the same type of "hyperdimensional grid" that had been successfully applied to Earth by Carl Munk and several others. These coordinates related ancient terrestrial sites (the Giza Plateau in Egypt, Teotihuacán in Mexico, etc.) to the geometry supplied by the *Martian* complex at Cydonia. Setting up a similar grid for the Moon, Hoagland reasoned, might narrow the location of potential lunar artifacts to a relative handful of locations out of literally millions—but that still left an awful lot of lunar real estate to try to examine in detail.

Now fate took a hand, and completely undercut all his previously outlined careful strategy for how to begin looking at the Moon for compelling evidence of intelligent design. It provided Hoagland, in late 1992, with one particularly egregious example of those previously cited, wildly unsupported lunar ET claims: Fred Steckling's someday sure-to-be-a-classic (because it is *so* bad) paperback *We Found Alien Bases on the Moon*. In Steckling's non-stop comedy of errors, misidentified photographic defects and general mish mash of mal-illustrated misinformation, one lunar image suddenly stood out: three reproductions (at various magnifications) of a single, hand-held Apollo orbital frame acquired by an astronaut "sometime during the mission of Apollo 10."

The badly reproduced images in Steckling's book, complete with arrows, purported to show several current lunar "domes" on the Sinus Medii plain, just east of the famous, aforementioned "Triesnecker." Further up in the image, on the outside of the shadowed southern slope of the nearby crater,

Ukert (over 100 miles away, yet also carefully marked by another prominent arrow)—the dark, foreboding "entrance to an underground alien base" was allegedly shown.

Hoagland looked hard, but could see nothing. Still, in looking at Steckling's reproduction, he was immediately intrigued by a pretty weird "coincidence"; Ukert was none other than the very same "triangular crater" he had seen in the lunar atlas, perhaps implying that this particular area, indeed, did have "special significance."

His eye then fell on a formation immediately adjacent to and southeast of "Ukert": a curious, rectilinear arrangement of small hills, dark grooves and suspiciously parallel lines in the middle of one representation of the photograph, apparently completely overlooked by Steckling.

Honed by almost a decade of grappling with the Cydonia geometry amid the jumbled mesas and dunes of the Martian sands, the blatantly obvious geometric order of this remarkably compact arrangement on the Moon—grouped, like Cydonia, in a very tiny geographic region—almost leaped off the page [Fig. 4-2].

Awestruck, he took out a ruler and began to measure—but even as he confirmed more and more regular geometry in this one tiny geographic region, he couldn't help wondering: Why didn't Steckling call the attention of his readers to *this* fascinating, highly anomalous, highly organized formation, instead of having all those silly arrows pointing at absolutely *nothing*?

Still measuring, he copied down the Apollo 10 frame number. The next step involved attempting to acquire, from sources outside NASA, Apollo 10 photographic maps and other data pertaining to this image. In this he had another stroke of luck; one of Hoagland's colleagues had carefully saved original documents from the Apollo missions, from over a quarter of a century ago. Soon, copies of the crucial Apollo 10 maps and other documentation were on their way.

When they arrived, it soon became obvious, from even casually looking at these quarter-of-a-century-old NASA "footprint maps," that someone, early on, had very much wanted the Apollo 10 astronauts to simply photograph the hell out of the entire Sinus Medii/Ukert region [Fig. 4-3]. Given that this was under consideration as an Apollo landing site, this in itself was not unusual—but many of the photographs, like AS10-32-4819 [Fig. 4-2], seemed to focus less on the flatlands where a landing might be made, and more on the anomalous geometric areas and mountains where such a landing would be well nigh impossible.

Was NASA looking for "something else"—besides a future Apollo landing site ...?

Then, another opportunity to gain some data on this central region of the Moon came from a most unexpected source.

During an autumn speaking tour, Hoagland was introduced to a high-tech inventor and industrialist who had founded a company based on his perfection of a unique military optical technology. The industrialist said he wanted to help (but wished to remain anonymous). Hoagland was led to believe that, with appropriate coaxing, this "industrialist" might be persuaded to adapt his remarkable optical technology for unparalleled terrestrial observations of *the Moon*—specifically now, the central Ukert region.

This strategy, as opposed to going straight to NASA for a negative of the intriguing Apollo 10 frame, was based on Hoagland's early suspicion (since disproved) that all easily accessible NASA images might have been "sanitized" in the intervening thirty years since they were originally taken, thus making Steckling's frame immediately suspect. It was Hoagland's original intention, upon finding the really intriguing objects on AS10-32-4819, to attempt *direct confirmation* of the largest geometric forms telescopically from Earth—provided his new "executive friend" cooperated by providing his "black" technology.

In the course of one of these discussions, held in his firm's quite impressive, richly paneled conference room on the seventh floor of his corporate headquarters in a major southern city, the company founder (who boasted repeatedly of "extensive NASA and military contacts"...) told Hoagland that he had been carefully following his Cydonia work for "several years." He also informed Hoagland that high-powered military brass involved in Pentagon "procurement" (later to lay the groundwork of the Navy's "Clementine" lunar probe) routinely gathered in *his* conference room, to watch Hoagland's latest "NASA-Cydonia Briefing" video (from the UN presentations ...) on the large-screen television console which dominated one entire end of the room.

Several times throughout the Fall of 1992, this individual would call Hoagland without warning from "that" conference room—in the company of "this Colonel," or "that General," visiting from Washington, D.C.—*while* they were (ostensibly) watching the "NASA-Cydonia" and "U.N." video presentations, so his distinguished visitors could ask questions directly regarding the Cydonia investigation. This didn't mean much at the time, but in hindsight, has taken on a significantly deeper meaning *vis-à-vis* recent developments around the Moon.

These meetings were briefly interrupted by an acceleration of events occurring on the "Martian front," and the hurried call and a flight to and from New York to debate Dr. French on *Good Morning America*. When Hoagland returned south, his newfound industrial friend demonstrated his support for opening "a lunar wing" to the on-going Mars investigation, by presenting him with the rare lunar atlas.

The Atlas provided the key to the entire subject of ancient artifacts upon the Moon—but before proceeding to the Atlas, Hoagland brought out his copy of the Apollo 10 image from the Steckling book, and asked if his "executive friend" could procure a negative on his behalf through his contacts at the National Space Science Data Center. Hoagland told him of his apprehensions regarding what might happen to the lunar data inside NASA, if he personally, visibly, involved himself in looking at the Moon, in the midst of the on-going controversy over the Cydonia investigation. He then pointed out the fake domes and bases Steckling had noted by his arrows on the NASA photograph, and his discovery of far more interesting *real* geometrical anomalies elsewhere in this same region.

His executive acquaintance suddenly looked very serious, and asked to take the image to the firm's copy room to make some duplicates.

Months passed.

Phone calls—lots of phone calls—were exchanged with this industrial "friend," and the conversations, which began with genuine excitement over the possibility of verifying a bona fide connection between Cydonia and a set of artificial structures on the Moon, now began to shift to the need for "large amounts of money"—a million dollars—for him to simply point his telescope towards the Moon!

After several months of this, through another friend, a real one this time, Hoagland quietly placed an order for the Apollo frame with NASA's Johnson Space Center in Houston.

It came within a week, but with a shock: it *wasn't* the photo that he'd seen in Steckling's book.

Hurriedly comparing the features in the image with the *Apollo 10* footprint maps, Hoagland now discovered a fascinating fact—the frame in Steckling's book was *not* frame "AS 10-32-4810," the photo he'd just received from Houston. So, what was the NASA photograph that Steckling reproduced, and carefully mislabeled not once, but *three* different times in the same volume? The Apollo 10 photomap said "AS10-32-4819," but if Hoagland hadn't had the official AS-10 image footprint map, he'd have been up the proverbial creek ... without a clue as to the real identity of the remarkable image in the Steckling book (we should probably mention at this point that Hoagland eventually discovered that Fred Steckling, at one time, also worked for the CIA ...).

The NASA 8 x 10 "glossy" he now held from Houston (AS10-32-4810), while turning out *not* to be the one he thought he'd ordered, quickly turned into another major stroke of luck. Hoagland now discovered it was the second frame (according to the footprint maps) of an entire hand-held Apollo 10 photo *sequence* acquired across the Sinus Medii region, taken by the astronauts orbiting just south of the craters Triesnecker and Ukert.

The frame that he'd received allowed, when compared with the image in the book, an almost 90° perspective on the curious "geometric square" he'd first noticed on the Steckling image. Further, if the first image (which he now knew was *really* AS10-32-4819) had the sun coming from the right, then this image (the real AS10-32-4810, no thanks to Steckling) had it coming from over the astronaut's shoulder, providing critical comparative surface reflectance information.

Using his enormous thirty-year digital technical advantage, Hoagland began computer analysis of the Ukert region of the Moon. Curiously, the location of the Apollo photograph he now had in his possession was also quite consistent with the "Cydonia grid" discussed earlier (within a few degrees of the center of the Earthside lunar disc at 00 latitude, 00 longitude). At times, this sixteen-mile-diameter crater is directly opposite, or under (if you're standing on the lunar surface) the "sub-Earth point" (the location on the Moon where the Earth would be seen directly overhead).

From that point, things developed very rapidly. Hoagland proceeded to order—from another "inside" source at NASA, then directly from NSSDC and (simultaneously) through NASA's Johnson Space Center—the entire 70mm, hand-held photographic sequence taken by the Apollo 10 crew across Ukert in May 1969. He also widened the investigation by ordering overlapping Sinus Medii/Ukert data from two previous classes of unmanned NASA lunar missions (the 1966-1967 Lunar Orbiters II, III and IV, and the late 1967 soft-landed Surveyor 6 spacecraft). This strategy allowed him to compare three different photographic technologies, from three different kinds of NASA reconnaissance missions, from a variety of different lighting angles, on what he now strongly suspected to exist in the Sinus Medii region.

When all assembled it made for a compelling case:

That—for the previous *thirty years*—"someone" inside NASA (probably, a lot of "someones" ...) had been hiding major segments of the *real* U.S. space program—not just on Mars ... but on the Moon.

The alternative hypothesis—that of all the thousands of scientists in and outside NASA who had examined this lunar information over the past thirty years, Hoagland was the *first* to notice anything "unusual"—was simply not credible.

This distasteful political conclusion was based primarily on his startling discovery of "blacked out" thumbnail images in official NASA catalogues—such as the NASA Apollo 10 "special publication" [Fig. 4-4]. The blacked-out thumbnails seemed to imply that the images themselves had been dramatically under-exposed, or even *not* exposed—and thus were useless.

So, of course, he ordered these blacked out frame numbers from the catalogs *first*. When he got them, not only were they *not* the hopelessly dark, non-detailed images the catalogs had deceptively (and obviously *deliberately*) made them out to be—they were literally full of new discoveries, and on scale he could not have imagined ....

Ohio State

On June 2, 1994, Hoagland presented the preliminary results of his two-year old, quiet investigation into the possibility of "alien" lunar ruins at Ohio State University.

The presentation—made at the largest land grant college in the United States, which in the 1950s hosted one of the first radio SETI projects in the world—spanned four hours, and was conducted before an audience of over 700 students, faculty and members of the general public. Anticipating the major interest that would follow, he arranged for a professional videotaping of the event, later produced and released as "The Moon-Mars Connection: Alien Ruins on the Moon?"

Unfortunately, several people in the audience recorded the presentation on home video. And, more unfortunate, inferior stills from these videos, frame-grabbed and uploaded to the Internet, instantly appeared around the world. These distorted versions of the highly controversial NASA data fostered a veritable firestorm of reaction and misinformation on the information superhighway, including heated discussions on CompuServe, AmericaOnline and other then-popular electronic forums. Due to the poor quality of these pirated copies of the original lunar images, the electronic discussions and arguments raging over the reality of the data and its meaning were not accompanied by anything approaching the quality of the original supporting evidence.

As a result, serious misinformation on the original analytical techniques applied and the official sources of this lunar "artifact data" persists ... even to the present.

Because of the diversity of lunar data used in this early phase of the investigation—a mix of electronic imagery, hard-copy photographic film and, in the case of the Soviet imaging, even half-tone reproductions from official NASA publications—it was decided to reduce all images to photographic prints from the earliest possible generation negatives (where available), physically enlarge these negatives at a variety of scales in a professional darkroom, and then electronically scan the resulting prints. Because each photographic print reproduces, on average, only about *one thousandth* of the full-intensity exposure range found on the original negatives, dozens of prints of selected features were often required to capture the range of detail and light intensity present on any one NASA "original." This process was continued until the right combination of exposure and f-stop were achieved to bring out subtle features.

The derived digitized lunar image frames were then imported into a variety of commercial and specialized software imaging programs for compositing, resolution and gray scale enhancement, analysis, and electronic and hard copy display.

The Shard

The first image presented by Hoagland at Ohio State was an oblique, medium-angle Lunar Orbiter frame acquired from a thirty-mile altitude, looking across Sinus Medii to objects located on its southwest perimeter [Fig. 4-5]. The Lunar Orbiter dataset provided unique challenges because of the novel way the data was obtained, sent back to Earth and processed.

The five successful Lunar Orbiter spacecraft utilized two telescopic lenses, with different focal lengths (magnifications), simultaneously framing two separate images on one long roll of ultra-fine-grained, high-resolution Kodak film inside each pressurized spacecraft Orbiter "bus." This created a "medium resolution" lunar image aboard each Orbiter (the "M" frames), and another, bore-sighted "high resolution" image (the "H" frames). After all exposures had been commanded from Earth, and the on-board Orbiter film totally exposed, each roll was remotely developed in lunar orbit in the on-board (pressurized) "mini photo lab." Then, each roll of developed film was scanned (section by section via a mechanical "flying spot scanner"—a very narrow beam of light, 6.5 microns wide, swept back and forth across the film by a rotating mirror system), thereby transforming the film image into a gridded "analog electronic code" corresponding to an array of varying intensity points across the original photograph. Subsequently, this grid code was then used to modulate the spacecraft analog radio transmissions to Earth, as the image was read out to NASA receivers on the ground, literally line by line.

(The Russians used an essentially identical film-based photographic/optical scanning/analog electronic transmission system in their early lunar probes—including ... *Kodak film*)

Once on Earth, this varying radio signal—a frequency-modulated analog version of the original scanned film image in the spacecraft (remember, no *digital* electronics then, or even "imaging computers"—in 1966)—was transformed *back* into photography by displaying the modulated signal output on an analog "cathode ray tube" (essentially, a highly primitive, miniature TV *picture* tube) and literally *photographing* the faceplate display.

The 35mm partial lunar *positive* film image strips thus produced were then *manually* pieced together by NASA employees, Army mapmakers and contractors, and then photographed again—to create "master original negatives" of each successfully re-assembled, individual Lunar Orbiter frame (remember, only a generational *copy* of the developed film images still *physically existing* aboard the lunar orbiting spacecraft itself ...), from all five *Lunar Orbiter* missions.

The first photographic image from this elaborate film/electronic process that Hoagland was able to acquire and examine (LO-III-84M) immediately revealed a number of striking, new "architectural" anomalies.

After optically scanning (again!) an analog photographic print made from this assembled NASA negative (printed and developed in a commercial photo lab, in New York City), and then running it through his own computer-enhancement process, Hoagland reviewed the results: one highly anomalous object immediately "popped out" above the distant lunar horizon on frame "LO-III-84M"—directly beneath a geometric calibration "cross" [Fig. 4-5] placed by Kodak on the original NASA film during manufacture.

Dubbed "the Shard" by Hoagland, the extraordinary object stood about 1.5 miles above the lunar surface (judging by the spacecraft geometry of the original image acquisition), instantly making it a truly inexplicable new wonder of the Moon, if not the universe.

On a bashed and battered lunar surface, exposed to billions of years of unshielded asteroid and micrometeorite bombardment, the Shard was a highly anomalous, defiantly upright, "bowling-pin" shaped structure—with an irregularly-pointed apex, a swollen middle "node" and a narrowing "foot."

Back toward the northeast (right) in the Lunar Orbiter III photography lay the equally-curiously-shaped *shadow* of this astonishing object—mirroring to a remarkable (and crucial) degree the Shard's unique morphology. The readily-apparent shadow was precisely consistent with the local geography on which "the Shard" was sitting (an ancient lunar ridge). In addition, the shadow was also consistent with the time of the month, the Lunar Orbiter camera angle, and the actual sun-angle illuminating this object when the image was initially acquired: mid-February 1967.

It was this *shadow*—more than any other single aspect of this object—which, in the eyes of many observers, solidified the Shard's reality as an extremely anomalous, potentially *manufactured* lunar structure.

The sheer height and shape of this object—a rough spire, *over a mile and a half tall*—also argued forcefully for an origin far more recent than the extremely ancient, rounded and eroded lunar landscape that surrounds it. Located on the southwest perimeter of Sinus Medii, across a large ancient crater called Flammarion, the Shard has somehow managed to defy the laws of entropy that have blasted those surroundings, according to geochemical and isotope age data returned by the Apollo missions, for eons. This "incessant meteoric rain," in the absence of "recent" geologic activity, should have reduced this object to rubble indistinguishable from its surroundings *billions* of years ago— but there it defiantly, inexplicably, still stood.

There were other, crucial clues that the Shard was, in fact, a *real* object—still standing upright against gravity on the lunar surface. One was its alignment with the local *vertical* rather than with the "grain" of the Lunar Orbiter film; the second, was a highly geometric *internal* "organization" [Fig. 4-6].

This striking pattern—which strongly supported the initial impression of "artificiality"—was composed of a repeating, complex, internal *crystalline* geometry, visible all throughout the object; additional Hoagland computer enhancements of additional enlarged prints, also made from NSSDC negatives of III-84M, revealed this stunning, internal pattern was made up of highly reflective, cubical—possibly *hexagonal*—geometric "compartments," greatly damaged but still visible and overwhelmingly redundant.

Resolution estimates, based on the Lunar Orbiter acquisition geometry, indicated the "cell size" of this geometry to be at least several times the resolution of the imaging enhancements (approximately seventy meters). Furthermore, this still- visible geometric pattern also was *not* orthogonally controlled (as if imposed as an aspect of the rectilinear pattern of computer pixels in the computer scanner, or on the display screen), but appeared to be determined by an internally consistent, *intrinsic*, architectural pattern *inside* the object itself.

Finally, a series of fine vertical lines appeared to overlay this astonishing structure—with bits and pieces of brighter material still suspended on these vertical "lines" off the main apex of "the Shard"—as if on a separate, rigid matrix entirely surrounding this highly exotic lunar structure.

The overall impression one was left with was that of a once-much-larger, complex, *glass-like,* artificial object—now extremely eroded by eons of meteorite impact processes.

Since it is virtually impossible to argue with the shadow cast by the Shard, and *since the object's* very *existence* (therefore established) argues forcefully for its artificiality, some critics have looked to alternate explanations for its presence in LO-III-84M. It has been suggested, for instance, that the "Shard" may be a transient "out-gassing" event, fortuitously captured by the Lunar Orbiter camera. While this cannot be totally discounted (based on Nikolai Kozyrev's cited, widely accepted detection of such a previous event, in fact not too far away, in the crater Alphonsus, in 1958), the absence of any diffusion around the Shard's sharply defined edges—and all that obvious internal *geometry*—works heavily against this explanation.

But there was an even more compelling argument favoring the Shard's reality, to be made right from the same LO-III-84M frame: the striking presence of "the Tower"....

The "Tower/Cube"

The "Tower/Cube" anomaly, seen to the left of the Shard in an enlarged portion of III-84M, appeared at first to be merely a smudge on the original full-frame NASA photograph [Fig. 4-7]. With a "tail" extending down toward the lunar horizon, it bore an eerie similarity to the Shard, although it appeared to be much more massive. Other explanations were immediately considered—perhaps it was a comet?

A quick historical search revealed no such passing solar system object in February, 1967.

OK, then, maybe a nebula: a cloud of gas and dust deep in interstellar space, serendipitously captured in the line of sight by the Orbiter's camera as it photographed the Moon; or, perhaps it was even a much more distant galaxy, likewise "coincidentally captured" on this one Lunar Orbiter III image beamed back to Earth in February 1967.

In other words, it was possible that this peculiar "smudge" could be one of a variety of mundane deep space objects—all far beyond the Moon.

A brief calculation, however, quickly eliminated *any* of those possibilities. NASA's published data on Orbiter photography showed that the f-stop setting for this image was 5.6, the shutter speed 1/100th of a second. Simultaneously capturing the bright lunar landscape and a faint background interstellar "nebula," on the same photograph, with those specific camera settings, on this type of slow-speed, *fine-grained* film, would be technically impossible. If the camera aperture was held open long enough to capture such an incredibly faint object, the brilliant, *sunlit* lunar landscape would have been completely blown out.

[This, of course, is a simple fact of optics and 1960's imaging technology that the lunar hoax proponents *never* seem to "get": imaging the daytime sunlit Moon, *and* incredibly dim background stars (or nebulae) simultaneously, was just *impossible* in 1966 from Lunar Orbiter ... and, in 1969-1972, with similar Apollo *film* cameras.)

With these possibilities eliminated, the other "mundane" explanation— that the feature was merely a photographic blemish—was also eliminated, through independent analysis by photographic experts at High Definition Image Photographic Laboratories in New York.

It was only after High Definition reproduced literally dozens of differently-exposed versions of this object (by printing the NSSDC negative over and over again at various exposure settings), with Hoagland repeatedly scanning the results until all the subtle details were brought out, then subjecting those raw scans to a series of extensive computer-imaging enhancements, utilizing a variety of digital filters within various commercial imaging programs, that the truly extraordinary nature of this remarkable "tower/cube structure" was fully revealed...

...as a massive, obviously *glass,* "megacube"— sitting more than *seven miles* above the Moon [Fig. 4-8].

More than a mile across (at an estimated distance of ~300 miles from Lunar Orbiter), the "Cube" itself appeared to be composed of literally *dozens* of smaller cubical (and/or hexagonal) sub-structures; the smallest of these details visible on frame III-84M measuring at least *fifty or sixty times* the size of the individual computer "pixels" of the imaging enhancements.

Comparison of the digitized, enhanced computer version of this object with the "raw" photographically-enlarged prints (viewed under a magnifier), confirmed the presence of the same, highly anomalous, internal geometric structure *on both* versions—analog and digital. (This was critically important as a "control," to rule out simple "computer artifacts" as the cause of the blatant geometric structure seen in the digital enhancements.

A different type of computer enhancement of the cube, this time specifically for "edges," revealed almost endless, repeating examples of much smaller, obviously geometric *structure* throughout its surviving mass. The constant meteor erosion was apparently abrading the structural detail from *the outside in*—with the most exposed "beams" and "girders" now visible on the exterior; the densest mass remaining was therefore concentrated toward the interior ... where the greatest light-scattering (roughly proportional to the surviving mass) was still occurring.

In other sections of this enlarged and enhanced image, long, vertical "transitions"—created by apparent "refraction-effects"—could also be traced, as if portions of the background object were being viewed *through* a heavily-distorting medium, located much closer to the spacecraft.

This glittering, long eroding, giant "mega-cube of glass" was, in turn—upon examination of the various computer enhancements—also confirmed to be perched atop a "glittering glass *tower*" ... stretching that "seven miles" down to the overexposed lunar surface [Fig. 4-9].

This "Tower" appeared to taper towards that surface, and simultaneously, to be leaning in a southerly direction in the photograph. This obvious departure from the local vertical (similar to the internal details in the "Shard") could be directly connected with the distance of this object from the Lunar Orbiter—cruising thirty miles above the Moon, looking downward toward the lunar. A "real object," connected with the surface by an actual *vertical* tower, could indeed appear to "lean," if (a) it was located closer to the spacecraft than the "Shard," or (b) was in the process of intrinsically "slowly tipping over" ... from more inexorable effects of constant meteor erosion, across literally millions of years.

The first possibility would be due to the relatively sharp curvature of the Moon's circumference, coupled with a sharper downward "look-angle" required for the Orbiter to view a closer object. The second was simply endemic to the obvious great age of this astonishing, formerly geometrically perfect, massive lunar artifact that Hoagland had discovered.

Uncertainty in deciding between these two different explanations for "the leaning tower" effect was keyed directly to a major uncertainty in the actual height/scale of the "Tower/Cube" combination itself—which, in turn, was critically dependent on the actual distance from the Lunar Orbiter camera and the observed structure's elevation.

The primary reason for the latter uncertainty was the additional discovery (on the computer enhancements) of a much fainter, surrounding "grid-like framework." This enveloping structure appeared to be composed of more highly reflective "debris," sparkling around the "Tower/Cube" itself, embedded in a much darker, vertical matrix Hoagland termed "rebar" (for reinforcing bars). The scattered, highly reflective fragments seen apparently clinging to a darker "grid" are consistent with this "two-component" matrix: micrometeorite sandblasted "glass," and a surrounding, darker "structural material."

This "prairie fire effect" stretched from south of the Tower/Cube, north past the "Shard" (which also seems to be embedded in this stuff), until the grid disappeared out of frame, still heading north. Somehow, these scattered, massive fragments of reflective "stuff"—like the Tower/Cube—were also suspended against airless space above the Moon. Thus, the implied necessity for some kind of extensive, darker "rebar" matrix, to hold them up.

It was apparently an intervening portion of this vertical glass "grid" which was causing the strong refractive effects seen in the close-ups of the "cube."

Not only is all this patently absurd within any current (or proposed) "geologic model" of the lunar surface, it totally supports the concept that the Tower/Cube (and the Shard, not too far away) are but remaining fragments of a once *far-larger*, also clearly *artificial* structure, apparently once composed of *glass* (a lot of it) ... and, attached to some kind of darker, vertical, overarching structural framework.

The existence of these few remaining "glittering glass fragments" from this former complete structure—viewed by Lunar Orbiter stretching north along the western edge of Sinus Medii—is what prompted Hoagland to propose the one-time existence of an artificial, very ancient "Sinus Medii Dome"—formerly *completely enclosing* this central "maria" on the Moon.

Analysis of Apollo images of this same western Sinus Medii region—AS 10-32-4854, AS 10-32-4855 and ASIO-32-4856 [Fig. 4-10] —confirmed the presence of the Tower/Cube, *and* from a completely different mission than Lunar Orbiter (Apollo 10). This confirmation came via a different technology (actual film, personally shot by a NASA astronaut himself and physically returned to Earth—as opposed to being scanned), and from a very different photographic azimuth (45°) from the original Lunar Orbiter III frame discussed above. The time of lunar day, and thus the lighting for this Apollo imagery, was also clearly different—just before local sunrise during Apollo 10's photography, as opposed to "mid-morning" for III-84M—but ... the shattered remains of some kind of "tower," extending upward from a shattered glass horizon—with a glittering glass cube still visible on top of it—*was still there.*

These multiple, independent appearances of the Tower/Cube on Lunar Orbiter and several, subsequent Apollo 10 photographs, not only confirms the surprising context in which it is embedded—a "grid-like" matrix, stretching far north along the western edge of Sinus Medii—these images, when magnified, confirm the highly structural details of both the Cube as well as the relationship of this amazing object to its supporting tower (which has the appearance in the Apollo images of multiple supporting arches), stretching in multiple, increasingly conspicuous layers down to the surface of the Moon: all apparently part of a once far larger, intact Sinus Medii "dome-like" geometric structure—now almost completely obliterated by an unknown (but certainly immense) amount of exposure to unshielded meteorite bombardment on an airless lunar surface.

As the investigation progressed from a handful of early NASA lunar images through to the current database of literally hundreds of sophisticated imaging enhancements, evidence steadily accumulated consistent with this "lunar dome hypothesis." This evidence now strongly argues, not only for the existence of a former, miles-high, multi-layered "Dome" once stretching over all of Sinus Medii, but increasingly for the existence of other "lunar domes" as well, constructed over other, widely-separated locations on the Earthside lunar surface—now visible from Earth as the classic lunar "maria."

As impossible as that may seem, the more NASA images that Hoagland studied, the more evidence in support of this seemingly outrageous idea inexorably accumulated ....

Surveyor 6–"The Sinus Medii Dome"

Surveyor 6 successfully landed on the Moon in November 1967—thirty or so miles west of "Bruce," a small five-mile wide crater near the center of Sinus Medii. From there, the unmanned Surveyor spacecraft took over 35,000 low- and high-resolution images of the surrounding lunar landscape by means of a 600-line analog television system. After local sunset, on November 24, an additional set of time-exposed images were acquired looking west, for purposes of studying light-scattering properties in interplanetary space caused by the solar corona, far beyond the lunar horizon.

One such image was intensively analyzed during Hoagland's initial lunar investigation. The analysis yielded a remarkable corroboration of a peculiar light-scattering phenomenon strongly correlated with the geometric optical anomalies observed by Lunar Orbiter III and Apollo 10 (the Shard and the Tower) approximately 100 miles southwest of the Surveyor 6 landing site—along the western edge of Sinus Medii.

This Surveyor 6 image differs, however, in one dramatic detail from the other NASA images previously analyzed: those photographs recorded these enigmatic geometric structures in back-scattered light (with the sun more or less behind the camera). The Surveyor image, with its remarkably brilliant beads stretching along the western horizon and an intensely geometric structure of scattered light seen against the airless lunar sky above it—was taken with the light coming from the front, from *beneath* the western horizon [Fig. 4-11].

The Surveyor 6 photography of "brilliant beads of light" along Sinus Medii's western horizon, recorded over an hour after sunset, was considered—even by NASA in 1967—as "a remarkable discovery." However, subsequent official efforts to explain this inexplicable phenomenon (on an airless world ...) focused exclusively on "forward-scattering, caused by small, electro-statically elevated particles of lunar dust ..." redirecting sunlight from beyond the visible horizon, as seen from the location of Surveyor 6.

However, later "in-situ" studies of lunar surface processes, accomplished by analyzing actual dust samples returned by the Apollo astronauts from elsewhere on the lunar surface, effectively eliminated "electro-statically suspended lunar dust particles" as a probable cause of the intensely brilliant "beading" seen in the Surveyor after-sunset images.

Which leaves a major mystery—what could cause such intense illumination, for more than six lateral degrees along the western horizon and *more than an hour after sunset*—so as to completely saturate (according to NASA's own studies of the images) –the Surveyor vidicon television system within the "beads" themselves?

It was Hoagland's contention that this phenomenon could *only* be caused by forward *refracted* sunlight: the myriad, tiny lens-like images of the actual *photosphere* of the sun, being literally bent (for over twenty miles) around the sharp curvature of the lunar horizon by the densest layers of his surviving "lunar glass dome architecture" ... just *above* the lunar surface.

The much fainter—but extraordinarily geometric lattice-like structure seen arching far above those mysterious brilliant beads along the horizon, in Hoagland's model—was crucial supporting evidence for the model's ultimate reality. This delicate geometry was not merely the sun's distant corona seen against space (as NASA claimed), but was caused by the still geometrically organized, still surviving bits of glass and rebar *physically* present at much higher altitudes, faintly scattering (as opposed to refracting) the brilliant light of the sun still directly visible at their high altitude down toward Surveyor's camera ... in these blatant, highly geometric patterns [Fig. 4-12].

Patterns which, of course, *couldn't* exist in this form in the sun's high-temperature corona—located literally millions of miles beyond the Moon.

The astonishing discovery of a highly geometric structure—just above the beads—on the original Surveyor after-sunset image is completely inconsistent with any radial, ray-like pattern expected for a true coronal photograph. It is, however, remarkable additional support for the presence of a former *geometric* lunar dome—now in ruins.

These collective optical phenomena, in Hoagland's analysis, could only be caused by the remnants of some kind of ancient, massive, glass-like, originally highly geometric architecture—still anchored in the lunar surface and extending literally miles above. This was precisely the same kind of lunar construction that was strongly implied by the independent Lunar Orbiter and Apollo 10 observations of the miles-high "Shard and Tower" (on approximately *the same* Sinus Medii meridian—but over 100 miles south of the Surveyor 6 landing site).

Eons long meteorite abrasion would work from the top down—stripping glass from its supporting rebar high above, but leaving greater amounts of protected, remnant glass structure closer to the horizon (from the vantage point of the Surveyor spacecraft)—precisely as the surface images portraying the anomalous horizon beading strongly implied.

Later in the investigation, on analyzing one of the Apollo 10 Hasselblad shots taken by the crew as they flew over Sinus Medii in May of 1979 (AS10-32-4816), Hoagland discovered some astonishing, totally *confirming* close-ups of this amazing "rebar" [Fig. 4-13].

It is now crystal clear from all of this official NASA imagery that there indeed, once existed a magnificent, incredibly sophisticated, geometrically complex, artificial *dome-like architecture*—stretching all across what today is seen by Moon watchers as "just another barren lunar lava flow" in Sinus Medii.

What it must have looked like in its "heyday"—and what it may one day tell us of its long-vanished, extraordinarily gifted architects and construction engineers—can only be imagined from its ghostly remnants, which still glitter—ancient and mysterious—literally *miles* above the Moon ....

Ukert–"Los Angeles"

Of all the evidence analyzed in the early investigation of potential lunar artifacts, none was more revealing, both scientifically and politically, than the acquisition and examination of Apollo 10 frame AS10-32-4822. What makes this all the more remarkable is the fact this frame supposedly "does not even exist"—as can be seen by examination of a portion of the official NASA mission catalogue for Apollo 10, published immediately after the 1969 mission, SP-232 [Fig. 4-14].

Hoagland's initial efforts to acquire even one version of 4822 began in 1992, in Houston at NASA's Johnson Space Center, where the Apollo photography acquired by the astronauts was initially processed on their return from the Moon. To his surprise, the commercial photo lab delegated by NASA to respond to the request (not in Houston, but in Dallas) promptly sent a high-quality version of this supposedly "missing image"—completely belying the *blank* frame printed in the official mission catalogue.

It was on this initial version of this crucial Apollo 10 image that a remarkable, highly patterned, geometric surface region of the Moon, just northeast of Ukert, first appeared.

Hoagland quickly realized that this pattern defied conventional explanation. One consulting geologist even nicknamed the region "L.A. on the Moon," because of its strikingly urban appearance.

In the photograph, over an area roughly equivalent to the real Los Angeles on Earth (hundreds of square miles), there appears a remarkably regular, rectangular, raised, repeating 3D geometric pattern. Large surface lineations—stretching for tens of miles, appearing remarkably similar to streets running across the actual L.A. basin in Southern California—crisscross this northeast Ukert region in a rectilinear pattern. Here and there, small round craters cut into the areas of sharply contrasting, remarkably rectangular relief, like mile-sized cookie cutters. In a close-up from 4822 [Fig. 4-16], this rectilinear, artificial block-like pattern, interspersed with a smattering of remarkably uniform impact craters, is even more apparent.

Overall, the overwhelming impression is that of finding a vast, ancient, bombed-out city on the Moon.

Images taken of this same region only moments earlier—such as Apollo 10 photograph AS10-32-4819 [Fig. 4-15], with a more "over the shoulder" illumination—underscore this city-like "L.A." comparison. Under both lighting angles, this region unmistakably resembles war-devastated city blocks and buildings, now lying in hundreds-of-square-mile ruins.

Within this general, very artificial-looking landscape, a series of smaller, very brilliant, horizontal, vertical and near-vertical features also appear—some clearly resolved as rectangular structures, others as possible "sky-scrapers" (on the scale of the images); other features are seen as merely brilliant, geometrically arrayed points of light—possibly specular reflections from surviving optically flat areas similar to "windows" or entire glass walls.

And, if you look very closely, you'll see that *all* the features are vertically distorted on a very fine scale ... as if the photograph had been taken looking at "L.A." through *wavy glass!*

Because—

That's *exactly* how it was taken—*through* the same vertical "glass/rebar" matrix demonstrated with the "glass cube and tower" earlier.

Years after the preceding analysis, Hoagland (with the help of Steve Troy) found a *second* set of Apollo 10 images of "L.A. on the Moon": the "4600 series."

One such image (AS10-31-4652) revealed an astonishing perspective: layer upon layer of obviously geometric *glass*, reflecting the brilliant light of the rising sun almost directly back toward the approaching spacecraft. One can clearly see multiple "layers," "floors," and innumerable "right angle" of what can only be geometrically-arranged "manufactured structures."

There are even what appear to be "suspended walkways" and "transportation bridges" clearly visible [Fig. 4-17].

There is simply no plausible explanation for these incredibly reflective, geometrically aligned, *transparent* structures except an artificial one...

Subsequent to the presentation of this data at Ohio State, some unofficial reaction by NASA officials attempted to dismiss the anomalous geometric pattern northeast of Ukert as nothing more than "standard lunar geological activity": lunar lava flows cooling to create geometric fractures, subsequently excavated and exposed by "random meteorite impacts."

This glib explanation completely ignored simple facts about such cooling processes familiar to geologists on Earth; such as the fact that on this planet, the size of such typically expected lava fractures is mere *inches,* or at best, feet ... not thousands of feet as observed in the city-block-sized pattern photographed, on *multiple* Apollo 10 images (and from almost a full 180-degree arc) near Ukert!

This early, semi-official explanation" for the astonishing surface geometry seen near Ukert also blithely ignored the fact that *billions of years* of meteorite impacts should have created impact debris (a lunar regolith) sufficient to bury any such original natural lava fractures. They would not reveal a rigidly-controlled network extending both laterally and vertically orders of magnitude (factors of 10, 100 and 1000 ...) beyond *any* known geological analogs on Earth.

When such trivial explanations to explain this mystery are firmly put to rest, we are left with a real scientific puzzle: what natural forces could have possibly created, and then preserved, such a stunning 3D geometric pattern? If it was formed billions of years ago, when lava last flowed across the Moon (according to Apollo's dating of the returned rocks), what could have maintained such a striking, regular geometry against the inevitable forces of erosion (even on the Moon) for those *billions* of years?

It is our contention that *no such natural model* can, even at this point, adequately explain "L.A. on the Moon." Thus, only an *artificial* model—which automatically incorporates tall, rectangular structures, a rigidly controlled, rectilinear horizontal layout, coupled with a relatively "recent" time frame (because of limited erosion)—is the only serious contender.

And, of course, no "geological" explanation can possibly account for the fact that the astronauts took these amazing Hasselblad images of "L.A. on the Moon" through *wavy glass*—located (impossibly) somewhere "between" the surface ... and the orbiting Apollo spacecraft.

A separate estimate, based on NASA's attempted measurements of the rate of past and present impact crater formation on the Moon (that smattering of "one-to-two mile craters" lying over "L.A.," mentioned previously), independently supports a relatively "young" geological age for this enigmatic pattern near Ukert—a "few hundred million years," at best. If that is the actual age of these remarkable geometric formations, this opens up literally astounding possibilities for "who" might have constructed "L.A. on the Moon," if not the "when" and "why."

But beyond "L.A."—which was bizarre enough—the *real* prize of 4822 turned out to be what, at first, appeared to be just a small "scratch" on the negative.

Triesnecker—"The Castle"

The Castle [Fig. 4-18] is another glittering, remarkably intact geometric formation Hoagland first identified on a special version of frame 4822— provided him by "a highly-placed" source at NASA's Goddard Spaceflight Center in 1992. Bearing a striking resemblance to "Schloss Neuschwanstein," built by King Ludwig II of Bavaria in 1869 (which served as the model for Cinderella's Castle in Disneyland) it is, in fact, another astonishingly real lunar artifact "hanging high" *above* the Moon.

The Castle's location on the lunar landscape is as remarkable as its appearance.

As judged by the geometry on frame 4822, this highly anomalous object is actually suspended some *nine miles* above the lunar surface, somewhere between the eighteen-mile diameter crater, "Triesnecker" and the well-known "Hyginus Rille."

A composite of three close-ups [Fig. 4-19]– a "raw" version on the left, and two enhancements (right, and bottom)—reveal additional extraordinary aspects of this object ... including the fact that, like the "cube/tower," the "castle" too is surrounded by a faint matrix of "sparkling geometric 'stuff.'" There is also the striking presence of a clearly *sagging* cable seen at the very top of this amazing artifact ... to which the obviously large and massive structure is *physically* attached!

Eventually identified on two separate NSSDC versions of "4822," the

Castle has presented Hoagland's critics with a series of extraordinary puzzles, beginning with the obvious:

What's holding it up?

Stereo analysis of these two "versions" [Fig. 4-19] confirmed the Castle's presence miles *above* the lunar surface—apparently just "hanging out in space." Around it, as previously noted, appeared a collection of much smaller, equally reflective geometric slivers—as if all were simply fragments of a once much larger, previously intact, somehow suspended structure.

A separate clue to the anomalous height of all these objects came from their remarkable surface brightness, particularly the Castle, compared with the lunar surface below.

The fragments all appeared highly reflective—much brighter than the actual lunar surface far beneath. For some reason, across all the 4822 frames, there was a curious "optical veiling" of this lunar surface (viewed from left to right [Fig. 4-20] This "veiling" extended from below the Apollo spacecraft's location, out across the Hyginus Rille to the horizon—and a twenty-five-mile diameter crater barely visible there, called "Manilius." This crater is hundreds of miles from the Apollo spacecraft, yet both it and the lunar horizon around it were inexplicably "fuzzy"—as if darkened by the same mysterious "obscuration" covering the lunar landscape much closer to Apollo.

In fact, this entire *airless* lunar landscape, lit by a mid-morning sun, should be brilliantly illuminated; instead, across more than half the frame, this striking scene was mysteriously shrouded by some kind of "dark, highly absorbent, apparently highly *directional,* optical medium" (think "window slats," or "blinds")—with only the Castle and a scattering of other, equally reflective debris apparently left sticking up *above* this "interfering layer" into unimpeded sunlight.

With no definable lunar atmosphere—rain, fog or clouds, or any other form of familiar terrestrial optical absorption mechanism—the only logical explanation for this obscuration and fuzziness (after the possibility of simple photographic defects to the original 4822 negatives had been eliminated—which it was) is that the Apollo astronauts actually photographed the remains of some kind of remarkable, constructed optical anomaly stretching over the eastern end of Sinus Medii—a semi-transparent, glass-like, mechanical medium, with remarkably focused optical properties when viewed from *key directions.*

In other words, the Apollo 10 crew, via 4822, apparently recorded another section (far northeast of the Tower/Cube remains) of this extensive "Sinus Medii dome."

The most visible, most blatant anomalous object on 4822 is, of course, the Castle—which, to be so strikingly visible, must be hanging *above* the heavy optical obscuration blanking out the lower levels of the Dome, and (from the angle of the photograph) the sun-drenched lunar surface underneath. In an effort to confirm his findings, Hoagland began ordering (through additional friends and acquaintances) more copies of AS10-32-4822 from various NASA archives. To his surprise he discovered that there was something even stranger about the image than the simple fact that it was blacked out in the catalogs—it seemed to exist, simultaneously, in a number of "different" forms—but all labeled with the *identical* frame number:

AS10-32-4822.

Comparison of "the Castle" in his original Goddard version of 4822 with a second copy provided by a student, Alex Cook (who ordered it separately through Goddard), revealed a *major* surprise: Alex's copy turned out, in fact, to be a completely *different* photograph, hitherto apparently unknown. Cook's "4822" thus immediately formed a natural stereo pair to the "4822" that Hoagland had obtained completely independently (from his NASA-Goddard source, in 1992). Setting up both photographs side-by-side for stereo viewing revealed several startling new details about "the Castle" ... and frame 4822 [Fig. 4-21].

In Hoagland's original version, the amazing drooping cable (apparently, *really* holding the Castle "in the grid!") is clearly visible; but in Cook's version of "4822," not only has the cable disappeared with the increased viewing angle, but the entire structure has visibly "foreshortened"—clearly the result of optical parallax, as the Apollo spacecraft moved farther to the west between the two exposures. From this fact alone, there is no doubt that—despite (inexplicably ...) sharing the *same* frame number—these are two views of the same object, present on two distinctly *different* photographic images!

In addition, formerly clear details of the Castle in the left-hand shot have been overlaid on the right with more of that "wavy shower-glass effect"—the result of the spacecraft moving to where its new line of sight was *through* an intervening, denser section of the remaining "glass grid matrix" overhead ... hanging "somewhere" *between* the spacecraft and the Castle. In close-up, the bulk of the material within the Castle looks exactly like *frosted* glass; like any originally transparent substance would look, after being exposed for countless millennia to hypervelocity micro-meteorite impacts in a vacuum.

In an effort to resolve, among other fundamental issues, why there were "so *many* different versions of 4822" (Hoagland eventually found over a *dozen* ...), in 1995 he led a team of diverse colleagues and associates on a unique visit to the National Space Science Data Center in Greenbelt, Maryland—NASA's chief repository for space imagery. The visit lasted two days, with one major objective being to show NSSDC officials some of the astonishing "anomalies" he'd found, and solicit their professional reactions (below). Another objective was to survey first-hand what *other* NASA data might exist, which could shed additional light on the physical nature of the astonishing objects he had found ....

Unfortunately, although much was accomplished in the course of those two days—including identification of several *additional* versions of "4822" also existing at NSSDC, ultimately revealing even more critical details of the "Sinus Medii Dome"—the nagging question "why are there 'too many versions of *this* key Apollo frame?'" was never satisfactorily answered. On the second day of the presentations and discussions, embarrassed NSSDC officials reluctantly revealed to Hoagland and his eight visiting associates that—apparently, overnight—there had been a mysterious "disappearance" of (presumably) *the one Apollo 10 negative* from which Hoagland's "special Castle print" had been produced and sent to Hoagland, at the request of that "high-level Goddard official," three years earlier.

Without that crucial negative—freely accessible to other scientists world-wide from an *official* NASA archive—any public claims made by Hoagland regarding the properties of "the Castle" *on that negative*—would be impossible for anyone to verify—including the officials at NSSDC to whom Hoagland had just presented the striking evidence for its existence.

Which, of course, seemed to be exactly what "someone" at NSSDC—after they'd seen what Hoagland had discovered on *that* NASA negative—had (overnight) carefully arranged.

This, however, was not the only "data integrity problem" Hoagland verified first-hand at NASA's official data archive ....

Shortly before the trip to NSSDC, Hoagland had been given (by a "source") a remarkable Apollo 16 lunar photographic print. The story that came with it was almost right out of "The Man from Uncle" (a popular 1960's "cloak-and-dagger" TV show, which ran during the timeframe of the Apollo missions).

The "source" was well-known to NASA and (according to his account), "regularly visited" at NASA Headquarters. He had been there, in the Administrator's office, a few days after Apollo 16 returned to Earth from its successful visit to "the highlands of the Moon"—April 27, 1972. For some reason he was left alone in the Administrators office between meetings; he looked over and saw a massive bunch of Apollo photographs lying on the Administrator's desk. Bored, he casually leafed through a few ... and was shocked by what he saw.

On impulse (he said later ...) he quickly slipped one of the prints into his briefcase and—before the Administrator could return—abruptly left.

Twenty three years later, he handed a copy of that "stolen Headquarters Apollo print" to Richard Hoagland.

What was *on* that purloined NASA photo was simply astonishing— unequivocal evidence for more "massive, ancient lunar engineering"—this time, stretching all across *Mare Crisium* (the aptly-named "Sea of Crises"). The centerpiece of that construction was none other than another several- miles-high *tower* [Fig. 4-22]—surrounded by a host of other glass-like stuff, all arrayed above a series of incredibly reflective lunar "craters" on the actual surface of Crisium—"craters" which, in this close-up, actually looked more like "miniature, *round* glass domes."

One of the prime objectives of Goddard trip was to find an *official* negative of this "hot" *Apollo 16* photograph (it had an ID number)—to see if the two versions were *the same*.

They weren't.

After putting in a request for a large print the night before, on the second day of the visit when he laid his second-generation enlargement of the ~20 year-old "stolen photograph" alongside the NSSDC version (produced from the on-site negative of AS16- 121-19438, kept in the NSSDC "vault") all the marvelous, "glittering detail" hanging in the sky *over* Crisium on Hoagland's version had *disappeared* ... replaced by a dead, dead "black"—which obviously had been airbrushed in on the copy negative, to represent what someone thought space above the lunar surface *ought* to look like.

And, of course, there was no longer any stunning "tower" ....

So, like the "unverifiable veracity" of 4822, the authenticity of Hoagland's version of AS16-121-19438 would forever stay unproven ... until the *real* negative decides to "stand up."

* * *

All in all, as Hoagland would be the first to volunteer, despite these *still* unanswered questions, the two-day Goddard visit in 1995 by Hoagland's lunar investigation team was quite productive; when they left, Hoagland was able to walk away with a number of additional Hasselblad negatives of *new* areas of interest on the Moon (nothing as spectacular as the Crisium Spire, unfortunately ...), as well as (surprisingly) several *reels* of 5-inch Apollo 16 *panoramic* camera negatives (the highest-resolution Apollo photographs currently available), comprising literally thousands of superb high-resolution negatives, which are still being carefully examined.

Politically, because of the scientific case he and his associates made over those two days to NSSDC officials, demonstrating first-hand some of the major anomalies he'd found on NASA's own source data, Hoagland was able to spark enough serious "inside" interest to insure a steady flow of *future* images from this official NASA archive.

And, there was more ....

The Russian Connection

During the NSSDC visit, Hoagland was also able to examine some official NASA publications containing images from one of the key USSR lunar missions of the 1960s, Zond 3. If his ideas about the Moon had any merit, then it seemed likely that the independent Russian probes would have encountered—and recorded—the same sorts of anomalies Hoagland was finding in the American data.

Unfortunately, much of the data gathered by a variety of Soviet space probes was (and still is ...) inexplicably off-limits to researchers from the West. Even after the Cold War officially ended, in 1991, and Hoagland was finally able to send personal "emissaries" to Moscow in months-long efforts to locate *original* versions (negatives or prints) of old Soviet Mars and lunar photographs—he had met with a virtual "Berlin wall" around the subject. At one point, one of his folks was actually pulled out of a long hallway and into an empty room at the National Academy of Sciences, in Moscow by the KGB— and told flatly to "stop looking!"

Eventually, at the Goddard library during the NSSDC visit, Hoagland was at least able to examine some of the NASA *published* photographs from the Zond-3 Mission, sent to NASA Headquarters in Washington shortly after the Mission first returned its images to Moscow, in July 1965. Those photographs, having somewhat lower quality than the American Lunar Orbiter series (Zond's images were also taken on film, then relayed back to Earth as a series of "facsimile scanned" negatives), would have at least some historical value ....

The first of the Zond series to actually make it to the Moon (Zonds 1 and 2 were aimed at Venus and Mars, respectively), Zond-3 was originally designed to go to Mars as well; the Soviets at that time either lacked the capability to safely place a spacecraft into orbit around the Moon, or simply thought Mars was more important ....

Somehow, the primary Zond 3 launch opportunity ("window") *for Mars,* in 1964, was "missed," yet the spacecraft was launched anyway ... months later, in 1965 ... towards "where Mars *would* have been," even if the actual planet was no longer a reachable target.

At least, that's what the Russians actually said in 1965!

Whatever the initial plans for Zond 3, its unique "delayed" mission took it past the leading hemisphere of the Moon as a "fly-by" ... ostensibly en route to "deep space and a 'phantom' planet Mars." It ended up, however, having unique significance for *lunar* investigation.

As the spacecraft passed thousands of miles "over" the massive Mare Orientale impact "basin," heading forever away from Earth, in looking back toward the Moon (see diagram - below) it serendipitously recorded two images of two completely different "anomalous lunar structures," that fit perfectly into Hoagland's evolving model of an ancient Sinus Medii Dome.

The first image [Fig. 4-24], published in TRW's "*Solar System Log*" in 1967, captured *another* "twenty-mile-high tower" (like the one in Crisium)— extending vertically straight up from the lunar surface, located somewhere in the western end of *Oceanus Procellarum* ("The Ocean of Storms"). The next Zond 3 image, taken thirty-four seconds later, had no "tower"—indicating it had, by that time, slipped over the lunar horizon due to the fast Soviet spacecraft motion and direction.

There were, in fact, a number of other intriguing "geometric anomalies" to be seen on the lunar Farside in this first Zond image—but the reproduction quality in the *Solar System Log* was not quite high enough to be confident of their true nature; the "tower," however—standing stark against *black space*, looking remarkably like the features seen on LO-III-84M *and* AS16-121- 19438 (although from a lot farther away)—was unmistakable.

The second amazing Zond 3 shot was in an official NASA publication, "Exploring Space with a Camera" (NASA SP-168, 1968)—also in the Goddard Library.

After the "tower" vanished (due to the continuing trajectory of Zond 3 past the Moon ...), it was replaced by an equally "anomalous feature," located on the lunar horizon about a thousand miles further to the south. This glassy-looking structure appeared strikingly like another lunar "dome" —very similar, in fact, to the "miniature domes" caught on that Apollo 16 shot of Crisium—a few miles high, and about the width of a medium-sized lunar "crater."

Again ... this "Zond 3 dome" extended several *miles* above the airless lunar horizon, against black space. And, like the earlier "Zond tower," neatly aligned with the local *vertical* [Fig. 4-25].

A close-up of this Zond 3 "dome" (rotated 90 degrees [Fig. 4-26]) reveals a significant amount of deterioration due, undoubtedly, to long-exposure effects from meteor erosion, but the outline is still strikingly clear: another giant, *glass-like* geometric structure ... extending *miles* above the Moon.

No wonder the KGB told Hoagland's guys to "stop looking!"

There is a postscript to all this, which instantly casts not only the Zond 3 Mission—on its unique "non-planet, planetary journey"—in an entirely new light (did the Soviets deliberately launch it as a "clandestine lunar mission," to actually take pictures of the largest artifacts they suspected on the Moon ... with "Mars" just as a cover story?). Whatever the truth, the mission also now raises the far more serious question—

"Just how much *did* Kennedy tell Khruschchev ... about what was waiting for *both* nations on the Moon ...?

Because—

The *date* when Zond 3 acquired its own explosive images of lunar artifacts was *July 20, 1965*—four years, *to the day*, before that date would be forever immortalized in history by Neil Armstrong and Buzz Aldrin's first human footsteps on the Moon.

As you shall see later in this volume, a date which also has bizarre "significance" far beyond that moment and event for the very "secret societies" that Kennedy publicly decried at his American Newspaper Publishers Association speech.

So, what was the final outcome of Hoagland's invited excursion into NASA's official repository for decades of accumulated lunar and planetary information?

The visit established that the lunar data sets on which Hoagland was finding truly astonishing, clearly artificial objects, was absolutely real. It further established that some in NASA (at least, in the beginning ...) were willing to look with open minds at what Hoagland and his associates were finding (this, tallying with the other official NASA invitations Hoagland had received earlier, to address other NASA Centers) ....

An alternative hypothesis is that NASA really needed to find out exactly *how much* Hoagland already *knew* ... and agreed to the "official NSSDC visit" as a clever means of getting him to show them—so they could effectively sanitize *that* data.

Regardless, the visit also demonstrated that "someone" in the Agency really was willing to make crucial evidence *supporting* Hoagland's claims— like, an official Apollo photographic record—simply *disappear* ... to effectively prevent further scientific investigation of key elements of Hoagland's research findings.

Hoagland's only lingering disappointment of the entire experience is the still unsolved mystery (and yes, it's still unanswered ... after 12 long years) what *really* happened to that critical Apollo 10 negative that night at Goddard—the one that holds the *proof* that "the Castle" is quite real?

Stay tuned ....

The Ken Johnston Collection

A few months after the NSSDC visit, in early 1995, Hoagland was on a lecture tour in Seattle. It was then that he met Ken Johnston—a Boeing engineer at the time, and a former fighter-jock and test pilot for Grumman Aerospace. After his tour of duty in the Marines as an F-4 pilot, Johnston had gone to work at NASA in the mid 1960s as the chief Lunar Module test pilot at the Manned Spacecraft Center in Houston. There he and his team subsequently trained all of the Apollo astronauts to fly the Lunar Module, while simultaneously being part of the extensive spacesuit development program ("I was 'capsule size,'" Johnston would later joke).

Johnston later moved across the center, going to work for Brown-Root Corporation and the Northrop Corporation in MSC's Lunar Receiving Laboratory (LRL)—the literal "eye of the storm" during Apollo. This consortium had the prime contract for the processing of the actual lunar samples coming back on Apollo, and Ken's key function was as "supervisor of the data and photo control department." This was the section of the LRL that handled all of the critical photographic and written documentation related to Humanity's first returned pieces of the Moon; after processing elsewhere in the Lab, the films and samples also went through Johnston's office for cataloging and long-term storage.

The specific "trigger" that brought Ken and Richard together, decades later, was Johnston's developing interest in Hoagland's latest line of research—what NASA might have actually discovered on the Moon ... and then kept *hidden*.

Having read *Monuments*, and seeing that Hoagland was coming to town as part of his national lecture tour, Johnston had written a letter of introduction and sent it to Hoagland. Earlier, while reading Hoagland's book, Johnston had recalled some disturbing "incidents" during his own time at NASA, events "that had always stuck in the back of my mind" he told Hoagland, which he could not easily dismiss even years later; reading *Monuments* had reawakened those old memories, and Johnston had began to seriously question if there might indeed be something to this whole idea of a "cover-up" at the Agency, after all. At Hoagland's Seattle presentation, during one of the breaks, Ken introduced himself and invited Richard over to his house the next day, to look at Ken's collection of "about 1,000 old NASA photos and other memorabilia" of his time at NASA. A new and lasting friendship was born.

However, the existence of this extensive number of Apollo photos, in a *private* collection—and how they got there, as he listened to Ken tell it later—really disturbed Hoagland.

Johnston explained that, as head of the LRL photo lab, it was his responsibility to catalog and archive *all* of the Apollo photographs. As part of the archiving process, the LRL eventually developed *four complete sets* of Apollo orbital and handheld photography, comprising literally tens of thousands of first-generation photographic negatives and prints.

Ken also had responsibility for managing the 16mm mission films from the on-board "sequence cameras" (modified military gun cameras), operating from the Command Module and Lunar Modules during various phases of the missions, including lunar orbit and descent/ascent. One of his duties was to frequently screen these "on-orbit films" at MSC, before members of the various scientific and engineering teams. Here is how he described one such screening on the popular *Coast to Coast AM* national radio program:

"Well, on that particular case—this was Apollo 14—after we had received the film, right after the astronauts had returned to the Earth, it had been processed in the NASA photo lab. It was my responsibility to put together a private viewing for the chief astronomer—that was Dr. Thornton Page and his associates and contributing scientists. I took the film over and set it up into what is called a 'sequence [projector]'; it's kind of like one of the gun cameras they use in the military [but in reverse—a projector]—where you can stop, freeze frame, go forward, back up and zoom in.

"And we were viewing the Apollo 14 footage, coming around the backside of the Moon as we were approaching a large crater. Now, due to the sun angle on the front side [of the Moon] that you would be looking at (you'd probably be looking at more of a crescent at that point on the backside) in the shadows in the craters, covering about half the crater, this particularly large crater showed a cluster of about *five or six lights down inside the rim.*

"And this column or plume—or out-gassing or something, coming up above the rim of the crater, where we could see that—at that point Dr. Page

had me stop and freeze, and back up; and go back and forth several times. And each time, he'd pause a second and look... and he finally turned to his associates and said: 'Well, isn't *that* interesting!' And they all chuckled and laughed, and Dr. Page said: 'Continue.'

"Well, I finished up that viewing and I was told to check it [the on-board sequence camera film] back into NASA bonded storage in the photo lab. The next day, I was to check it back out and show it to the rank-and-file engineers and scientists at the [Manned Spacecraft] Center.

"While we were viewing it the second time—and, several of my friends were sitting next to me—I was telling them: 'You can't believe what we saw on the backside of the Moon! Wait until you see this view.'

"And, as we were approaching the same crater... and we went *past* the crater—*there was nothing there!*

"I stopped the camera, took the film out to examine it—to see if anything had been cut out—and there was no evidence of anything being cut out. I told the audience that we were having 'technical difficulties,' put it back in and finished.

"That afternoon, I ran into Dr. Page over at the Lunar Receiving Laboratory and asked him what had happened to 'the lights and the out-gassing or steam we saw,' and he kind of grinned and gave me a little twinkle and a chuckle and said: 'There were no lights. There is nothing there.'

"And he walked away. And, we were so busy... I didn't get a chance to question him again."

* * *

This is typical of stories we have heard over the years from other former NASA employees.

Johnston had also observed various and sundry oddities with the still camera images. Once, while passing through a classified building on the Center he normally didn't frequent, Ken observed artists airbrushing the "sky" in various photos. That in itself wasn't unusual, as press release prints were regularly cleaned up. What bothered Johnston in this case was that these weren't prints that were being airbrushed, but rather *photographic negatives*— meaning that, after that drastic process had been applied, the original data could *never be reproduced in the form it had originally been taken.*

(Shades of what Hoagland had personally observed, with frame AS16-121-19438, at NSSDC).

All of this took an even more sinister twist in 1972, near the end of the manned lunar program. Johnston was called into the office of Bud Laskawa, Johnston's lead at the LRL records division. At the meeting, Laskawa told Johnston that orders had come down from NASA Headquarters (through Dr. Michael Duke, Laskawa and Johnston's NASA boss) to destroy *all* of the copies of the original lunar photography that he had been protecting and archiving for the past several years.[71] Johnston was dumbfounded that anyone could order the *destruction* of the official record of Mankind's first venture beyond the earth. He protested, and begged to be allowed to donate the photographs to various universities or foundations, but was told there was "no chance." The orders were explicit—he was to destroy all four sets of the literally tens of thousands of Apollo lunar photos taken by the astronauts.

Johnston found this situation unconscionable. Eventually, after further protests, he relented and destroyed three full sets of the data—but with his guilt eating away at him, he decided to save one complete set "elsewhere." Some of the images and negatives he kept for himself. However, since the collection was so vast, he eventually decided to donate the rest to his alma mater, Oklahoma City University, where the data quietly resided—out of NASA's oversight—for over thirty years...

Hoagland was saddened and repelled, all at the same time, by the whole tragic affair.

That the extremely limited number of high-quality, first or second generation copies of the unique photographic record of Man's first voyages *to the Moon* could be so blithely—so deliberately—destroyed, and by an *official* NASA order, infuriated him almost beyond words. Obviously, after years of looking at the remaining database, with its "blacked-out catalog images, and endless, mysterious 'discrepancies' and 'disappearances'" ... and wondering ... Hoagland finally had his "smoking gun":

He was now totally convinced of a *deliberate* Apollo cover-up of "ET artifacts" —coming *officially* from NASA headquarters, in Washington.

So, what had NASA been trying so desperately to *hide*...?

A close examination of Ken's surviving photos revealed overwhelming evidence that Hoagland's darkest fears—about "deliberately concealed lunar artifacts"—were definitely well-founded; many of Johnston's photos were full of the same sorts of anomalies Hoagland had spotted on the unmanned spacecraft lunar frames at the beginning of his investigation, and on other frames quietly gathered prior to his personal visit to NSSDC a few months earlier.

Hoagland, you will recall, had been slipped a remarkable 8 X 10 original Apollo 16 print just before his NSSDC visit—revealing an unmistakable "miles-high, spire-like structure" in Mare Crisium [Fig. 4-28]—astonishingly similar to the "Zond 3 tower" plainly visible on the other side of the Moon. The existence of other "miles-high, tower-like structures" was now clearly confirmed, by the "untouched" Apollo data still in Johnston's possession, after over thirty years.

Hoagland, after seeing the Zond 3 "tower" in the library at Goddard, and witnessing the result of the blatant "air-brushing out of existence" of the equally revealing vertical anomalies on the official NSSDC version of "AS16-121-19438"—which included the extraordinary "Crisium tower" itself—had modified his evolving "lunar dome model" to include *all* the visible "lunar mare" on the Earthside of the Moon. Interestingly, his slowly accumulating photographic data re a possible "Crisium Dome" had been *visually* confirmed (and recorded) by NASA, from astronaut Al Worden, in his official Apollo 15 debriefing:

> "... It is very strange the way the ejecta... from Proclus crosses Crisium.It is almost like flying above *a haze layer* and looking down *through the haze...* It looks like it is suspended *over* it [emphasis added] ...."[72]

In fact the "haze layer" Worden describes over Mare Crisium was/is obviously not "haze" or an "optical illusion," but rather another in-situ confirmation, over a completely separate mare (from Sinus Medii), of Hoagland's lunar dome hypothesis. Craters, like neighboring Proclus and Picard (below), simply *can't* have "overlaying haze" on an airless Moon. The only way for Worden to have observed this "haze" is if he is looking down from orbit *through* a partially transparent *intervening medium*. And the only "intervening medium" we have any photographic evidence of is one of Hoagland's "shattered, glass-like, ancient lunar domes" ... this one lying over Crisium.

If the "pie wedge" partial dome over Picard is a real lunar feature [Fig. 4-28], then it would logically follow that there might be other, similar domes of this "watch-crystal-type" (rather than the "scaffolding type"—as seen over larger regions, such as entire mare ...), placed over similar smaller areas. Structurally, this makes complete sense: a large, translucent, box-like "scaffolding dome" over the entire area, reinforced by smaller, rounded "pressurized domes" over specific features, like small "craters," would be a perfect "back-up system" to possible disaster ... on an airless world.

A dome ... within a dome ... within a dome [Fig. 4-29].

This would also have the effect of providing additional, layered shielding from the harmful effects of solar radiation and catastrophic meteor decompression.

Fortunately, in ferreting out additional observational details which would flesh out and substantiate this model, we have had the able assistance over the years of a plethora of associates who were interested in extending and confirming Hoagland's original work. One of the best of these is Steve Troy—an amateur astronomer/geologist and professional mural artist/painter, with a degree in fine arts, from South Dakota.

Steve took an interest in Hoagland's lunar work early on, in 1996. As it turns out, soon after "joining the team," Steve made an important, independent discovery which significantly reinforced Hoagland's earlier observations re Mare Crisium.

Troy, like Hoagland, had obtained copies of the official photographic catalogs from the Apollo missions. Also like Hoagland, he'd gotten in the habit of ordering negatives of the images that were "blacked out" or "very dark" in the catalogs. While going through a set of Apollo 10 negatives in the mid 1990's, he struck gold.

In studying frame AS10-30-4421, he noticed a very bright area at the far right. After making a series of sectional enlargements, he soon spotted the source of this strange illumination: the anomalous brightening seemed to be reflecting off a fully preserved, rounded ... *lunar dome* [Fig. 4-30].

Seen edge-on, and up against the mountainous background of the north "shore" of Mare Crisium, this bright, clearly dome-like structure appeared *translucent* against the basin's background rim; in addition, there seemed to be a second, also semi-translucent curved structure (with a higher slope angle) *behind* and to the right of it.

Steve soon discovered that this frame was part of a photographic sequence taken as the Apollo 10 Command/Service Module traveled "right to left" on its orbital path across the visible face of the Moon, and the "southern shore" of Mare Crisium [Fig. 4-32].

But, in looking at the other images, they showed that this area had already been covered in previous photographs, and showed no unusual brightening. Steve quickly surmised that whoever was taking the photographs out the Command Module window must have been looking forward along the spacecrafts' orbital track, snapping picture after picture. Then "something," probably a flare of sunlight off the dome itself, had enticed the astronaut to turn back to his right and *re-photograph* the area he had already covered.

It was fairly easy to recreate the geometry of the photo itself and discover that the phase angle—the geometric relationship between the sun light, the dome and the spacecraft—had been perfect at that time to create just such an anomalous "flare."

Since Steve didn't have a computer at that time, Hoagland suggested that he take his negatives and prints to one of the authors (Mike Bara) for enhancement and processing. The first thing Bara noticed was that the image map for AS10-30-4421 showed that a pair of Mare Crisium craters, Cleomedes F and Cleomedes F-a, were placed just "outside" the right edge of the photo "footprint" (above). Noting that the larger crater, Cleomedes F, would have been in front of the smaller crater (F-a) from the camera's perspective, it didn't take long to connect it to the two "translucent domes" in Steve's image. The larger, wider and shallower dome was "in front" while the smaller, narrower and steeper dome was behind.

Bara quickly realized that the official NASA "footprint map" was simply *not* accurate.

Eventually, false-color enhancements confirmed that the domes were definitely there. High resolution scans showed that the Cleomedes F dome had *two* distinct edges, in essence proving that it was not part of the mountainous background, but sitting separately out on the Mare Crisium plane [Fig. 4-31].

So Steve's image represented a completely independent confirmation of the astonishing architectural details Hoagland was seeing in the Apollo 16 "Crisium tower image," both on the floor and high above Mare Crisium— slipped to him almost a year before.

Most of the prints in Ken's priceless collection at his home (kept carefully in thick, three-ring-binder notebooks, each photo in a separate glassine envelope) turned out to be from earlier Missions—Apollo 12 and Apollo 14. The 12 and 14 landing sites were unique in the entire Apollo Program—in that they were closer together than any other landings (later, some scientists complained they were "too close"). Only 122 miles separated the two Apollo touchdowns in Oceanus Procellarum (the "Sea of Storms"). In hindsight, this in itself is suspicious, since the scientific value of rock samples from two sites in such proximity (on the non-geologically active Moon, subject to only external meteor bombardment) would have been dubious.

In *To a Rocky Moon* (one of the "Bibles" of lunar geology), planetary geologist and author Don Wilhelms describes the canonic rationale behind this "curious" site selection—arguing that the Fra Mauro location of Apollo 14 was in a more "clearly defined maria region," and might therefore lend more insight into these mysterious dark features of the lunar surface than Apollo 12's site, some 120 miles to the west.

But while the unique proximity of the two landing sites may not have made much sense from a purely geological perspective, it did serendipitously enable Hoagland to make an invaluable photographic comparison—utilizing Johnston's secret Apollo stash—that would ultimately be key to validating his entire "glass dome, lunar ruins model."

NASA's obvious political efforts to provide only *altered* versions of the Apollo images for the public and the press, after 1972—by summarily ordering Johnston to destroy all *"superfluous"* copies of the existing Apollo photographic archive—clearly backfired; by driving Johnston to sequester away one untouched set in a hidden "time capsule." Hoagland was then able to discover two untouched panoramas of the Apollo 14 landing site—each a mosaic composed of individual Hasselblad images, printed on two 8x10 glossies. When these 8x10 prints were scanned, by simply adjusting the gamma and contrast in the resulting digital images Hoagland was stunned to see an enormous amount of previously invisible *geometric* detail suddenly appear ... hanging in the supposedly pitch-black, airless lunar *sky*.

Overwhelming *surface* confirmation of his "glass-like lunar domes" hypothesis, in *another* "mare location" on the Moon—and in exquisite detail (see Color Fig. 4).

This geometry—for those who might try to argue that Ken's hidden NASA prints had, somehow, simply "lightened" over those thirty intervening years— is strikingly "aware" of where the Sun was when the images were taken; as can be seen in this 360-degree enhanced version of the panorama. Most of the light above the lunar horizon is being scattered directly *opposite* the Sun (hiding behind the Lunar Module, in the center)—precisely consistent with sunlight being scattered by the faint remnants of a former glass-like lunar construction, arching high over the landing site.

Among the treasury of additional prints Hoagland found in Johnston's files were copies of the *individual* Hasselblad images making up this amazing Apollo 14 panorama, which possessed the much higher resolution needed to discern the real nature of the remarkable geometric detail hanging in the sky. One particularly spectacular frame stood out—what has come to be known as "Mitchell Under Glass" (AS14-66-9301)—a shot of astronaut Edgar Mitchell deploying the TV camera just north of the A-14 landing site, taken by Apollo 14 Commander Alan Shepard early in EVA-1, as part of the preceding panorama sequence.

Mitchell is apparently completely oblivious to the awesome, layered, glass-like geometric architecture rising overhead (see Color Fig. 5).

Or, another heavily refractive, multi-leveled, *glass tower* rising to the north.

As Hoagland made further enhancements and enlargements of this astonishing full-resolution Apollo 14 image, he discovered multiple *layers* of breathtaking "structural construction" embedded in the NASA frame; multiple surviving "cell-like rooms," three-dimensional "cross-bracing," angled "stringers," etc... all following logical *structural* patterns for a massive work of shattered, but once coherent, glass-like *mega-engineering*.

A close-up of this section revealed astonishing additional details of this complex, three-dimensional layered architecture—including the totally unexpected discovery that at the base were *sharply angled, massive buttresses*, apparently anchoring the whole towering engineering framework deep beneath the surface.

This steeply-angled intersection of the vast geometric superstructure with the lunar horizon, strikingly echoed engineering techniques used by engineers on earth for centuries, in fact, techniques going back to the construction of Europe's most impressive architectural forms—the cathedrals and their flying buttresses—and, obviously, for the same reason. The cathedral buttress was designed to spread vertical compression forces caused by massive gravity loads on the roofs and walls, which ultimately would have shattered the relatively fragile construction materials used in the Middle Ages (cut stone), over a wider "footprint."

The buttresses thus provided resistance against lateral forces, permitting much higher construction.

Extreme close-ups of these amazing "lunar buttresses" [Fig. 4-33] confirmed a beautiful set of at least three parallel constructions in this narrow-angle view to the north, each inclined at about a 45° angle—and anchored, apparently, well beyond the visible horizon.

These unique architectural and engineering forms—but on a *lunar* photograph—were solid confirmation that the additional, small-scale constructional details Hoagland was seeing on this "hidden" NASA print, were far beyond mere photographic defects; they were, in fact, solid *proof* of amazing, ancient lunar engineering ... but on an almost *unimaginable scale*.

After all, why would photographic "blemishes" hold such a consistent set of angles to the visible horizon, and, simultaneously, be so rich in internally consistent geometric and logically *constructional* details? And why would such "photo defects" restrict themselves to just the airless lunar sky *above* the lunar horizon?

They wouldn't.

After returning to the East Coast with boxes of Ken's pristine photographs in tow, and after his astonishing discoveries on his initial scans, Hoagland decided to begin an intensive search again through NASA's public database at NSSDC, for similar *untouched* Apollo images still under NASA's control. As a scientist and former consultant to CBS, he realized that only if he could find publicly-accessible confirmation of the amazing details on Ken's "hidden stash," would other scientists, the public ... and, most critically, the press ... even begin to take seriously the overall hypothesis that NASA had found, and then *deliberately* suppressed, the remains of an almost inconceivable ancient, extraterrestrial civilization—one which had left miles-high, incredibly geometrically-precise ruins on the Moon.

And, now, after seeing what was on Ken's carefully preserved images, he had a new idea of exactly where to look.

At the height of the Apollo Program, when new lunar landing missions were coming every few months, NASA's Public Affairs Offices were beyond "busy"; their job was to try to communicate the on-going success of Apollo to the American people and the Congress through the press (remember, this was long before the Internet). Their primary tools—in an era of only the printed word (mainly newspapers and four-color, "glossy" magazines—like Life and national Geographic) and television—were high-quality still photographs, and a series of short films focusing on each new mission as it successfully ended. (These historic NASA films can now be seen, endlessly repeated, on NASA Television, for those who have satellite or cable access to the Space Agency's own television channel—caution, they're usually shown at three or four a.m.)

Hoagland, in a burst of inspiration, realized that these "classic" 16mm Agency films—produced in NASA's frenetic haste to "get the word" out around the world—offered just the possibility for *uncensored* original lunar data. That, in the rush to get photographic and other materials from the just-returned missions into the films, the rushed images might not have been completely "sanitized" before being recorded in these films, to get them produced by a private contractor in time and then distributed (literally) worldwide ... before the *next* Apollo mission.

This extraordinary time-pressure on NASA Public Affairs to get the word out, might really have allowed photographic details of what the crews *really* saw and photographed upon the Moon to "slip through the cracks." It was certainly worth a shot.

Hoagland decided to focus first on the 16mm film that NASA had released just after Apollo 12, called *Pinpoint for Science*.

The name came from one of the engineering objectives of Apollo 12, in the wake of the historic and overwhelmingly successful *Apollo 11* mission. One of the (small) "failures" of Apollo 11, according to NASA, was their inability to know precisely "where" Neil Armstrong had actually touched down after he took manual control and flew the LM over "a crater surrounded by rocks" to a landing site about five miles further "downrange" in Tranquility than the intended landing site. Because this official "landing uncertainty" had dragged on for several weeks, well after the crew had returned and been debriefed, NASA was determined to refine the lunar landing procedures for the next Mission, Apollo 12.

Besides, since the second landing was being optimistically targeted within walking distance of the two-year-old unmanned Surveyor 3 (which had successfully soft-landed in April 1967), NASA had to come up with a fool-proof means for real-time "pinpoint navigation"—if it was to successfully set down Apollo 12 anywhere near Surveyor. Ergo the name of that official mission film, after Pete Conrad and Alan Bean's incredibly successful "pinpoint" lunar landing.

The full story of how Hoagland got access to early generations of these NASA PR films at NSSDC, and simultaneously managed to convince a photographic engineering friend and colleague in Los Angeles, John Stevens, to load up a massive, one-of-a-kind "moviola/telecine" machine Stevens had invented—an original version of a standard screening device used in Hollywood for editing feature films, *without* introducing scratches—would fill another book.

Suffice to say, at Hoagland's request Stevens rented a U-Haul, loaded up his "floating bed film viewer/projector" in L.A., and *personally* drove the equipment 10,000 miles (round trip) from Los Angeles to the Goddard Spaceflight Center, just outside Washington, D.C., and back—just to transfer key Apollo mission films to video for Hoagland's research.

They immediately struck gold.

After having an NSSDC 16mm print of *Pinpoint for Science* transferred to video, via John's "miracle machine" set up in the Goddard parking lot, Hoagland made selected screen grabs from the resulting video in the computer. He was focusing on original Oxberry Animation Stand (another Hollywood device, used to animate still photographs in movies) animations that NASA's "contract producers" had incorporated in the film— pans and zooms of "stills" from the original Apollo 12 Hasselblad shots taken on the Moon by the Apollo 12 crew, Pete Conrad and Alan Bean.

Hoagland's reasoning was simple: in scanning Ken's priceless Apollo 14 C-prints, he'd discovered that the computer could "see" what the human eye could not—incredible geometric detail in the pitch black areas, like the lunar sky. The sensitivity of modern "CCD" imaging technology, in even commercially-available image scanners, coupled with the amazing enhancement capabilities of state-of-the-art commercial software—like Adobe's Photoshop—allowed the invisible detail buried in these supposedly black layers, of these thirty-year-old emulsions, to ultimately be revealed—a "democratization" of technology that no censor at NASA could have *possibly* foreseen over more than thirty years.

Taking this discovery one step further, Hoagland reasoned that if the *Pinpoint for Lunar Landing* producers had taken rush 8x10 hardcopy prints from the original Hasselblad Moon shots, placed them under the intense lights on a professional Oxberry Animation Stand (to add a few pans and zooms to the final film, to give it more of a feeling of "being there"), then on those frames of the 16mm release print, even after several generations of copying, some vital, invisible detail in the otherwise pitch black lunar skies over the Apollo 12 landing site might remain—still detectable by the computer.

The results of this entire chain of logic [Fig. 4-34] far exceeded Hoagland's cautiously optimistic expectations, even coming off his recent astonishing experience with Ken Johnston's pristine NASA prints.

The enhanced 16mm frames (captured by John Stevens' machine from NASA's official *Pinpoint for Science* print) unquestionably revealed more "reflective, glass-like ruins" over the horizon from the Apollo 12 landing site, and the *same* massive inclined buttresses, slanting down beyond the lunar horizon in the distance. Bright star-like objects sparkled amid the amazing, steeply-pitched, stair-stepped lunar ruins; these could not be *real* stars, of course (the exposure times of the original Hasselblad frames were far too short to enable stellar photography, even on the airless lunar surface). They had to be more remaining "bits of the lunar glass matrix" that Hoagland had previously established—intensely reflecting sunlight, *still attached* to the visible but shattered geometric framework of the once vast "lunar dome" that stretched miles overhead.

The Lunar Module Intrepid was visible to the left, partially obscured by the near-by lunar horizon, but it was the striking, unmistakable appearance of those slanted buttresses, seen on a much bigger scale from the Apollo 12 landing site, which gave Hoagland independent but vital confirmation for his entire model of the "ancient lunar domes."

Given that two Apollo missions—14 and now 12—had photographed the same crystalline geometry, and apparently in generally the same location (given the spacing and location of the two landing sites—only 122 miles apart), Hoagland was more convinced than ever that his lunar dome theory was correct.

That these inclined buttresses were actual foundational components of a single mega-engineering structure—apparently once covering a huge portion of Oceanus Procellarum—Hoagland was now virtually certain; even a superficial comparison between the "domes" appearance from the relatively nearby landing sites, revealed striking engineering commonalities.

But the icing on the cake was still to come.

Coincidentally, Hoagland learned of a series of interviews, conducted by *Discover* magazine in the summer of 1994, about a year before his pioneering film experiments at Goddard and the NSSDC, with eleven of the original Apollo astronauts. One of those interviewed was Alan Bean—the Lunar Module pilot on Apollo 12.

In the course of his interview, *Discover* asked Bean—an astronaut-turned-accomplished-artist after his historic November, 1969 Mission to the Moon—a pretty simple question, certainly not one to evoke any unusual answer, especially after all the "sanitized" official NASA photographs everyone's seen.

Discover: What did the sky look like from the Moon?

Bean: It looks like a *shiny* black. It doesn't look like Earth black at night. Up there, space has a real *shiny* look. It reminded me a little bit of *patent-leather shoes*. And I kept asking myself, as I looked at it, "Why does it seem so *shiny*?"[Emphasis added.][73]

The core prediction of Hoagland's entire lunar dome hypothesis is that what we (and the astronauts) are seeing in the sky—in all these *untouched* lunar images—is *glass*. Lots and lots of *glass*... that *this* is what Alan Bean, the astronaut-turned-artist, was subconsciously remembering from his own epic journey to the Moon, long after he returned home from the most unusual environment to which any artist could ever be exposed. That it was *this* combined effect, from innumerable jagged pieces of softly shining *glass*, sparkling across the supposedly empty vacuum of a lunar sky, which caused Bean to *still* wonder ... decades after his experience—

"Why does it seem so *shiny*?"

Now we know.

Hoagland's confirmation of his "shattered domes" hypothesis lies not only in the stunning, original lunar images he'd been able to uncover, beginning with Ken Johnston's "secret stash," but also in the anomalous "shiny black space" that Alan Bean still so vividly *remembers*. This is now photographically confirmed as having been caused by a billion bits of meteor-shattered glass still catching sunlight against what should have been a pitch black lunar sky.

But Hoagland would have to wait more than a decade after this initial photographic evidence for the final, full official confirmation of his "lunar dome hypothesis."

It is only in the last year, since 2006, that a flood of literally thousands of equally untouched Apollo lunar images have suddenly, mysteriously, begun appearing all across the Internet, including on *official* NASA websites. These recently released Apollo lunar images—containing overwhelming evidence of precisely the type of "vaulting, crystalline, *geometric* lunar architecture" that Hoagland first confirmed on Ken Johnston's carefully preserved *original* images from NASA's LRL, from over thirty years ago, have now made it easy for *anyone*, armed with a computer, an Internet connection and commercially available software, to *independently* confirm *everything* Hoagland has been saying all along.

These current official websites range from the Apollo Lunar Surface Journal—an Apollo history site, maintained by former Los Alamos National Laboratory astronomer Eric Jones[74] for NASA headquarters itself—to the Project Apollo Archive, a companion website maintained by one of the chief image suppliers for NASA's Lunar Surface Journal, Kipp Teague.[75]

As specifically noted on both sites—the Apollo lunar images presented are "digitally scanned by Johnson Space Center directly from the original Hasselblad film roll[s]."[76]

Obviously, "someone" inside NASA—perhaps following Ken Johnston's courageous, patriotic, pioneering lead—has *finally* decided to tell the truth about what Apollo *really* found. Hoagland, of course, is continuing to pursue additional, scientific confirmation of the "ancient dome hypothesis." The newest releases of thousands of high-resolution (~2,500 line) scans of the original Hasselblad film rolls has now provided overwhelming evidence of the existence of the "lunar glass" he first predicted over a decade ago.

In addition to reflection, the other "hallmark signature" of glass is its ability for refraction—to bend light by slowing it down, as the light passes *through* the glass—but light doesn't slow down (bend) uniformly when entering or leaving a transparent medium, like glass; it bends *selectively*—depending on the wavelength and the angle. This is the basic physics of a prism: higher frequencies (shorter wavelengths—blue—violet) are slowed down *more* (thus, they are deflected through a greater angle of "dispersion") than lower wavelengths (longer wavelengths—orange—red).

This results in the classic "rainbow" seen after light has passed through anything transparent... glass, water droplets, ice, quartz crystals... *any* refractive medium capable of differentially bending (slowing down the velocity of) light, compared to its velocity in a vacuum.

Based on these fundamental laws of optics, Hoagland made a second flat prediction vis-à-vis his "lunar domes" years ago: Hoagland forecast the existence of "a billion prismatic rainbows," sprayed across the desolate lunar landscape from a billion shards of glass suspended in the gridded network of the awesome lunar domes still arching overhead... all *differentially* refracting sunlight (see Color Fig. 8 - Apollo 17 frame AS17-134-20426 [Apollo Lunar Surface Journal]).

This impossible "prismatic color dispersion," from a supposedly matterless lunar vacuum above the lunar surface, is found on an amazing , new NASA image—AS17-134-20426. It shows Apollo 17 astronaut/geologist Jack Schmidt holding a collection rake during one of the Apollo 17 EVAs, while behind him, suspended in the visible "grid" (texture pattern) above a distant lunar "hill," is apparently a relatively large (or nearby) piece of glass (the enlarged inset)—reflecting (and then differentially *refracting*) sunlight at just the right angle to be prismatically captured in this official Apollo 17 photograph! The original Hasselblad image was taken by Apollo 17

Commander, Gene Cernan, at the Apollo 17 Taurus-Littrow Lunar Landing Site, looking west.

Sharp eyed readers will notice an even greater "anomaly" about this image—the entire western lunar horizon beyond Schmidt seems to be *layered* above the horizon, and in bands of *reddish light* ... grading upward to deep blue. The full frame of AS17-134-20426 reveals the true extent of this equally remarkable (and impossible!) optical phenomenon—if the lunar sky is truly empty. The physical explanation—given the "shattered glass dome hypothesis"—is as simple as it is (at first) unbelievable.

What Cernan captured on this spectacular Hasselblad color image was nothing less than the backscattered *reflection* of "a refractive sunrise on the Moon," coming through the literally "hundreds of miles of shattered bits of glass and rebar" still suspended in the remains of the lunar dome arching over Mare Serenitatis ("The Sea of Serenity")—directly *behind* Gene Cernan as he took this picture (AS17-134-20426).

Terrestrial examples of exactly the same phenomenon can be seen side by side with AS17-134-20426—but in reverse (see Color Fig. 10) This comparison is of a sunset occurring over San Francisco, with the scattered "refractive rainbow" occurring in the west (above the setting sun *behind* the photographer) being *reflected* back into the camera lens from the east—from the earth's atmosphere scattering the "prism of atmospheric sunset."

The situation on the Moon seems to be essentially identical, except that the material scattering the sunlight (and causing the color refraction) is not a "lunar atmosphere of gases," but a rigid, grid-like matrix of "rebar"—on which is still attached sufficient shards of glass to differentially *refract* sunlight (at relatively low angles—like near lunar sunrise or sunset), *exactly* like the earth's atmosphere does on this planet.

Imagine if those Surveyor 6 "post-sunset shots" taken in 1967 of "the glass-like lunar dome rising over Sinus Medii," refracting sunlight *forward*, had been in *color* (and, who's to say it wasn't ... and we just never got to see them).

Another specific prediction implicit in this intensely *refractive*, glass-like superstructure model, was that the lunar surface would be anything but "a dull, grey monotone"—if properly photographed in color. When the refracted "rainbows" from two (or more) prisms optically combine, the result is an array of additional, unfamiliar hues, including pastel pinks and shades of indigoes.

In the case of "towering glass lunar buttresses," and hundreds of miles of intervening, layered "rebar"—all studded with trillions of additional fragments of transparent and highly refractive glass—the combined effect of sunlight, optically bent *through* this geometric array of literal "mega-prisms" and projected onto the lunar surface below, should border on the *psychedelic* in some areas!

In fact, another original Apollo image from the newly-released (on the Internet) official NASA archive blatantly confirms this "colorful" prediction of the lunar dome hypothesis. This specific image—AS17-137-20990—which includes a standardized "color chart" for calibration (see Color Fig. 9), is anything but the well-known "NASA grey," showcasing, instead, a range of vibrant hues—including "pink" mountains in the distance, and nearby rocks ... in vivid *blue!*

In our opinion, this (and other, equally blatant examples of "psychedelic colors" on the Moon) is unquestionably the direct result of "overlapping, prismatic beams" created by sunlight *refracted* through a vast network of suspended glass—in this case, located literally miles above the Taurus-Littrow landing site, overlapping exactly like multiple laboratory prisms (see Color Fig 13)—but on the actual surface of the Moon.

All of which returns us to the one human witness to have physically experienced these astonishing phenomenon *with his own eyes*—and wants to tell us ... *in his art:* Alan Bean.

Since returning from the Moon and retiring from the space program, Bean has spent a great deal of his time developing his obviously innate talents as an artist. He has painted lunar images of both real events—as well as "imagined depictions" of his fellow astronauts doing things they didn't get to do on the "real" Moon.

As part of developing his art, Bean has worked on a number of color studies over many years. These evolving techniques and forms are readily apparent on his impressive "Alan Bean Gallery" website.[77]

Invariably, when he would paint a vision from a mission other than his own, the sky would appear as the absolute black that it "should" be; but, when he took to painting the Moon from the depths of his own memory, the sky suddenly became the odd bluish tone we have come to expect from all the "shattered, geometric glass" in the un-retouched surface images to which we've now had access. Additionally, Bean tends to paint the lunar surface not in the dull, uniform gray tones that are "canonical" (by NASA standards)—but with a wide variety of curiously pastel shades and colors.

Interestingly, from our perspective, this is *exactly* the way we would *expect* the lunar surface to actually appear—if sunlight was reflecting and *refracting* through the remaining fragments of once vast transparent structures overhead. In fact, simple color "stretches" of key Apollo 14 surface images (remember, taken just over a hundred miles from the Apollo 12 landing site), show that Bean is *not* depicting the lunar surface as "he imagines it," but clearly, as he either consciously ... or unconsciously ... *remembers* it (see Color Fig. 11).

Of all Bean's fascinating paintings, one in particular stands out above the others. Titled "Rock 'n Roll on the Ocean of Storms," it depicts Bean and his Mission Commander, Pete Conrad, horse-playing on the surface of the Moon. Not only does it display the bright, *refractive* color scheme which has become the hallmark of the Bean-interpretation of the lunar surface—the sky above the astronauts unmistakably depicts not only Hoagland's "battered lunar dome" but its specifically "inclined buttresses" as well.

A simple, side-by-side comparison shows that Bean is clearly operating from memory here of what he actually *saw* on the lunar surface, back in 1969. The match is simply too uncanny (see Color Fig. 12).

Keep in mind that the "blue shift" in the sky color of the *Enterprise* enhancements is partially a consequence of the color stretching technique of the computer enhancement process, and partially a consequence of a possible "dye shift" in Ken Johnston's private collection of Apollo C-prints, from over the last thirty years. This is likely also why the stunning scattered sunrise colors, seen in the newly-released Apollo 17 images on the NASA websites, scanned from original, frozen rolls of film, did not previously show up in enhancements made from Johnston's preserved C-prints; again, the dyes in his uncooled color photographs have inevitably shifted over time.

The only question remaining in our minds is whether this was some kind of intentional, but oblique "disclosure" on Bean's part—as a way of legally getting around his responsibilities to stay silent under "Brookings"—or, if this is a sign that his unconscious mind has been "remembering things" his conscious mind had been trained to effectively forget.

As far as the astronauts go, this is simply the most specific in a long line of not-so-subtle hints over the past three decades that something is *radically* amiss with the "official" story of Apollo.

As far as the cover-up itself goes, Bean's highly personal colorful and geometric Moon is one more building block in a growing case arguing for literally *decades* of deliberate NASA deceptions and disinformation vis-à-vis the Moon—a case that is only getting stronger with each new discovery and revelation.

Interestingly, Bean's "colorful view" of the Moon was reconfirmed—and from orbit—just as we were going to press.

As we were examining the latest images posted to the Apollo Lunar Surface Journal (as noted earlier, an *official* NASA website "dedicated to preserving the original Apollo photographs and transcripts"), one of the "team" (Ken Johnston) found yet *another* image that ultimately confirmed the "prismatic colors" of the Moon; the image was taken from the Apollo 17 Lunar Module, looking down upon the Taurus-Littrow Valley landing site, just before the A-17 Landing itself. The view included the mysterious "South Massif" (to be discussed in detail later on). This newly scanned and posted Hasselblad frame (AS17-147-22465) was copied (according to the blurb on the NASA site) "directly from an original Hasselblad negative, stored in the vault in Houston."

The web image, part of an entire pre-landing series, was crystal clear ... and very large (~16 MB)—but, curiously, *didn't* show the colors we'd expected from Hoagland's "prismatic" dome model, now *confirmed* by the striking Apollo 17 surface color photographs from Cernan, showing Jack Schmidt against a *lunar sunrise reflection* (presented and discussed previously).

Something was wrong ....

It was only when we spotted the Command and Service Module, "America," also in the shot (located in orbit, in the literal line-of-sight between the Lunar Module, where the camera was located, and "South Massif") that we realized *its* colors (particularly, the vivid gold-mylar thermal insulation around the SPS engine "bell"—turned toward the camera in the photograph) could act as a "color calibrator"—much like the gnomon in the orange-soil image we saw earlier; using accurate color images of this mylar covering, from the CSM during "stacking" activities at the Cape (Canaveral), Hoagland was then able to correct the color balance in the entire Hasselblad scene ... to reflect the *true* "colors of the Moon" at Taurus-Littrow that December morning in 1972.

The results were stunning (see Color Fig. 14).

Once again, just as we had seen from the surface, we are suddenly witness to the *real* "pink, purple and blue mountains" of the Moon—another striking confirmation of Hoagland's prismatic light refraction model of the ancient lunar domes.

And, we also learned something: that NASA is "still up to its old tricks."

That, regardless of the blurb currently on the Apollo Lunar Surface Journal website, the newly scanned photographs *cannot* be the "originals"— having all been curiously "desaturated" before posting, in terms of *calibrated, real lunar colors* ....

A completely independent confirmation of this continuing political reality came from another, unexpected source.

In perusing some amateur astronomy web sites, Hoagland discovered that this *same prismatic color pattern* could be found in high-resolution, multi-color CCD studies of the Moon taken from Earth. When compared side-by-side with a terrestrial sunset (or sunrise) seen from Earth orbit (from the shuttle or space station), the results confirm that *both* worlds—impossibly—possess the same *refractive* color properties!

But only the Earth has *an atmosphere.*

Since we know the Moon *does not*, the only scientific explanation remaining is that there is, indeed, "something" active *optically*—suspended above the lunar surface—which is performing the *same* extraordinary, prismatic light refraction and scattering role as the atmosphere on Earth (see Color Fig. 15).

And the only logical possibility for that is "a vast surface area of ancient, shattered glass ..." covering *most* of the visible lunar surface ... and to an altitude of "miles."

Earthrise

The 16mm Apollo "sequence camera" footage that Hoagland obtained from his NSSDC visit (part of his "agreement"), turned out to be another stunning treasure trove from the whole "Goddard expedition." Transfer of the Apollo NSSDC out-the-window film to video, and then electronically restoring it—via John Stevens' previously mentioned, state-of-the-art floating bed film viewer/projector system—revealed a number of major confirmations of his earlier assessments about the Moon.

After reviewing hours and hours of this, hitherto unseen "raw" mission footage, all graciously processed for the investigation by John Stevens, Hoagland became very interested in a particular sequence from Apollo 10—taken over the Mare Smythii region as the spacecraft "came around the hill" from the Farside of the Moon; on this particular "magazine" (as NASA called a reel), the crew had taken images of the Earth rising dramatically over the lunar horizon, as they returned to the Earthside in their two-hour circumlunar orbit.

As Hoagland watched this one sequence over and over again, he was increasingly baffled by the less-than-obvious fact (because there was also a lot of "jiggling" with the camera going on, even as this important event was taking place—the camera obviously being "hand-held" and not fixed properly in its window bracket, as it was supposed to be) that "something" was not right with the Earth ... rising above the visible "edge" of the Moon.

The planet—when he finally "freeze-framed" it on John's tape [Fig. 4-35]— was strangely, radically, *distorted;* it was both flattened *and* asymmetrically "angular" on one side—as if the whole globe had been subjected to some kind of bizarre "compression forces" as it rose ....

Hoagland immediately recognized that this newest "lunar anomaly" might just be the most significant to date.

On Earth (or, from a spacecraft orbiting above it), it is very common to see the Moon distorted as it rises or sets over the horizon [Fig. 4-36.] This is caused by the dense atmosphere of the Earth (which rapidly gets less dense with altitude ...) differentially refracting (bending) the light of distant objects (like the Moon or sun) seen at the horizon. As the Moon or sun rises higher in the sky, the distortion lessens (because the atmospheric density/refraction decreases with height ...)—and these objects optically resume their correct shape.

The only problem with applying this optical analogy to the "Earthrise" sequence captured over Mare Smythii, is that the Moon *has no atmosphere* to bend the light. And thus, *nothing* to distort the shape of the Earth as it rises over the *lunar* horizon; without such an intervening, "refractive medium," what could be so *dramatically* mangling the Earth (as seen in Fig. 4-37, taken from the Apollo 10 "Earthrise" film)?

(And before anyone thinks "Ah, come on, it's the *window* in the spacecraft!": the Command and Lunar Module windows were constructed of specially ground, *very* expensive, *optically flat* glass—specifically designed by NASA to transmit *undistorted* exterior views ....)

As Hoagland ran the film out to its last few frames, he noticed that there, in fact, seemed to be a visible hint of "something" at the very end ... indeed, positioned *between* the window ... and the rising Earth; in other words, something ... "out over the Moon [Fig. 4-38]"

Some important background:

As power is switched off in those old-style 16mm "gun cameras" (like the ones used on Apollo), the film motion takes a few seconds to wind down and come to a complete stop. So, as the film slows down, the last few frames are increasingly overexposed (because the shutter speed is *mechanically* linked to the film-transport mechanism - above), essentially paralleling—via longer exposures times for those last few frames—the "brightness enhancement process" Hoagland had used in the computer on Ken Johnston's earlier Apollo still photography.

Thus, on those last "Earthrise frames," the *identity* of the much fainter, distorting "intervening medium" could finally, vividly, be *seen* [Fig. 4-39]. It was—analyzed literally "frame-by frame"—none other than *another* "massive, battered, glass-like *lunar dome*" — stretching over Mare Smythii.

Complete with obvious *layers...* and a jagged, *meteor-bombarded* upper surface—

Identical to the model Hoagland himself had sketched (from the various versions of 4822 and the other frames in the Apollo 10 "4800" series ...), anciently arching over Sinus Medii.

The fact that the Earth showed a distinctly non-linear *geometric distortion*— when viewed directly *through* "this stuff"—Hoagland felt, was a tantalizing clue as to the potential *architectural* make-up of the "intervening medium" itself; it was clear, however, that some serious engineering "constraints" would be needed before any definitive conclusions, as to the precise structural nature of the dome's mechanical or optical elements, could be scientifically advanced.

On a hunch that he already had the "right person" for such a big job, Hoagland turned to Award-winning architectural student, Robert Fierteck— studying for his degree at the prestigious Pratt Institute in New York (below). Bob had been working with Hoagland for years, trying to figure out the original layout of "the City" at Cydonia, but recently had also been pressing "to do something with structures on the Moon"; he had been with him on the key trip to Goddard.

Based on his first-rate Cydonia work, Hoagland felt Fiertek was definitely "up to the task": of constructing a *valid* architectural model of "the Mare Smythii dome" using as the end point, the remarkable "optical calibration" now serendipitously provided by the A-10 Earthrise film ... the Earth itself.

Using a Pratt structural CAD program (computer assisted design), and adding some "optical ray tracing" capability, Fierteck soon built a general, simulated "engineering dome structure" in the computer—a generic "mare-type lunar dome," based on—to start with—a photographic analysis of the many individual "dome elements" seen *above the horizon* on the Sinus Medii Apollo 10 photographs—especially, the many variations of "4822" [Fig. 4-40].

Without precise stereo measurements (extremely difficult to make, with photo elements composed of heavily eroded, fuzzy *glass*), the distances (and thus scales) of various structural features seen on the Apollo 10 imagery —such as the equally-spaced horizontal and vertical "layers" and "supports" [Fig. 4-40]—could only be estimated within certain limits.

To reduce this potential error, Bob now tried to fold into his on-going Sinus Medii analysis the extremely clear and striking *identical dome geometry* seen "hanging" in the background above Ed Mitchell—in the infamous "Mitchell Under Glass" Apollo 14 shot. That image provided a set of critical numerical constraints on the scaling of both the vertical and horizontal "dome elements" [Fig. 4-41] ... again, on the (logical) assumption that there was really only *one* workable way to build a lunar dome—covering an entire lunar mare.

Now, Fiertek was ready to fold those numbers *back* into his other dome model, stretching over Sinus Medii [Fig. 4-42] ... to see if the two, independently arrived at sets of numbers *fit*.

They did.

With due regard that they were, in fact, only "estimates" (with about a factor of "two" uncertainty) these numerical parameters were then inserted back into the original CAD program—to create a limited, three-dimensional "slice" of the vastly larger structural "lattice-like" dome that one could assume covered the entire region over Sinus Medii and Mare Smythii ....

The next step was to take this 3D *physical* model of "the dome," and insert the "ray trace" 1 assumptions [regarding "the refractive indices of the glass elements" (assuming the "rebar" was, indeed, still holding up a lot of glass ...)].

This would be the penultimate step in this entire, long, elaborate math-ematical sequence—to an actual computer view of a "simulated Earth" ... seen *through* this massive, reconstructed "rebar and glass matrix" over the Moon.

Fiertek now inserted the "simulated Earth" *behind* all this geometric glass, added some "fractal" meteor erosion ... and waited to see if the effect was anywhere close to what the Apollo 10 crew had actually seen , as the Earth rose—so incredibly distorted—over Mare Smythii ....

The results were astonishing [Fig. 4-43].

With all this extraordinary evidence piling up, Hoagland was under serious new pressure to make some kind of "new public statement."

With some trepidation (would NASA, in response, publicly acknowledge *any* of these scientifically inexplicable features on (and above) the lunar surface ... or would they continue to "stonewall"—behind their decades-long facade, he scheduled a March 1996 press conference at the National Press Club in Washington, D.C.

Yet, even stranger events were yet to come...

Peer Review and Press Conferences

As preparations began for a March press briefing, Hoagland was keenly aware that he needed some kind of "independent analysis" to present at the event. The McDaniel Report had served as an excellent, independent "peer-review" of the methods and findings of the early Mars investigations. What Hoagland needed now was a similar review for the *lunar* data.

At Ohio State, Hoagland had brought along Dr. Bruce Cornet, a geologist who had independently reviewed his initial lunar findings earlier—but in the intervening months, Cornet had wallowed in personal problems and was not up to publicly handling the press on such a controversial subject. Further, because of these domestic problems, Hoagland had been unable to consult with him about the subsequent stunning discoveries stemming from access to the Ken Johnston collection, so it was reluctantly determined that Cornet would not appear.

Ron Nicks, another consulting geologist, had volunteered his services to the investigation and was more than willing to discuss his first impressions—but Hoagland still hoped to have another "voice" to add, as "inside the space program" support at the briefing.

He got it in the personage of Marvin Czarnik—a NASA veteran who had been quietly reviewing Hoagland's claims about the Moon ever since Ohio State; Mr. Czarnik had been an employee of McDonnell Douglas for over thirty-five years, and had worked with NASA on guidance and mission planning throughout the manned spacecraft program.

After seeing a tape of Hoagland's Ohio State presentation, Czarnik decided to try and confirm some of the lunar claims himself. He formed the "Lunar Artifacts Research Group" (L.A.RGE) with five other Douglas engineers, and made several trips to the NASA archives in Houston, St. Louis and other sites for Lunar Orbiter and Apollo negatives and prints.

L.A.RGE validated most of the findings from Hoagland's presentation at Ohio State, including, the existence and optical properties of the Shard and the Tower; due to the fact that it appeared *only* on that one (now missing) Goddard Apollo 10 negative, they were, however, *unable* to confirm the Castle—from Houston's versions of the same frames Hoagland had obtained from NSSDC.

The final three presenters were Ken Johnston, architect Robert Fiertek and Alex Cook, the latter a student from Bellingham Washington, whom Hoagland had flown to Washington DC specifically for the event. Cook had been the first to obtain his own mysteriously "different" copy of the now-infamous 4822 frame from NSSDC—and Hoagland felt he should be the one to tell his own, extremely puzzling story to the national press.

The press conference came off on March 21, 1996. At least sixty members of the national and international media, including CNN, C-SPAN, Telmundo and NBC attended. There were also major print media represented—including reporters from some of the major news magazines and "national" newspapers, such as the *Washington Post*.

The proceedings began with Hoagland recounting the Mars data and some of the earlier lunar images he'd presented in Ohio, then, gave way to Czarnik, who explained that he had confirmed many of the earlier observations Hoagland had made around Sinus Medii, which he had presented at Ohio State—including the two separate views of the "Tower," from completely different missions.

Ken Johnston followed, describing his remarkable experiences while at the Lunar Receiving Laboratory. He discussed the Thornton Paige incident, the airbrushing of negatives, and how he came to be in possession of the hand-held Apollo images being shown at the briefing.

Johnston was followed by Ron Nicks, who said that—as an "engineering geologist" (thus, being familiar with both geology and man-made engineering)— he simply could not come up with any "geologic" explanations for the objects he had seen on Hoagland's NSSDC images.

After that, Alex Cook described his search for the "real" version of frame 4822, and how—after repeatedly ordering that one frame and persistently getting *different* versions, yet all under the *same* ID number—he had reluctantly come to the conclusion that there were (at least ...) "ten *different* pictures" with this designation, and that they were likely all part of a "power-winder sequence" taken seconds apart, albeit all hidden behind "4822."

Next, more confirmation was provided by Brian Moore, Ph.D., an expert in lunar construction techniques, who'd worked at Kennedy Space Center for two years. He stated that he had also enhanced the same frames as Hoagland, and had achieved the same results. He also discussed the feasibility of such a lunar dome, and confirmed Hoagland's assertion that the images presented were consistent with Hoagland's proposal ... that such a "dome" over Simus Medii existed.

Yet all of this assembled expertise didn't translate into anything close to "fair" press coverage.

Some reports in the national press went to extreme lengths to minimize the data presented. One particularly harsh critic was longtime *Washington Post* science reporter Richard Leiby, whose scathing article minimized the decade-long NASA career of Johnston (calling him merely "a contractor") and made no mention of the supporting testimony of NASA experts like Czarnik and his team. Fiertek and Nicks were likewise dismissed as "fans of his [Hoagland's] book."

Leiby simply dismissed Hoagland as a "kook," and misled his readers into thinking that Hoagland had been up at the podium without any expert support.

He then sought to get his own expert opinion from Paul Lowman at NASA. Lowman was the author of the infamous "*Technical Review of the Monuments of Mars*" document that NASA had distributed in the 1980s, and which Stanley McDaniel's report had dismissed as a piece of unscientific propaganda. Leiby called Lowman an "expert in orbital photography," which he is not. Lowman quickly ventured his opinion that the objects in the images were merely "photographic defects," despite the fact that he was a geologist, with no experience in photo processing or image enhancement of the type that Hoagland and the others had presented.

In other words, to bolster his pre-formed bias against Hoagland, Leiby had enlisted the expertise of a *known* critic of Hoagland's, who had already demonstrated a willingness to distort the facts on the subject at hand, and then took his opinion on subjects he knew little or nothing about as gospel— over the opinions of experts who did have such knowledge.

Other outlets were more generous, with several radio and TV stations across the nation running video from the conference and reporting on it fairly. Despite showing up and shooting video of the proceedings, CNN and C-SPAN did not broadcast any footage.

If the U.S. media response was somewhat tepid, the international media was more enthusiastic. Immediately after the news conference, Hoagland was sought for interviews with news outlets from Mexico, France and Brazil, among others. He and Ken Johnston also made an appearance that evening on *Coast to Coast AM* with Art Bell. Several astronauts came out and immediately denied that there was anything like the structures Hoagland had alleged on the Moon, or that NASA had covered it up. Buzz Aldrin, asked by a caller on a C-SPAN program about the charges, stated that there *were* artificial structures on the Moon, namely the LEMs and equipment that the astronauts had left behind, and that any attempt to investigate the types of ruins that Hoagland had alleged was a waste of time.

Aldrin's response is particularly interesting in the way he parsed it. He never said flat out that there *were not* towering glass-like ruins on the Moon; merely that investigating further was a waste of time. His response could be taken as a statement to the effect that NASA had indeed found such ruins, but could gain no practical use out of what they found.

One other side note: Aldrin and Czarnik had been friends since the mid 1960s, when they worked together to develop NASA's rendezvous and docking mission rules. In the late 1990s, Czarnik related a story through a mutual friend. Aldrin had joined this friend for a round of golf, and the friend had decided he wanted to ask Aldrin about the rumors of UFOs and the strange ruins that they had supposedly seen on the Moon. Around the fourth hole, Czarnik's mutual friend put the question to Aldrin, who responded by picking up his clubs and stomping off the course. The two men have not spoken since.

The news conference had one additional impact in that former astronaut Edgar Mitchell, who knew Hoagland from his NASA days, agreed to debate the issue of artificial structures on the Moon with Hoagland on *Coast to Coast*. The debate was really more of a broadcast discussion between the two men, as it was clear Mitchell was not as well versed in the subject as he could have been.

The program ended with Hoagland promising to send Mitchell additional materials to review, and Mitchell agreeing to another mutual program to review the new material. Hoagland did send the pictures and other data, but Mitchell never responded, and the follow-up program never took place.

Yes, Virginia, We Really Went to the Moon

Before we leave the Moon phase of the investigation, the authors feel we must comment on a particularly damaging urban myth that has taken hold of the popular consciousness in the last few years. As we discussed in the Introduction, this idea (most recently advanced by a few well-known self-promoters such as David Percy, Bill Kaysing and the late James Collier) had its origins as far back as the Apollo 11 mission itself. This myth is based on the simple—albeit naïve and absurd—notion that the Apollo missions and subsequent Moon landings were "faked."

If NASA is eventually forced to admit that there is more to the Face on Mars than meets the eye, that maybe they missed something the first three times around, or that there is truly something ancient and extraordinary on the Moon, then it will be crucial to have thoroughly discredited the "conspiracy theorists" out there (by the deliberate promotion and then effective debunking of obviously bad conspiracy theories like "We Never Went to the Moon.") If enough people can be convinced by this deliberate disinformation campaign that there are no NASA conspiracies then it will be much easier to sell the idea that NASA just missed a couple of things on those Mars pictures all those years ago.

Let us be clear:

We are uniformly, unabashedly "conspiracy theorists," and have—against our wills—become truly convinced that there has been a conscious, carefully planned and permanent cover-up by NASA of some of its most extraordinary discoveries made in the course of the Agency's forty-year history.

That said one thing they did *not* do, unquestionably, was *fake* the Moon landings.

In fact, most of the charges made by these "Moon Hoax" advocates are so absurd, so easily discredited and so lacking in any kind of scientific analysis (and just plain common sense) that they give legitimate conspiracy theories (like ours) a bad name (which is more than likely now the *real* objective—see below). The comedy of errors and willful ignorance represented by the Moon Hoax advocates is too extensive to detail here. We instead refer our readers to the "Who Mourns for Apollo?" series on Mike Bara's Lunar Anomalies web site for a detailed, claim-by-claim dismantling of the whole Moon Hoax mythos.[78]

To reiterate our own political position: we have come, slowly and reluctantly, to believe that the entire "Moon Hoax myth" was, in fact, *professionally* instigated—and is currently being kept very much alive—by none other than *NASA!* That, the myth forms the "perfect cover" against an entire (growing ...) segment of the population, who *do* increasingly suspect "the government" is not telling them the truth about a *lot* of things ... including ... going to the Moon.

So, as Hoagland observed *first-hand* at JPL all those years ago, NASA carefully planned, from the beginning, to give those folks a "conspiracy," alright, but—a *fake* conspiracy ... to cover up the *real* one.

How else do you explain NASA formally commissioning James Oberg, a leading space authority, just last year (2006) to finally write an official NASA book on "why the Moon Hoax crowd is 'out to lunch.'" Only, to have *another* part of NASA suddenly *cancel* Oberg's contract ... abruptly withdrawing from publishing the one book which would have had a chance to finally lay out "why" and "how" the Agency *really* landed humans on the Moon—still, after over thirty years, its most astonishing achievment?

Because "someone" wants—*needs*—the endless ambiguity of "did they *really* do it" ... to cover up what NASA *really* found .....

The Golden Fleece

When bringing up the, admittedly, "totally unbelievable" lunar data you have seen presented in this Chapter, we inevitably get two questions objecting to our premise: "Why didn't the *astronauts* see these ruins, and then tell us all about them?" and "Why would you make domes out of *glass,* on a place like the Moon—with so many meteors and asteroids bombarding it all the time?"

The second question is actually the easiest to answer: all materials, including glass-based minerals like quartz silica, take on different properties in the hard, cold vacuum of space than they do on Earth. One of the most abundant substances of this world is water; in liquid, gaseous or solid form. Space however, has an appalling lack of water, so much so that it can be easily stated that water is one of the *least* available resources in the vast expanse of emptiness we call 'space,'" at least, within the solar system.

As it turns out, this is one of the qualities of the vacuum that makes glass not just a desirable, but an *ideal* material for building structures on an airless world like our Moon. Glass on Earth is well known to have little tensile strength, meaning it doesn't stretch easily (because it is brittle) and will not withstand even a very weak impact from a hard object (shear). When you throw a baseball at a glass window, it fractures and cracks easily, and with little resistance. However, if you attempt to crush a glass sphere, you'll find that it has a great deal of strength under "compression" stresses.

The reason for these properties on Earth is that it is pretty much impossible to extract the water from glass as it is forming under typical terrestrial conditions. Water is all around us, even in the most arid deserts. It is in the ground as a liquid, frozen in the arctic as a solid, and even in the air around us as humidity. All this water causes a phenomenon called "hydrolytic weakening" when glass is being manufactured on Earth, meaning that at the molecular level, the bonding of silicates and oxygen is resultantly weakened. This produces a transparent, brittle and fragile material we call common "glass." Manufactured under Earthly conditions, we have found that glass is a very useful and artistically pleasing medium for a variety of uses—but structural construction is *not* one of them.

In short, on Earth, we don't build glass houses.

But the Moon is a completely different story. It is airless, with no humidity to interfere with the molecular bonding of the silicates that make-up the glass that is omnipresent. The hard-cold vacuum enhances the strength of lunar glass to the point that it is approximately *twice as strong as steel* under the same stress conditions. In fact, several papers from scientists at Harvard and other universities have suggested that lunar glass is the ideal substance from which to construct a domed lunar base.[79]

All we are proposing is that somebody else came up with the idea long before we did.

The reality is that if Alan Shepard's famous (low velocity) golf shots on the Moon had actually struck some of this glass material, it would have simply bounced off the structures like a rubber ball off a battleship. Only the incessant, high velocity "meteoric rain" over millions of years could have reduced these once magnificent feats of mega-engineering to the ghostly relics we see them as today.

Imagine the irony of one day returning to the Moon, only to use the same shattered shards of glass to *rebuild* a truss-work lunar dome that someone, eons ago, had originally erected. Talk about literally "touching the face of god ...."

As to the second question—the issue of what the astronauts might have seen or not seen from the surface—this is a problem we have debated since the time of Hoagland's earliest suspicions of the glass-dome model. Originally, Hoagland was willing to give the astronauts the benefit of the doubt. Early on in the investigation, during the 1996 "Hoagland-Mitchell debate" with Apollo 14 astronaut Edgar Mitchell on *Coast to Coast AM,* Hoagland went out of his way to consider that the distant ruins might have been very faint, and might have even been further filtered by the gold visors that the astronauts wore on the surface.

Ostensibly added to the helmets to protect the astronauts eyes from the harmful effects of raw ultraviolet light, Hoagland had yet to check on the specific optical properties of the gold visors when he discussed the subject with Mitchell on the air.

Given that most of the astronauts initially denied seeing anything like what Hoagland had found in the lunar images, Hoagland initially assumed that the visors might be "tuned" to keep *out* the kind of light scattered by the lunar glass arching overhead; that perhaps the astronauts were literally "in the dark" about the distant lunar ruins. Under this assumption, the Apollo crews themselves would have been given photographic tasks without being told *why* they were being asked to shoot certain scenes (like, the panoramas), using special film that could be properly filtered or enhanced later in the Manned Spacecraft Center photo lab to reveal what their NASA overseers knew to be there.

But, over the course of time, Hoagland was able to find the specs on the gold visors ... and they painted a *completely* different picture.

Far from filtering out the areas of the spectrum in which the glass ruins would have been most prominent, it turned out that the gold visors were designed specifically to *enhance* this blue-violet portion of the spectrum! And, by a factor of over *twenty to one*[80] (see Color Figure 6).

What this meant was that there was *no way* the astronauts could have avoided seeing the massive bluish-violet glass structures that were all around them. Their denials aside, could they still have been telling a partial truth? Could they have seen the actual ruins with these amazing gold film "amplifiers" ... but somehow, then *forgotten* it?

Early in his low-key lunar ruins investigation, a NASA insider (an MD directly involved in the medical aspects of the Program ...) confirmed to Hoagland and Johnston anecdotally that during their "debriefings" by NASA, all of the Apollo astronauts had been *hypnotized*—ostensibly to help them remember more clearly their time on the Moon. In actual fact, it seems more likely that these sessions were used to make them *forget* what they had seen, as evidenced by the astronauts own post-Apollo behavior. Neil Armstrong, for one, basically fell off the edge of the Earth, all but becoming a hermit. Besides the case of Alan Bean, which we covered earlier in this Chapter, there are at least two other cases of astronauts struggling to remember their experiences.

In his autobiographical second book, "Return to Earth," Buzz Aldrin recounts an experience he had during an interview in the early 1970's. While speaking to local Kiwanis Club in Palmdale California (home of Edwards Air Force Base and many NASA contractor firms), Aldrin was asked a simple question: "What did it *feel* like to be on the Moon?" Almost immediately, Aldrin was overcome with a panic attack, and the harder he tried to remember, the worse it got. Finally, he was forced to abandon the interview and rush from the stage, followed by his then-wife Joan. In the alleyway behind the Club he then became physically, violently ill.

Aldrin has continued to struggle with his inability to remember certain parts of the Apollo 11 mission right up until recent years. In a 1999 interview with Salon.com, he put it this way:

"I try to answer," he admits wearily. "I say, 'It felt terrific. Tremendously satisfying. The mission was going well, and our training had prepared us perfectly.'

"But then people say, 'No ... how did it feel? How did it *really* feel?'" He bristles. "For Christ's sake, I don't know! I just don't know. I have been frustrated since the day I left the Moon by that question."[81]

Later, he describes watching a video tape of the TV coverage of his own spacewalk with Armstrong while they were in quarantine.

"And as we watched," Aldrin says quietly, "I remember turning to Neil and saying: 'Look: We missed the whole thing.'"

And Aldrin is not alone in his struggles to recollect his experiences on the Moon. The Apollo 12 commander, the late Pete Conrad, got so frustrated with his inadequacy at answering the same question that he resorted to giving the same, banal answer each time: "'Super! Really enjoyed it!'"[82]

These stories are absolutely consistent with a deep hypnotic suggestion to *not* remember certain things they may have seen there. It is also consistent with a story told by another Apollo astronaut, Apollo 14's Edgar Mitchell. During the "Hoagland-Mitchell debate," Mitchell came forth with this little nugget:

"Basically, when people asked me, "What did it feel like to be on the Moon?" being a super rationalist and a Ph.D., and all of that, I didn't think it was a germane question. I thought if you ask me what did I *do* on the Moon, or what did I *think* about on the Moon, I could have told you. But what did I 'feel?' I didn't know. And so I set out to... I started thinking about that question. First of all, it irritated me because I didn't have an answer to it, and eventually I asked myself, "*Should* I know what I felt like on the Moon?"

"So I went to some good friends of mine, Dr. Jean Houston and her husband Bob Master, and said "help me find out what I *felt* like on the Moon," and that began the investigation of inner experiences for me back in 1972 and led to the approaches that I have taken in understanding experience and the psychic experience and all this whole subject matter of consciousness that we've been looking at for 25 years."

In his book, "The Way of the Explorer," Mitchell recounts the same desperate search for his "true feelings" while walking on the Moon—a time which *should* have been indelibly imprinted on his psyche—on *all* the astronauts psyches— if not on their lasting emotions, for the rest of their entire lives.

It was obviously not.

These are amazing admissions from Mitchell, coming from a man whose original NASA photograph we actually, physically possess (thanks again to Ken Johnston)—clearly standing on the Moon underneath *an immense canopy of shattered, deep blue, geometric lunar glass ...* towering above the airless, cratered landscape; a man who obviously *can't remember a thing* about what it must have *felt* like just to gaze up in sheer wonder and engineering admiration at "the awe-inspiring magnificence of what was arching overhead, and stretching into the distance all around him (see Color Figure 5) ...."

Mitchell is thus providing a third, specific source for the astronauts not remembering key parts of their unique, once-in-a-lifetime experiences on the Moon. Mitchell struggled so much with the issue that, on his own, he eventually sought the help of a professional hypnotist and psychologist, to basically "deprogram" himself!

Incidentally, it didn't work.

According to a Hoagland source with a long professional association with Jean Houston, when Mitchell came to that part of his (remember, *personally* requested) sessions in which Houston would repeatedly prod him with "Now, try to remember what it *felt* like when you were walking on the Moon ..." Mitchell would repeatedly deflect the question, saying "That's not important ... let's move on."

In fact, in all of these cases, the astronauts are able to recite practically "chapter and verse" from all the mission tasks that they performed—literally, even *decades* after these intricate and complex mission plans, down to minute technical details—but *little else*. In other words, the "scripted" portions of their duties are still crystal clear, but the broader "feelings" that inevitably go with observation and recollection—and are a fundamental part of any integrated personality– are strangely (and tellingly ...) completely absent from their narratives.

So, even though we now *know* that there was no way they could have missed seeing the Crystal Towers of the Moon, rising all around them, we must stop short of calling the Apollo astronauts themselves blatant "liars." For, as you will read in subsequent pages, it is now clear that they themselves are struggling to "remember"—and to "somehow" tell the rest of us, based on the same official NASA images that we are seeing—the literally overwhelming truth of all that they experienced ... and then somehow "forgot."

Chapter Four Images

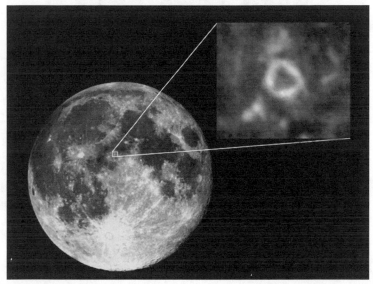

Fig. 4-1- Close-up of Ukert crater (inset) from North American catalog

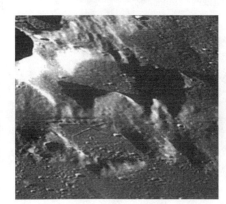

Fig. 4-2 – "Geometric Square" near Ukert Crater.

Fig. 4-3 – Apollo 10 photographic map of Sinus MedII.

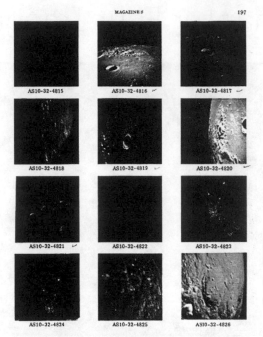

Fig. 4-4 – Apollo 10 photographic catalog showing blacked out and dramatically darkened images.

Fig. 4-5 – The "Shard" from Lunar Orbiter frame LO-III-84M. Note shadow being cast by shard along the lunar surface (and at proper sun angle). White cross above "Shard" is a photo registration mark, placed on the original unmanned spacecraft film before insertion into Lunar Orbiter by Eastman Kodak. Light "smudge" above and to the left of the Shard is far distant "Tower."

Fig. 4-6 – Close-up of The Shard, an anomalous vertical lunar feature with elaborate internal geometric detail (see inset), extending at least 1.5 miles above the airless lunar surface. Computer-enhancement and enlargement from NASA Lunar Orbiter frame III-84M.

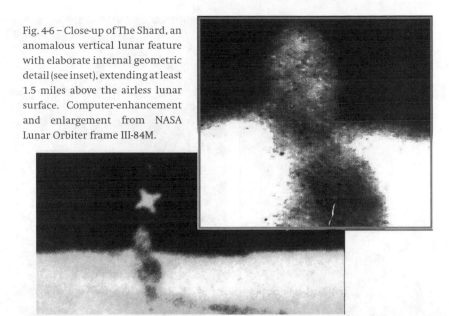

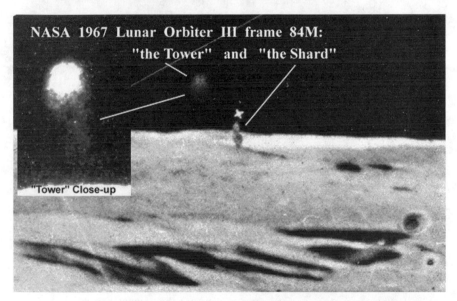

Fig. 4-7 - This wide-angle enhancement from III-84M reveals a second "artificial anomaly" above the Moon, on the same NASA frame —a peculiar "Tower/Cube." Located just to the left (south) of the "Shard," somewhere beyond the lunar horizon (~260 miles distant), this "faint smudge" is not a comet or a galaxy, accidentally captured by the unmanned lunar spacecraft in 1967. Rather, it is revealed to be another glass-like lunar structure—one that on analysis turns out to be even more remarkable than the "Shard" itself. Apparent is a distinct, geometric, leaning "column" of fainter optical material – "the Tower" – extending from the "Cube" (inset) down toward the lunar surface at least seven miles below.

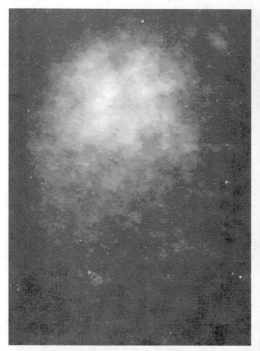

Fig. 4-8 – Close-up of the "Cube."

Fig. 4-10 – Side view of the Tower and Cube from NASA frame AS10-32-4856. This second confirming image, taken on another mission and using a different photographic medium, proves the Tower is a real object on the Moon and not a photographic defect or enhancement artifact.

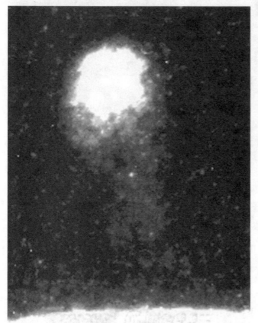

Fig. 4-9 – The Tower in close-up. Note supporting "filaments" emanating from the Lunar surface.

Fig. 4-11 – Surveyor 6 Post sunset "brilliant beads of light" post sunset image.

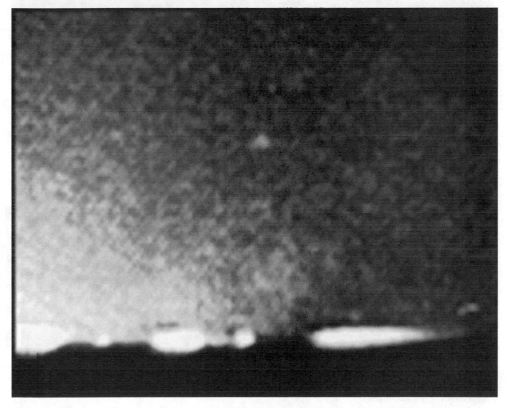

Fig. 4-12 – Close-up of geometric structure within coronal refraction.

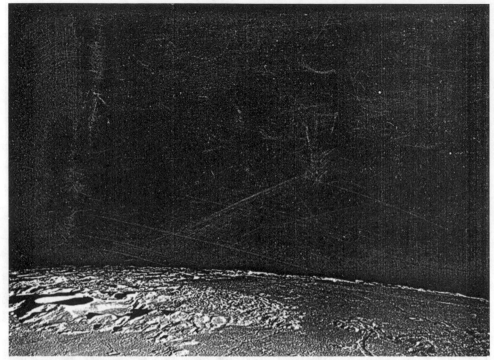

Fig 4-13 - Lunar architecture. This wide-angle Apollo 10 Hasselblad image reveals a stunning array of blatantly architectural forms suspended over Sinus Medii - Obviously impossible in any "natural" model of the Moon. (NASA AS10-32-4810)

AS10-32-4821   AS10-32-4822   AS10-32-4823

Fig. 4-14 – Section of Apollo 10 photographic catalog with Blacked out frames, including AS-10-32-4822. From NASA publication SP-232.

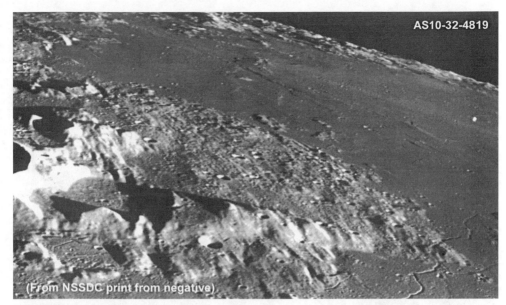

Fig. 4-15 – AS-10-32-4819

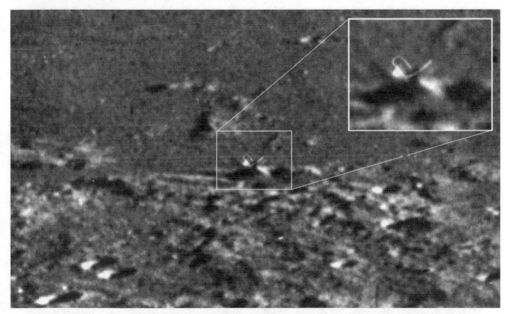

Fig 4-16 - Close-up from different Apollo 10 frame of "L.A." – AS-10-32-4822. Object resembling a "twisted paper clip" appears suspended over same geometric surface features (inset). This object does not appear on other versions of the same frame number – revealing that NASA secretly acquired several different images, all now maintained under *the same frame number*.

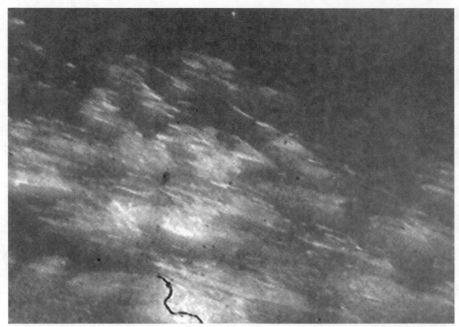

Fig. 4-17– Sectional of recently discovered Apollo 10 "4600 series" frame AS10-31-4652 showing close-up of "L.A..". Note angular, geometric nature of structures as well as anomalously reflective material making up most of the image.

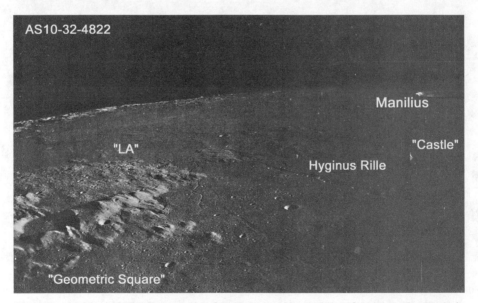

Fig. 4-18 – Version of AS10-32-4822 showing "L.A." and the "Castle" (far right). The Castle is actually suspended some nine miles above the Lunar surface.

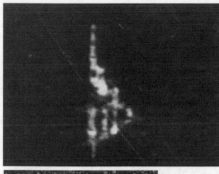

Fig. 4-19 - Sectional enlargement from Apollo 10 frame 4822 shows enigmatic, highly geometric object called "the Castle" above the lunar surface. At least eleven additional versions of this photo have been identified, in various NASA archives around the world – all inexplicably under the same ID number. Evidently part of a "power winder" sequence of photos taken by the astronauts, the changing perspectives allow positive verification of the 3-D reality of the object. Note sagging "suspension cable" – independent evidence of a massive object attached.

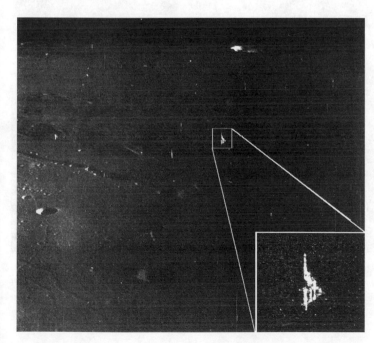

Fig. 4-20 - The Castle (inset), a brilliant, glass-like, complex geometric object estimated to be about nine miles above the darkened lunar surface in this Apollo 10 frame, 4822. The Castle is surrounded by a host of other, equally bright, but much smaller "sliver-like" objects – apparently also surviving members of a former "Sinus Medii Dome" that Hoagland is proposing once covered this entire central region of the Moon. Note the strangely "veiled," though brilliantly sunlit, cratered lunar landscape underneath – which is mysteriously, increasingly, obscured from left to right, as the viewing angle approaches closer to the sun. What's causing that ... on a totally airless satellite?

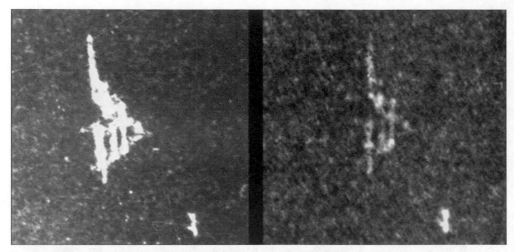

Fig. 4-21 – Stereo pair of the Castle from two different versions of 4822.

Fig. 4-22 - Richard C. Hoagland points out to NSSDC officials at NASA's Goddard Space Flight Center, in 1995, two photographic enlargements from AS10-32-4822 – one containing "the Castle," the other which did not. And, he's asking the critical question: "How can one official NASA image verifiably contain two different sets of data?"

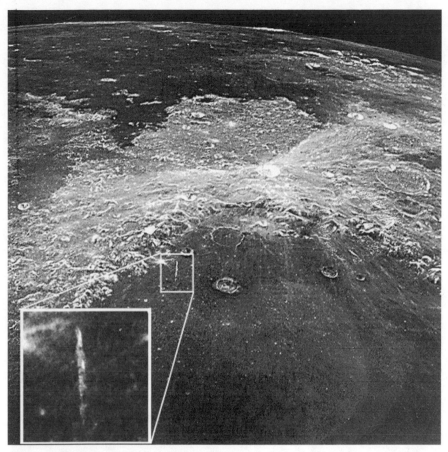

Fig. 4-23 – The "Mare Crisium Spire."

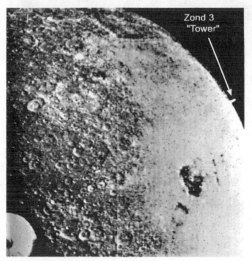

Fig. 4-24 – The Russian "Zond 3 Tower" published in TRW's "Solar System Log" in 1967.

Fig. 4-25 - Another lunar "dome" (right side, lower limb) – from Zond-3 (July 20, 1965)

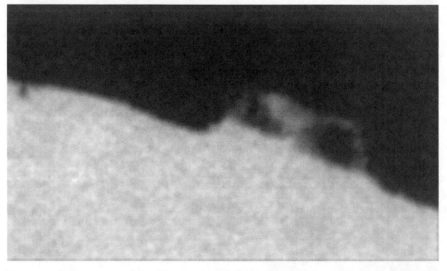

Fig. 4-26 – Enlargement of the "Zond 3 dome" from official NASA publication, "Exploring Space with a Camera" (NASA SP-168, 1968)

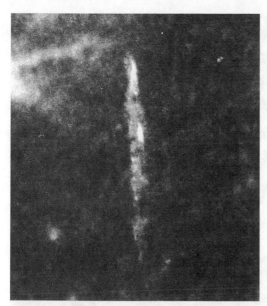

Fig. 4-27 - Close-up of the "Crisium Spire" from official NASA Apollo print AS16-121-19438. Note cross-beam supports meeting at the "spire" at 90° angle, part of obscuring geometric "matrix" surrounding this enigmatic object.

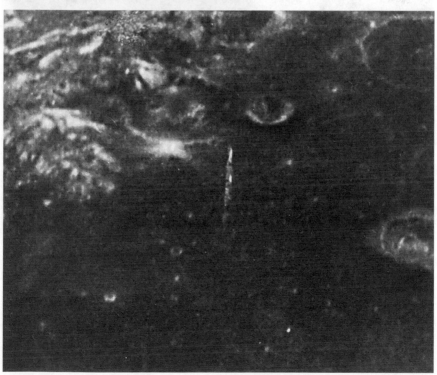

Fig. 4-28 - Partial dome over Picard crater in Mare Crisium (far right). "Crisium Spire" is in center of frame. There is no conventional explanation for Picard's anomalous reflectivity in this image. Enlargement from AS16-121-19438

Fig. 4-29 – Artists concept of glass lunar dome over a crater, identical to photographic image of Picard crater (Fig. 4-28).

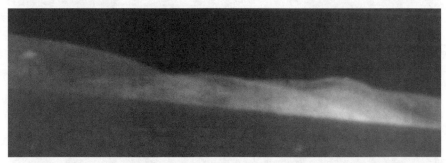

Fig. 4-30 – Steve Troy's glass domes in Mare Crisium, possibly covering the craters Cleomedes F and Clomedes Fa. From NASA frame AS10-30-4421.

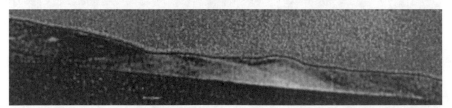

Fig. 4-31 - Grayscale version of false-color enhancement of Cleomedes F and Cleomedes Fa "domes." Note two distinct edges of major foreground dome, and how steeper-angled background dome can be seen through it. (Bara)

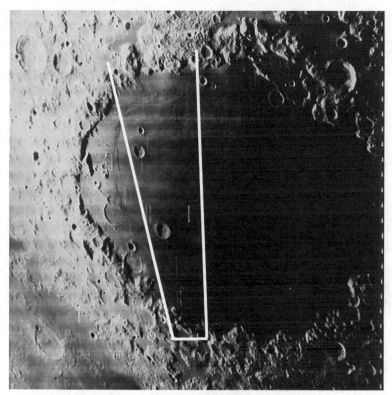

Fig. 4-32 - AS10-30-4421 image map. Picard is the crater just to the lower center of the A-10 "footprint" (rectangular outline). Cleomedes F and Cleomedes F-a are just to the right of the image edge at the top. Mark-up by Steve Troy.

Fig. 4-33 – Annotated "Mitchell Under Glass" frame from the Ken Johnston collection.

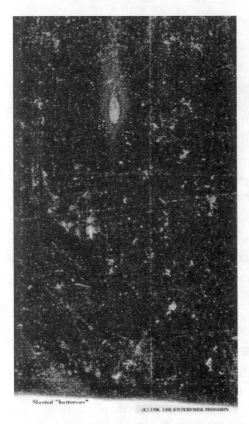

Fig. 4-34 - Inclined buttresses and cross-beam constructional "filaments" cover this astonishing enlargement from AS14-66-9301. Such redundant geometry is, of course, totally absurd in terms of any "geological" explanation for these features.

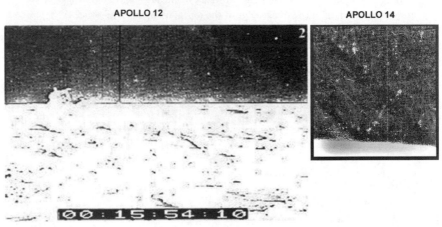

Fig. 4-35 – Screen cap from Apollo 12 film "Pinpoint for Science" (L) and from AS14-66-9301 (R) showing the same inclined buttresses from two different landing sites – separated by just 122 miles.

Fig.4-36 – Earthrise from Apollo 10. Note flattened, distorted shape of Earth's globe, a condition which is totally impossible on an airless satellite like the Moon.

Fig. 4-37 – The Moon distorted by atmospheric refraction as it rises.

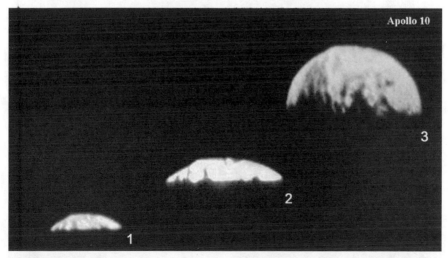

Fig. 4-38 – Earthrise from Apollo 10. With no atmosphere on the Moon, what is causing this notable distortion?

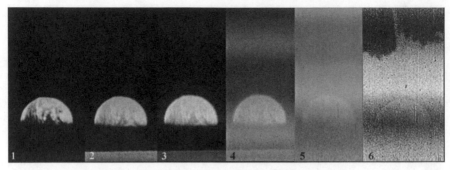

Fig. 4-39 - Apollo 10 16mm "Earthrise" film over Mare Smythii. Note flattened, angular shape of globe (due to refraction), and increasing exposure as film transport and shutter slow down after being turned off (frames 4, 5 and 6) – revealing a layered, light-scattering "glass dome matrix" over Mare Smythii ... with a jagged, meteor eroded upper surface (6).

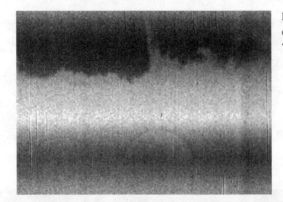

Fig. 4-40 – Close-up final overexposed frame of "Earthrise" film.

Fig. 4-41 – Four versions of 4822 used by Robert Fiertek to model his theoretical "Sinus Medii dome."

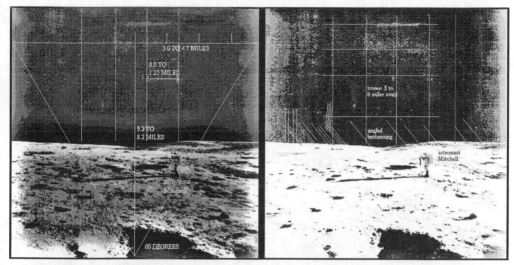

Fig. 4-42 – Calculations made by Fiertek to create a 3D CAD model of the dome.

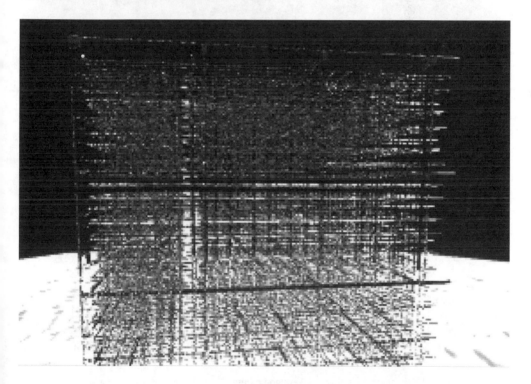

Fig. 4-43 - Reconstruction of box-like "dome" over Sinus Medii by architect Robert Fiertek. Note similarity to "scaffolding on the Moon" image from Color Figures 2 and 3.

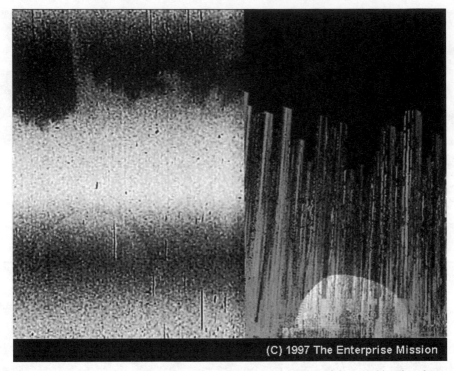

Fig. 4-44 – The last frame of Earthrise film side by side with 3D model created by Fiertek.

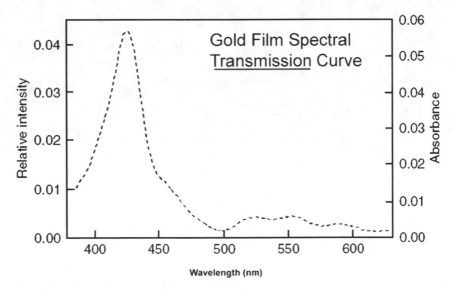

Fig. 4-45 – Spectral transmission curve for the gold visors worn by Apollo astronauts on the Moon. Visors were specially tuned to be sensitive in the visual spectrum that would make distant lunar ruins plainly evident.

Chapter Five

A Conspiracy Unfolds

"The very word 'secrecy' is repugnant in a free and open society; and we are as a people inherently and historically opposed to secret societies, to secret oaths and to secret proceedings. We decided long ago that the dangers of excessive and unwarranted concealment of pertinent facts far outweighed the dangers which are cited to justify it." –President John F. Kennedy, April 27, 1961

With the press conference now behind him and a steady source of new lunar NASA imagery to examine from his new "friends" at NSSDC, Hoagland now had time to turn his attention to an entirely new (and potentially, far more sinister) aspect of his investigation. Increasingly, he was haunted by the question: What did NASA know ... and *when* did they first know it?

During the National Press Club preparations, he'd had occasion to go back over Ken Johnston's collection of first generation Apollo hand-held photography. As he and Ken were thumbing through the pictures in Ken's vast photo albums, they had come across an official Apollo mission patch from the early days of the Program[Fig. 5-1]. In looking at the patch, it struck Hoagland just how incongruent it was with the stated goals and mythology of NASA's official Apollo Program to the Moon.

The patch depicts the Earth and Moon, with a flight path tracing from Cape Canaveral to a landing spot on the Moon. The lunar disk is overlaid with a mythological deity—presumably Apollo—and there is a large "A" in the center of the patch. The design was built around a very curious feature—the constellation of Orion, with its three distinctive belt stars comprising the bar in "A."

According to Agency lore, NASA Director of Space Flight Development Abe Silverstein had consciously tapped Greco-Roman mythology in the naming of the Mercury, Gemini and Apollo projects. The projects' names were approved at a NASA conference in July 1960. Yet, in examining the mythology further, each of the Program names seemed to have a second, alternative interpretation...

As far as the mythology goes, Mercury, the gods' messenger, seems appropriate for the single man capsule and its quick missions. Yet the NASA symbol for Project Mercury is a representation of the *element* mercury, rather than the Greek god Mercury, and is similar to a stylized Egyptian ankh[Fig. 5-2]. Alchemists considered mercury the "first matter" from which all other metals were derived. So was this a dual, *alchemical* meaning for the name "Project Mercury?"

Gemini, which means "the twins" in Latin, could not only refer to the two-man capsule, but also to the rendezvous and docking procedures which were perfected on Project Gemini and which were crucial to the later Apollo missions to the Moon. Gemini is also frequently referenced to the constellation that borders Orion and boasts the twin stars Castor and Pollux as its most prominent features; other myths link Castor and Pollux with the morning and evening stars (the planet Venus).

Certainly, there is an inherent "duality" with all of these associations that well fits the "twins" designation of the NASA Program. Two astronauts ... two vehicles rendezvousing in space ... the twin stars Castor and Pollux.

Then we have Apollo.

A semi-official book on the space program published in 1985, called *All We Did Was Fly to the Moon*, describes the Apollo insignia this way:

"The Earth and the Moon flank a large stylized letter 'A' against a background of stars. The Constellation Orion, 'The Mighty Hunter,' is positioned so that its three central stars, known as Orion's belt, form the bar of the letter 'A.' These stars are Mintaka, Alnilam and Alnitak. The star shown above the Moon is Orion's shoulder, the red star Betelgeuse, and his other shoulder on the right top side of the 'A' is the white star Bellatrix. Under the right side of the 'A' is Orion's foot, the blue-white star Rigel and under the left side of the 'A' is Orion's other foot, the blue-white star Saiph. Rigel is one of the Apollo Astronaut's thirty-seven navigational stars. Between the lines of the 'A' is Orion's sword and in the center of the sword is the Orion Nebula...

"*World Book Encyclopedia* notes that 'Orion was a mighty hunter in Greek mythology. He was the son of Poseidon (Neptune), who gave him the power to walk through the sea and on its surface.

"'The goddess Artemis (Diana) fell in love with the handsome Orion. Her brother, Apollo, did not like this, and plotted to destroy Orion. One day while Orion was swimming, Apollo walked by with Artemis. Apollo challenged her to hit the target bobbing in the water. Artemis did not know it was the head of her lover, and killed him with her arrow. Her sorrow was great and she placed Orion in the sky as a constellation.'

"The face on the Moon represents the mythical god [sic], Apollo."[83]

Even this description, full of line after line about "Orion" compared to a single mention of Apollo shows how much more important Orion is to the patch than Apollo. So why call the Program "Apollo" at all? And why name the Program after a petulant "god" who tricks his sister into murdering her lover? Is this the image NASA wanted of its first space pioneers, rather than the mighty hunter Orion? Why not just call the Project "Orion?" No matter how you look at it, the Greco-Roman designation of Apollo just doesn't fit such a noble and risky venture of Man's first foray to another world.

With the Greek interpretation of Orion seemingly incompatible with the mythology of the space program to this point, Hoagland looked deeper. He knew from his studies of ancient religions and myths that Orion had a special significance in the beliefs of many ancient cultures, especially the Egyptians. The constellation of Orion, according to most Egyptian mythologists,[84] was the celestial representation of the central figure of the Egyptian pantheon of gods, the god of "resurrection" revered by the ancient Egyptians as "Osiris." Indeed, the story of Orion/Osiris, his murder at the hands of his evil brother Set/Taurus, his magical resurrection by his wife and sister Isis (represented by the star Sirius and sometimes associated with the Moon) and his vengeance against Set through the hands of his son and heir Horus/Leo, is the most ancient and sacred of the Egyptian origin myths. To call Orion/Osiris the "king" of the Egyptian pantheon of gods would not be far-fetched.

Isis, Osiris, Horus and Set

In the ancient texts, Isis and Osiris are children of Nut, the wife of Ra, who ruled upon the Earth. Nut also had two other children, Set and Nephthys.

Set and Osiris were brothers and rivals for their father's throne, and after Ra left the Earth to return to heaven, Osiris and Set engaged in a competition to succeed him and rule over the Earth. In the end, Osiris won the struggle because he was able to teach the men of the earth farming and agriculture, science, music and poetry. He was much loved and praised throughout the kingdom of Egypt, which only made his brother jealous. Osiris then set out to spread the wisdom of Egypt throughout the rest of the world.

While he was away, Set plotted his demise and sprang a trap upon his return. Osiris was dismembered and the pieces cast around the kingdom. Isis eventually found all the parts (save one) and resurrected Osiris in order to mate with him. Osiris then returned to Heaven in the West, where he would rule as judge of the dead, and his son Horus was born to Isis. Isis hid Horus from his evil uncle amongst the reeds of the Nile.

When he became a man, Horus confronted his uncle Set, who had murdered his father and usurped his kingdom. During the course of this war, Horus is temporarily blinded by Set, only to regain his sight and then vanquish Set and restore his father's kingdom. The texts make clear that Horus must return to battle with Set in the future to posses the souls of men. The stories go on to tell us that Horus will triumph in this war, and when Set is destroyed in spirit form Osiris will return to Earth to rule over a new "Golden Age." According to Egyptian belief, Horus was the first "man-god," and all of the human Pharaohs that followed were direct descendants of Horus.

This story is full of astronomical references, including hints (via key, "coded" numbers in the narratives) that the Egyptians fully understood the concepts of precession, lunar cycles and stellar motion. There are also strong hints that these beings were not mythical "gods" at all, but rather real flesh and blood beings from an advanced civilization that visited Earth in antediluvian times.

The Egyptians developed a stellar religion that retold this story in various forms. In this religion, the constellation Orion came to represent Osiris, Taurus represents Set, and Leo, Horus. The Star Sirius was the living embodiment of Isis, the goddess of life and nurture, sister, wife and consort of Osiris.

As Above, So Below

Hoagland had long been interested in the possible connections between the Giza Plateau in Egypt (home of the Great Pyramids and the Sphinx) and the Martian Pyramids and Face (termed "The Martian Sphinx" by Russian researcher Vladimir Avinsky) in Cydonia. During his U.N. presentation in 1992, a possible mathematical and geometric linkage was presented, specifically connecting the location of these two distant sites. Hoagland had noted that the arctangent (.865) of the "D&M Pyramid" geodetic Martian latitude (40.87 N) is the same (within one part in a thousand) as the cosine of the Sphinx's current latitude on Earth. While this may seem a bit tenuous, Hoagland found it curious that two such base mathematical constants could be the same on two such key features—and, on two completely *separate* worlds. These identical "shared" latitude relationships are, in turn, reflected redundantly in the *internal* geometry of the D&M Pyramid on Mars, and in the base slope angle of the Great Pyramid of Giza (using a sexagesimal base numbering system).

In the same 1992 U.N. presentation Hoagland went on to demonstrate an additional level to this possible Earth/Mars connection. He showed that the right half of the Face, when flipped symmetrically over the left half, appeared distinctly *feline*. This symbolic parallel to the Great Sphinx of Egypt seemed to complete the "terrestrial connection" to Cydonia—via the same, haunting hominid/feline fusion evident in the Face's famed terrestrial counterpart on Earth.

As he delved deeper into the mysteries of the Giza Plateau, he discovered some contemporary work that linked the pyramids, Sphinx and Nile even more deeply with the Orion/Osiris epoch of Egypt's ancient past. In 1994, civil engineer Robert Bauval demonstrated a remarkable geometric linkage between the relative layout of the three major pyramids at Giza to each other, and the celestial geometry of the belt stars in the brilliant constellation of Orion [Fig. 5-3]—all relative to the Milky Way itself (represented by the Nile).[85]

An additional surprise: this reconstructed Orion pyramid geometry coincided with the position of the Milky Way (the Nile)—*not* when most Egyptologists believe that the Pyramids were built, in 2500 BC—but in *10,500 BC*.

This implied that the Pyramids and possibly the Sphinx (which was already dated to least 5,000 BC by geologist Robert Schock) were built at a time that reflected what they saw in the skies above them. If the Egyptian tenet "as above, so below" was indeed applied during the building of the Giza wonders, then certainly they must be *far older* than any known or acknowledged organized human civilization.

Previous work (Badawy and Trimble, 1965), as well as Bauval's own 1994 refinements, also demonstrated that key internal passages within the Great Pyramid itself actually point directly toward "Osiris" in the sky—toward Orion's key belt stars when they cross the Meridian.

Other astronomical connections calculated by Bauval (and confirmed by robotic measurements made in 1993 within the Pyramid) also directly link the monuments at Giza, particularly the Great Pyramid and Sphinx, to additional specific stars, and two additional crucial constellations. The first was the so-called "dog star," Sirius (Isis in the mythology) in the constellation Canis Major. The second was to the feline constellation Leo, commonly linked to Horus, Osiris' avenging son.

These additional celestial alignments are most significant when appearing either directly in the center of the sky (at the "meridian"), or when rising or setting on the horizon. They also converge on the same 10,500 BC time frame [Fig. 5-4].

So certainly, Orion/Osiris, Sirius/Isis and Horus/Leo had great significance to the ancient Egyptians. But, what is a mythological *Egyptian* stellar deity doing representing an official *U.S. governmental* exploration of the Moon? And, in a program known under the designation of a Greek "sun god," Apollo. Why not, for example, "Diana"—the Greek goddess of the Moon?

The initial answer is quite simple: "Apollo" is actually "Horus"—if one examines in detail Greek mythological literature and its derivations from the earlier Egyptian.[86] Horus, like Apollo, is the Egyptian "god of the sun" (and also, curiously, ruled over Mars). Thus, it turns out that the $20 billion NASA Apollo Exploration Program of the Moon was, in fact, nothing less than a disguised "Osiris/Horus Lunar Landing Program" ... straight out of ancient *Egypt.*

This doesn't make the selection of solar mythology, to represent a lunar mission, any less mysterious—but, even then, why didn't NASA just openly call the lunar program by its earlier, Egyptian designation "Orion?" Why hide the underlying Egyptian persona under a *Greek* designation, only to reveal its true identity within the official Program logo?

Indeed, it is distinctly possible (and Hoagland certainly believes it ...) that the massive "A" on the program patch does *not* stand for "Apollo" at all, but rather the Greek derivation of Orion/Osiris—

"Asar."

Perhaps the answer lay in the patch itself, with its many hints and links to a far earlier mythological epoch, the era of Osiris/Asar and his successor, Horus/Apollo. The three belt stars seemed to represent the three astronauts as "sons of Orion/Osiris"—but, as Osiris was a god of "resurrection," it seemed odd that the first attempt by Man to visit the Moon would be associated with a rebirth.

A rebirth of *what?*

It was only after Hoagland noted the name of another space vehicle—the Shuttle Atlantis—that he began to put the pieces together. What if this *wasn't* the first time Man had reached across space to touch the Moon? What if the ancient Egyptian "Zep-Tepi" (literally, "the first time") Era of Osiris/Isis/Horus was not a myth ... but a *reality?*

Based on this crucially important, repeating and converging Orion/Leo/Sirius connection for the Sphinx and its associated Giza Pyramids, Bauval and his new co-author Graham Hancock had contemporaneously advanced a major new interpretation of key sections of the Fifth Dynasty Old Kingdom "Pyramid Texts."[87] These ancient, written records (based on much older, pre-historic oral traditions) repeatedly cite Orion ("Osiris") and its surrounding celestial regions ("the Duat") as the crucial linking symbol for not only the Giza Plateau structures, but for other pyramids built along the Nile—echoing on Earth, in limestone, the most fundamental Egyptian religious and mythological traditions apparent in the sky. Other texts reinforced an equally strong "leonide connection," indicating that the "celestial lion," Leo (according to Bauval, the precise skyborn counterpart of the enigmatic Sphinx on Earth), was also part of this comprehensive (though still profoundly mysterious) celestial mythology.

Bauval and Hancock's new interpretations also took seriously, for the first time, repeated, emphatic references found within the texts linking this Pyramid/Orion/Sirius/Sphinx/Leo connection to a literal "First Time." References to this extremely ancient, enigmatic and apparently catastrophically-ended civilization (circa 10,500 BC) still apparently echoed in the surviving celestial identifications with these particular "key" regions of the sky [Fig. 5-4].

This pre-Egyptian and potentially high-tech global terrestrial civilization (with mysterious extraterrestrial connections, also identified within the texts) was what was known later in the descendant Greek tradition—through Plato's own Egyptian sources—as "Atlantis."

This new data reinforces earlier interpretations, by myriad additional researchers, of increasing world-wide evidence for the existence of a pre-Egyptian, remarkably sophisticated, truly global culture—which flourished long before current "mainstream" archaeological interpretations, and which in some manner ended abruptly about 12,500 years ago.

So Hoagland asked himself: what if the reason for the hidden Egyptian motif of the Apollo patch *wasn't* merely "homage" to a long forgotten myth, but a literal acknowledgment that in Apollo Osiris' children were returning to their rightful place among the stars? Perhaps even to the very worlds where Osiris and his followers had left the signs and wonders of their previous inhabitance of the solar system? What if NASA had placed Osiris on the patch, not because they wanted to ... but because they felt *compelled* to?

What other signs might there be that "Project Osiris" had a very special place in human history, above and beyond that which was publicly acknowledged?

Tranquility Base

Faced with these larger questions, Hoagland began to look back at the Apollo missions for clues. He quickly found numerous references to Orion/Osiris in the Apollo mythology. He noted, for instance, that the Apollo 15 Lunar Module had been named "Falcon", seemingly a reference to the Air Force Academy mascot. Since Mission Commander David Scott was a graduate of the Academy, the name made sense—but as he studied up on Egyptian religion, Hoagland quickly found that the Falcon was also associated with Horus, the avenging son of Osiris.

The Lunar Module *"Horus?"*

As he went down the list of names, Hoagland found more of these fascinating double meanings. The Apollo 16 LEM had been openly named "Orion," an obvious reference to Osiris. And, the Apollo 11 Command Module, Columbia drew its name from St. Columba—a sixth-century monk who, in

Masonic lore, brought a "sacred stone" (said to be the stone upon which Jacob rested his head when he received the vision of his heavenly ladder) to Scotland from Egypt.

Also, Apollo 13's Lunar Module, which served as a lifeboat—literally saving the lives of Lovell, Swigert and Haise—was named "Aquarius." According to the Mission Commander, Jim Lovell:

> *"Contrary to popular belief, it was not named after the song in the play* Hair, *but after the Egyptian God Aquarius. She was symbolized as a water carrier who brought fertility, and therefore, life and knowledge to the Nile Valley, as we hoped our Lunar Module,* Aquarius, *would bring life back from the Moon."*[88]

All this was certainly intriguing, and fit with the recurring "Egyptian motif" Hoagland had uncovered, but it was when he read the personal accounts of the astronauts themselves that he began to gain a much clearer understanding of the "hidden pattern" he'd uncovered.

Edwin B. "Buzz" Aldrin was the second man to walk on the Moon alongside Neil Armstrong. After returning from the Moon and retiring from NASA, he penned an autobiography called *Men From Earth* about his experiences. In his book, he describes a "small religious offering" he made a short time after their landing on the Moon:

> *"During the first idle moment in the LM before eating our snack, I reached into my personal preference kit and pulled out two small packages which had been specially prepared at my request. One contained a small amount of wine, the other a small wafer. With them and a small chalice from the kit, I took communion on the Moon, reading to myself from a small card I carried on which I had written the portion of the Book of John used in the traditional communion ceremony."* [89]

Aldrin also made it clear that Armstrong did not share his enthusiasm for the ceremony. According to *Men From Earth*, Armstrong looked on "with an expression of faint disdain (as if to say, 'what's he up to now?')." This little-discussed event was dramatized in the 1998 HBO miniseries *From the Earth to the Moon*, and portrayed as Aldrin describes it.

Hoagland next discovered that Aldrin's ceremony (which was taken from Webster Presbyterian Church rituals, in Houston, which, in turn, "borrowed" it from the much older Catholic communion ceremony), in fact, had its real roots in *ancient Egypt*—as an offering to Osiris (naturally).[90] Further, he discovered that the *date* of the *Apollo 11* landing—and this mysterious offering to Osiris—was a sacred one in ancient Egypt. July 20 was the date of the annual inundation of the Nile Valley, marked by the so-called "helical rising" of Sirius around the time of the (generally believed) building of the Pyramids at Giza—2500 BC. July 20 was not only the date of the Egyptian New Year, but also represented the return of Isis from her period of exile with her son Horus, who would one day avenge his father's wrongful death at the hands of Set.

Sirius, along with Orion/Osiris was, of course, at the heart of not only the Egyptian mythological system—in the Isis/Osiris/Horus triumvirate—it was literally at the heart of the entire ancient Egyptian calendrical system. The helical rising was a crucial celestial coincidence that governed (for literally thousands of years) all Egyptian life, physical and spiritual, along the Nile [Fig. 5-5].

Because of Earth's annual orbit of the sun, Sirius "disappears" from the night sky at Giza for about seventy days each year, and has done so for hundreds of thousands of years. Its reappearance, just before sunrise in the east, is where the term "heliacal rising" comes from. Heliacal is from the Greek "Helios"—the rising sun—which is just another form of the Egyptian term for the same phenomenon: Horus. Our English word "horizon" comes from the Egyptian root, and literally means "Horus-Rising" in the ancient tongue.

Because of the ever-present effects of precession (the ~26,000-year "wobble" of the Earth), this annual disappearance has been systematically sliding through the seasons for all of Egypt's history. In modern times this event now takes place on August 5. 2,000 years ago, the heliacal rising was on July 20, the date of the landing of "Eagle" at Tranquility Base.

In ancient times, circa 3300 BC, this heliacal re-emergence took place on the summer Solstice (June 21, Gregorian). Around this same time, the melting snows of the mountains in Central Africa would flood the Nile and provide much needed irrigation for the crops of this otherwise arid land. This conjunction of events led to the marking of the New Year coincident with this magical rebirth of Sirius, the Nile and Isis. The New Year was considered to have started not at midnight, but when Sirius actually reappeared in the sky, at dawn, with Horus.

In his book *Echoes of Ancient Skies*, the archaeo-astronomer Dr. Ed Krupp writes about these events:

> *"After disappearing from the night sky (for seventy days), Sirius eventually reappears in the dawn, before the sun comes up. The first time this occurs each year is called the star's heliacal rising, and on this day Sirius remains visible for only a short time before the sky gets too bright to see it. In ancient Egypt, this annual reappearance of Sirius fell close to the summer solstice and coincided with the time of the Nile's inundation. Isis, as Sirius, was the 'Mistress of the Year's Beginning,' for the Egyptian new year was set by this event. New Year's ceremony texts at Dendera say Isis coaxes out the Nile and causes it to swell. The metaphor is astronomical, hydraulic and sexual, and it parallels the function of Isis in the myth. Sirius revives the Nile just as Isis revives Osiris. Her time in hiding from Seth is when Sirius is gone (seventy days) from the night sky. She (Isis) gives birth to her son Horus, as Sirius gives birth to the New Year, and in the texts Horus and the New Year are equated. She is the vehicle for renewal of life and order. Shining for a moment, one morning in summer, she stimulates the Nile and starts the year."*

So, what were the chances that "Project Osiris," carrying the veiled banner of Orion/Osiris/Asar on its patch, could land on another world on such a sacred date in the Egyptian year, only to then mark the event with a thinly-disguised ancient ceremony that paid homage to the chief Egyptian god himself, Osiris ... all purely by *coincidence*?

This sequence became even less explicable (in terms of a secular, "scientific" Agency ...) when Hoagland looked back at the Apollo mission profiles.

Apollo 1 was a ground test vehicle that was lost in the infamous fire that took the lives of astronauts Gus Grissom, Ed White and Roger Chaffee. Apollo 4 was the first test of an unmanned Command and Service module in space. Apollo 5 was the first test of the Lunar Module in space. Apollo 6 was a full-up test of the Saturn V launch vehicle. Apollo 7 was the first manned test of the CSM in Earth orbit. It was with Apollo 8 that things began to get interesting.

*   *   *

Apollo 8 was originally intended to be a full-up test of the CSM and LM in high Earth orbit—simulating a real mission to the Moon, by looping tens of thousands of miles out into space. But the Lunar Module wasn't ready yet for flight. Facing a deadline of 1969 to land on the Moon, and with the Russians breathing down NASA's neck, it was daringly decided that Apollo 8 would, instead, become NASA's first manned lunar *fly-by* mission—*without* the LM.

The Apollo 8 crew of Borman, Lovell and Anders became the first humans to leave the gravitational influence of the Earth and circumnavigate the Moon, proving that the voyage and return to the Moon was possible.

All that would be left to do was make the lunar landing.

Apollo 9 accomplished the next step toward this ultimate objective, back in Earth orbit.

On this Mission, the LM "Spider" was finally, successfully, flown for the first time, detached, docked and re-docked with the CSM all over a ten-day period. Apollo 9 was the first "all up" testing of all the components that would be used to land the astronauts safely on the Moon and return them to Earth. The Mission went off flawlessly, with all of the Mission parameters—including a manned LM being flown thousands of miles away from a lone astronaut remaining in the Command Module, and then back again—being met or exceeded.

So this left Apollo 10.

With all of the primary technical and mission planning components having been tested and proven out, the next mission was a full-up dress rehearsal for Apollo 11. Launched on May 18, 1969, Apollo 10 acted as a pathfinder for Apollo 11's "Eagle," following the same descent path that Apollo 11 would two months later. Eventually, Thomas P. Stafford piloted Snoopy to within 8.4 miles of the lunar surface (about 44,000 feet), prompting Lunar Module Pilot Gene Cernan to comment ominously; "Man, we's getting down among them."

Given that their altitude was nearly 50,000 feet above the lunar surface, we can't help but wonder what Cernan was talking about. At that altitude, lunar surface features, even mountains, would be obscure and far away. However, given where Stafford and Cernan were at that moment, passing through Sinus Medii and heading on toward Mare Smythii, the only thing they could have been "down among"—at 50,000 feet—would be Hoagland's theorized miles high glass domes. Undeniably, the location and the altitude would be correct for that to be what Cernan was talking about.

This bizarre comment also raises the other strangely incongruent aspect of Apollo 10; while the spacecraft was theoretically fully capable of landing on the Moon, inexplicably, it was *not* given the capability to do so.

Not only was the Mission denied the fuel to make a safe lunar landing (the tanks on-board were literally only half-filled), but the LM "Snoopy" was a crippled version of the "real" vehicle, unable to physically land on the lunar surface.

Politically, this really makes no sense.

Still in a fierce race with the Soviet Union to be the *first* to land a man on the Moon, Apollo 10 had everything necessary to accomplish this long sought after political goal—*except,* the tools to do so. The Saturn V, the LM and the CSM had all been tested on previous missions, and the NASA long-distance (lunar) communications network was tested on Apollo 8. There was no practical, canonical reason *not* to land Apollo 10. With only two more shots at making the goal before Kennedy's "end of the decade," the question is, why *wait*?

As he delved ever deeper into the arcane, *Egyptian* mysteries surrounding these supposedly secular, "scientific and engineering" NASA missions, Hoagland finally found his answer:

Because it wasn't "time" yet.

It finally occurred to Hoagland that there had to be a "hidden reason"—a *ceremonial* reason quite likely—why Apollo 10 was prevented from carrying out the Mission it was so capable of accomplishing, and thus achieving Kennedy's Goal with plenty of margin for error if something went wrong.

Perhaps it had something to do with the date of July 20th, or the project patch, the odd "communion ceremony," or maybe (as he found out with Alan Shepard and America's first manned sub-orbital flight, back in 1961 ...) the men themselves—Armstrong and Aldrin—had, for some reason, been *pre-selected.*

Or maybe, it was *all* of those things.

Recognizing that the ancient Egyptian religion was a stellar one, and reasoning that the stars themselves may have had a role to play in this odd series of "coincidences," Hoagland purchased some astronomy software and started to look back in time to the lunar landings.

Maybe Orion was more important in the Apollo Program than even he thought.

When he first plugged in the date, time and coordinates of the lunar landing, he found some curious stellar alignments that *almost* fit. He expected to see perhaps the belt stars of Orion or maybe Sirius/Isis herself either on the horizon or at the meridian, as was custom in the Egyptian stellar religion. Instead, he found that Sirius was hovering almost due east over the landing site, nearly 20° above the lunar horizon—but then he wondered, what if the landing itself was not the most important event, but Aldrin's "religious offering" was?

The first thing he had to do was pin point the time of the mysterious communion ceremony. Using transcripts from the Apollo 11 mission logs, he and Johnston painstakingly reconstructed the post-landing lockdown procedures that Aldrin and Armstrong had gone through on that historic day. In doing so, they determined that the ceremony took place precisely *thirty-three minutes* after touchdown on the lunar surface.

When he plugged *that* time into his astronomy software, he got quite a shock: Sirius was hovering over the Apollo 11 landing site at precisely 19.5° above the lunar horizon.

Hoagland was stunned. There was now no question in his mind that Aldrin's "communion ceremony" to Osiris was carefully planned for this *precise* moment and location on the Moon: with Osiris wife, Isis, his unique savior and consort, looking down as Sirius from her exact "*19.5°*" elevation in observation of the "sacred moment" [Fig. 5-6].

But the fact that Sirius was at 19.5°—not on the horizon or meridian as a truly pure interpretation of the ancient stellar religion of Egypt would call for—had far deeper implications. It suggested that as far back as 1969, NASA was *fully aware* of the significance of the Cydonia "*tetrahedral geometry*"—and perhaps the physics behind it—even though this was more than half a decade before the first images of that crucial Martian region would be taken. A further clue was found in the hieroglyphic symbol for Sirius: an *equilateral* triangle, the very 2D representation of the 3D tetrahedron [Fig. 5-6].

So Hoagland was left to ponder a ceremonial NASA "offering" to Osiris, made when Osiris' consort Isis was exactly at 19.5° above the landing site— represented by a redundant "tetrahedral" hieroglyph *and* at the "tetrahedral" altitude of 19.5° (and, at 33 minutes after the landing)—

All flying under an official NASA Program patch that symbolically— secretly—paid homage to "Orion/Asar/Osiris" ....

On a hunch, he decided to check another Apollo landing site—the "future" Apollo 12 landing site. If what was going on was what he *thought* was going on, this could be an excellent candidate location for *another* "ritual surprise." At the time of the Apollo 11 landing and Aldrin's "Osiris/Isis ceremony" back at Tranquility Base, NASA had already ritually marked the site by landing the unmanned Surveyor 3 probe there. Just four months after Apollo 11, in November, 1969, NASA set Pete Conrad and Alan Bean down (for their "pinpoint" lunar landing ...) right beside Surveyor 3—literally within walking distance [Fig. 5-7]!

Sure enough, he promptly found another astonishing alignment:

Orion's middle belt star, Alnilam, was on the east-southeast horizon, and rising; it was a hairs breadth away at the Apollo 11 Landing (only twelve arc minutes below the zero horizon), even closer to dead band "zero" (~three arc minutes) as Aldrin's "Osiris" ceremony, a thousand miles to the east, commenced [Fig. 5-8].

And *dead-on* as it concluded.

In the ancient Egyptian stellar religion, this would have been a most significant alignment, with Osiris/Asar dead on the horizon, "rising to life" just as his consort (and rescuer) Isis was being honored at Tranquility Base. Literally, the ceremony seemed to mark a re-enactment of the ancient legend of Osiris' *resurrection* at the hand of his consort Isis—with the stars *simultaneously* playing out the scene at what would become the two most sacred "temples" of Man's first lunar explorations.

Hoagland had to research the ancient mythical texts of Egypt extensively to understand the symbolism at work here. He found that certain positions in the sky had great significance, with the horizon and the meridian easily the most significant. To the Egyptians, the horizon represented a sort of netherworld between dimensions. By "dimensions," the Egyptians meant "life and death." To them, death was simply the next step on a spiritual quest to be reunited with Osiris. When a stellar object, like the star Sirius, was on the western horizon (and setting), it meant that the goddess Isis herself was moving from "the world of men" to "the world of the gods" (or vice versa, if it was on the other horizon ...).

The meridian marked an object's traverse from east to west in a nocturnal rising and setting that symbolized the daily birth and death of the sun. When a nighttime object crosses the Meridian, it attains its highest point in the night sky. It was at this moment that the object, be it Sirius, Orion's belt, or the heart of Leo the Lion, was most alive. From that brief moment, and for the rest of the evening, it would descend to the west, slowly decaying toward death.

Curious as to the validity of his observations and understanding their astonishing implications for the Apollo Program as a whole, Hoagland consulted Marv Czarnik, the twenty-five-year NASA veteran who had confirmed his lunar artifact findings before the 1996 DC press conference. Hoagland now asked

Czarnik if "the positions of the stars" were of "any great significance in mission planning?" Czarnik replied that they were more than "significant"; they were absolutely *crucial* to any hope of successful celestial navigation to and from the Moon, as well as landing in a precise location on its moving surface.

But—

Not *these* star positions.

In order to pull *these* "non-functional" star alignments off—and, simultaneously, at two *different* landing sites over a thousand miles apart—Czarnik confirmed the Orion/Osiris symbolism *must* have had the *highest* mission planning priority, over all *other* publicly stated goals of the Apollo Landing Program. They would have to be given top priority over all other objectives, be they political, general mission science, specific lunar geological sampling and even crew safety.

Going to the Moon (or any other planet) with current propulsion technology (such as Apollo's Saturn 5), requires a tremendous amount of careful and prior mission planning. Because fuel and rocket thrust are very limited, leaving for a specific planetary destination with a precisely-timed, pre-planned arrival time and specific landing site in mind, requires an immense amount of detailed knowledge of key "celestial mechanics." This encompasses all relative planetary and spacecraft motions, ranging from precise planetary orbits, to planned spacecraft departure and arrival times, to the individual planetary rotation rates themselves. The latter critically impact mission departure and arrival times, if not the intended landing site geometry itself [Fig. 5-9].[91]

This is because *everything* in space is constantly in motion—spinning on its own axis, orbiting the center of the solar system, or orbiting another object as it's moving around the sun (like another planet). With current rocket technology, if you wish to arrive and land at a specific place, at a specific time, on a specific object, that priority—and that priority *alone*—determines everything else about that specific planetary (or lunar) mission.

This includes any and all secondary considerations—such as science, operational planning, decisions related to landing-site geology, angle of sunlight at the time of landing, communications geometry to Earth, etc.

And, if you want to land on another planet so that the stars above that planetary landing site conform to some kind of "ritual celestial pattern"—say, the configuration of Orion and its associated constellations over the Pyramids at Giza, as they would have appeared several thousand years ago—that decision, and that decision *alone*, will determine (override!) *every other aspect* of the extremely complex mission planning just described. Not to mention, as Hoagland already knew and Czarnik independently confirmed, these considerations have zero *scientific* relevance to such a landing, at least in a secular context.

These severe constraints also shed new light on just why the Apollo 11 Landing had been so "dramatic" in its last few minutes.

According to the official NASA history, Neil Armstrong took over manual control and flew the Eagle "five miles further west of the intended landing site," supposedly because of "rocks." In doing so, he nearly exhausted the LM's supply of fuel, and thus risked the entire lunar landing; just a few seconds more, and Armstrong would have had to abort the landing and return to orbit.

But really, why did he have to fly so far, and take such a risk; how realistic is it that he couldn't find a boulder-free landing site to set her down for over *five miles* (in other words, for over *25,000 feet!*)—in a vehicle capable of landing *vertically* ... with a width of less than 30 feet?

Interestingly, if Armstrong had managed to set down at the original intended landing site, then the stars would *not* have been right for Aldrin's little "Osiris ceremony" thirty-three minutes after landing. The place they eventually did land—"Tranquility Base," as it would forever become known— was the one and only place in the entire pre-determined "Apollo Landing Zone," stretching along almost the entire visible equator of the Moon—where Sirius would be hovering at 19.5º ... thirty-three minutes *after* Apollo 11's Touchdown, the night of July 20th, 1969 (GMT).

As far as Hoagland was concerned, there could no longer be any question that this *entire* sequence of events was meticulously planned and executed flawlessly— by *someone*. But "who" within NASA would have had such power—and the desire— to literally hijack the First Lunar Landing ... and in such an arcane fashion?

He soon had his answer.

In looking at just who officially picked the Apollo Landing Sites, he discovered yet *another* "Egyptian connection."

Dr. Farouk El-Baz is an Egyptian geologist who tutored the Apollo astronauts in the developing science of "lunar planetology." Born in Cairo and educated in the United States, El-Baz taught in Germany and the U.S. before becoming involved in the U.S. space program in 1967. In that year, he applied to "Bellcom" (a subsidiary of AT&T)—which up to that point had been (as you would expect) handling communications for the Apollo Program. However, after El-Baz came aboard, Bellcom's job description underwent a significant change; they were now suddenly in charge of selecting the actual landing sites for the Apollo Program. In fact, his own biography shows that, curiously, El-Baz was *solely* in charge of the site selection process:

> "... *From 1967 to 1972, Dr. El-Baz participated in the Apollo Program as Supervisor of Lunar Science Planning at Bellcomm, Inc., a division of AT&T that conducted systems analysis for NASA Headquarters in Washington, D.C. During these six years, he was Secretary of the Landing Site Selection Committee for the Apollo missions to the Moon, Principal Investigator of Visual Observationsand Photography, and Chairman of the Astronaut Training Group.*"[92]

What all of these titles amount to is that El-Baz was the guy who picked the landing sites, controlled the dissemination and analysis of all the photography, and directly managed and oversaw the astronauts' geological training, preparing them for what they would actually *observe* on the lunar surface. In short—he was *the most powerful single individual* in the American space program.

And, he was in a unique position to manipulate the landing site selection in just the way that Hoagland now suspected.

Still, this did not *prove* that he had done it; only after examining El-Baz's fascinating family background, and the severe scientific objections raised to some of El-Baz' "curious selections," did Hoagland become convinced that he was, indeed, the "trigger man" for the strange events of July 20, 1969 and after.

Hoagland discovered that El-Baz was just a bit more than an Egyptian geologist who needed a job in the space program, and happened to be in the right place at the right time, with all the power necessary to manipulate subsequent events as he saw fit. Upon further investigation, it turned out that El-Baz's father was an expert in *Egyptian religions*—including the stellar religions of the ancients.

El-Baz himself was called "The King" by the astronauts he trained—literally "Pharaoh" in the ancient Egyptian tongue—revealing a degree of reverence by the hotshot test pilots in Apollo rarely lavished on "mere" field geologists. He also came from a very influential Egyptian family politically (which was how he came to be accepted into the program over the severe *objections* of Richard Nixon's own brother—who was a senior geologist at Bellcom).

His post-Apollo years have been spent studying terrestrial satellite images, looking for archeological ruins in the Egyptian desert (similar to the landscape one would find on Mars) and helping to manage and preserve the Pyramids and Sphinx at Giza. He was also science advisor to Egyptian president Anwar Sadat. So, El-Baz certainly had the necessary background and pedigree to be the point man in NASA's "Osiris" Program for the Moon ....

Now more than a little curious, Hoagland began to analyze the other missions and significant events in the Apollo Program for key alignments." He started with Apollo 8.

By far the most significant achievement in human space exploration up to that time, Hoagland had personally witnessed the historic events of Apollo 8 first-hand. From the fiery launch of the Apollo 8 crew atop the amazing Saturn 5 from Cape Canaveral, to a monitor-rich CBS Control Room in New York, where he saw Frank Borman read from Genesis live while orbiting the Moon with the cratered lunar landscape literally rolling by outside the window, Hoagland had always appreciated the historic impact of these special moments in the mission. But the most dangerous and important single moment in the mission came when the crew had engaged a rocket burn (called Lunar Orbit Insertion) to place man into Lunar orbit for the first time. Had the burn failed, the spacecraft would have spun around the backside of the Moon and been sent on a "free return" trajectory back to Earth, never having achieved a single complete orbit.

When he rolled the clock back to the Apollo 8 Lunar Orbit Insertion [Fig. 5-10], he found that Orion/Osiris had, indeed, "observed" that seminal Apollo event.

He started by "placing himself" at Tranquility Base, on the assumption that this might be some kind of anchor, or "master temple" in the "lunar ritual grid system"—that he now strongly suspected NASA had established *long* before Apollo.

Had he been standing at Tranquility Base, right where Armstrong would set the Lunar Module "Eagle" down just seven months later, he'd have seen the Orion's Belt star Mintaka *dead on the horizon* [Fig. 5-10], hovering in the netherworld between life and death—at the very moment that Apollo 8 fired its lone SPS engine, and decelerated into lunar orbit on the Farside of the Moon, that December 24th in 1968 on Christmas Eve.

He then rolled the clock ahead to the Apollo 12 landing site—at the precise date and time of the Lunar Module Intrepid's landing, November 24, 1969. Once again, he found a *major* alignment above the landing site: Orion's belt star "Mintaka," hovering right at 19.5° ... over Oceanus Procellarum [Fig.

5-11].

Satisfied that he had discovered an amazing, if not almost unbelievable "ritual pattern" in the most significant events in NASA's Lunar Program, Hoagland decided to check another possibility ....

Ever since he'd found this astonishing "Isis/Osiris pattern," he'd wondered about the first images of the Face on Mars from Viking.

If "they" could plan out (and then execute) an "alignment lunar landing" with this degree of astronomical precision—what *else* had "they" been able to accomplish over the years? Holding his breath, he punched the coordinates of "the Face" into Red Shift (the commercial astronomical program he was using), and then rolled the clock back to July 25, 1976—just as Viking 1 snapped the first image *looking down on* that mysterious, formation ... nearly twenty years before [Fig. 5-12].

There, at *19.5°*—precisely where he suspected it *had* to be—was another of Orion's belt stars, in this case "Alnitak."

Osiris had, indeed, been watching ....

The implications of this particular ritual alignment were even more staggering than the initial discovery of the alignments on the Moon. If the pattern he had found was indeed real (and, at this point, how could there be any reasonable doubt; the odds against all this happening by chance were literally "*trillions* to one"), then NASA *must have known about the Face even before the first Viking image had been taken.*

This strongly implied that NASA—somehow—had known about Cydonia's existence, if not the existence of *the Face* ... before any spacecraft had actually been there!

What other conclusion could there be?

One other possibility loomed: perhaps images taken by Mariner 9 (in 1971) had revealed enough of Cydonia as "a place of future interest" for the subsequent 1976 Viking Mission to take a closer look. If NASA *knew* from Mariner (though, no such images have ever been officially admitted to, or found ...) that the Face, and rest of the monuments at Cydonia were there—at 41°N x 9° W—then they obviously had gone to a great deal of trouble [Fig. 5-12] to ensure that Osiris "oversaw" the *next* major NASA Program to explore them—the first unmanned landing of Viking "to search for Life on Mars." All the while, of course, denying vigorously all these years that *anything* of interest was there at all.

What could be *so* compelling as to make NASA plan *the entire Viking Mission* around a *single* image ... and, to the extent that they *obviously* had?

One of Hoagland's tests for the reality of any of these findings involved the very genesis of NASA's Lunar Landing Program; just how far back, he wondered, could this "ritual hijacking" of Apollo actually be traced?

Since all that is required to determine "an alignment" is a precise location (on *any* moon or planet) and an exact time, he decided to go back to the very beginnings: Kennedy's historic, joint session of Congress, May 25, 1961—where the President issued his ringing call for placing "a man on the moon, *before* this decade is out, and returning him safely to the earth ...."

From NASA and Congressional records, Hoagland ultimately determined that the President had moved to his "Apollo announcement" section of the speech at about 12:50 p.m., local daylight time in Washington DC. From this, he was able to do a Red Shift survey of all of the future Apollo Lunar Landing Sites—to see if there were any significant alignments during Kennedy's address.

And, to his astonishment, he *found* one!

A crucial one—again, involving "Osiris," in the act of "celestial resurrection"—with Mintaka, the same star in the belt of Orion which would stand 19.5º over the landing site when Apollo 12 actually made it eight *years* in the future ... rising *precisely,* now, dead on the eastern horizon of the *future* Surveyor 3/Apollo 12 landing site [Fig. 5-13].

This unquestionable "hit" had some extremely far-reaching implications—again, not only for just how early NASA's "ritual conspiracy" had been set in motion, but for "who" was actually controlling it. In particular, the fact that this lunar site was somehow considered "special"—six years *before* the unmanned Surveyor 3 would touch down, and eight years before Pete Conrad and Alan Bean would follow—strongly implied that *this* was the "anchor point" of the entire "lunar ritual network" ... *not* the Apollo 11 landing site.

And that discovery, in turn, would provide a crucial clue as to just "who" was running this entire "hijack operation."

*   *   *

Hoagland knew that a great many of the Cydonia researchers, who had been willing to go quite far with him over the years, would have "serious problems" with this astonishing new information. The idea that NASA had been secretly planning *entire missions*—manned and unmanned, both to the

Moon and Mars—since its inception—*solely* around something as absurd as "key ritual alignments," would, he imagined, cause a *huge* rift between not only him and NASA ... but with many of the more cautious members of the "anomaly community."

Even he could not bring himself to totally believe it, yet ... and he had the incontrovertible celestial mechanics evidence to *prove* it in his hands.

Objections

Feeling he had a strong enough case to finally take this to the public, and in hopes that more "inside whistle blowers" like Ken Johnston would come forth—if they found out what had been *really* going on in their beloved Agency—Hoagland announced his new "ritual alignment model" on *Coast to Coast AM* in the spring of 1996, and simultaneously, on the newly formed www.enterprisemission.com.

However, when Hoagland presented these almost unbelievable "ritual findings" to the Cydonia research community, as expected, he didn't find a lot of enthusiasm; over the years, many objections have been put forth—from a simplistic "it's all just coincidence," to more complex epistemological arguments. One of his strongest critics was Dr. Tom Van Flandern, a friend and colleague who had stood with Hoagland on the dais all the way back to the Mars Observer press conferences in 1992.

Van Flandern is an expert in celestial mechanics, having obtained an astronomy Ph.D. from Yale in 1969. He spent twenty years at the U.S. Naval Observatory, where he became the Chief of the Celestial Mechanics Branch. Among other controversial ideas, Van Flandern is a strong advocate of the Exploded Planet Hypothesis (EPH), which argues that there is compelling evidence (which NASA, strangely, is also totally ignoring) of a former "exploded planet" once orbiting between Mars and Jupiter, where the asteroid belt is now located.

Tom had been an early advocate of the artificiality hypothesis for Cydonia—yet he objected to Hoagland's new data regarding a repeating pattern of "ritual alignments"—including at the time of acquisition of the key "Face" image, from Viking in 1976—on several counts. For one, he believed that no matter how many "hits" the Red Shift data had produced, it was all subject to selective testing by the observer. How had Hoagland decided that the Apollo 8 lunar orbit insertion was the most important moment of the Mission for instance, rather than trans-lunar injection or even when the astronauts had eaten breakfast on the morning of the launch? Further, he felt that testing anything that had taken place in the past (*a posteriori*, he called it) could not produce reliable test results in the area of probability. He even tackled the argument on his website (www.metaresearch.org) in a column titled "On Improbable Claims."

> "... In general, we tend to be deceived because our minds often do not recognize how truly vast is the number of possible coincidences that can occur [sic]. So when a few of them do occur, as they must if the odds are right, we tend to be amazed simply because the odds against that particular coincidence were very great. The odds against a flipped coin coming up tails ten straight times are 1,024-to-1 against. But if we make several thousand attempts, the odds become pretty good that it will happen one or more times.
>
> "In science, an improbable event that has already happened is called a posteriori (after the fact), and generally is taken to have no significance no matter how unlikely it might appear. By contrast, if we specified a certain specific highly improbable event in all its detail a priori (before the fact), and it happened anyway, that would be significant, and we would be obliged to pay attention ...."

Hoagland understood Van Flandern's position, but simply could not logically agree with it. For one, he had not applied his Red Shift technique to "thousands" of events, or even hundreds, but just an historically obvious *handful;* these were among the *most* significant events in the history of the entire space program. Certainly, the date and time of "the astronauts eating breakfast" was *not* a "significant event," and had no bearing on the significance of the other events in such a mission. To accuse Hoagland of "cherry picking" was to ignore the fact that several highly improbable stellar alignments—in *precise* context with *specific* mythical names and symbols, attached specifically to key NASA missions—had inexplicably taken place, in a supposedly "strictly scientific" Agency.

That simply should *not* be happening, under any normal operational scenario—and *had* to be explained ....

As to the second objection—that a past event cannot be the basis of an improbable future prediction—this too is intellectually fallacious. The fact that an event took place in the past is not the same thing as knowing the result of the test. Van Flandern would argue that drawing a royal flush from a randomly shuffled deck at some point in the past is not improbable at all. It has been done many times. But *predicting* that a randomly shuffled deck will dole out a royal flush on the next hand and having it come true *is* highly improbable, and therefore an event of significance scientifically.

However, he's simply wrong.

Just because the event took place in the past is no reason to disqualify it from consideration for the database. Even if the landing or "communion ceremony" took place thirty years before, it still represents an *a priori* prediction, because the *results* of the test are unknown before they're carried out. As long as the "significant event" to be tested is declared in advance, along with the expected outcome, then the test is just as valid as if the observer were predicting a future alignment around a significant event.

But there was little time for such abstract epistemological debates with colleagues. We were about to make a discovery that would truly illuminate the discussion in a way that was most unexpected.

The Occult Space Program

As Hoagland tried to grapple with the implications of this new discovery— that NASA somehow had manipulated the entire Apollo Program (and perhaps its other programs as well) around a bizarre set of "ancient religious rituals"— he was faced with very big questions: just "who" had been involved, and how much did they know? What could make them stay silent for so long? How many people at NASA would have to be "in on it" to pull off such an enormous coup, right under the noses of the literally tens of thousands of "honest, rank and file" NASA scientists and engineers?

That Dr. Farhouk El-Baz was one of the major architects of this arcane agenda he had little doubt. But who else had been in on the plans, and how were they recruited?

Once again, it was Ken Johnston who provided a key insight.

After discussing with Johnston the now infamous "communion ceremony" that Aldrin had conducted, Ken pointed out that Aldrin—like Johnston himself—had at the time been a *32° Scottish Rite Freemason*. He also noted that a recent book by two Masonic scholars (Christopher Knight and Robert Lomas) had concluded that virtually *all* of the Masonic rituals were derived from the story of Isis and Osiris.[93]

Their book *The Hiram Key* showed that, contrary to Masonry's own lore, the Craft was *not* founded in London in 1717, but in fact traced its roots all the way back to ancient Egypt. They followed a trail back through time, to the Templars, to Jesus and the Temple of Jerusalem, then on to the builder of the first Temple of Solomon, Hiram Abiff. They concluded that the ritual of the third-degree of Freemasonry was a re-enactment of Abiff's murder for refusing to reveal the high secret of the Craft, and that this same ritual was in fact derived from the ancient Pharaohaic rituals that paid direct homage to Isis and Osiris. They also asserted that Jesus himself was an initiate of this quasi-Masonic order, and that his real teachings had been usurped and distorted by the Catholic Church millennia before. They viewed Jesus as a martyred prophet, but not a divine being as the Church came to ultimately insist. None of this made them very popular with either the Christians or their own fellow Masons.

As strange as this evolving conspiracy theory had become, there was now a direct linkage between one of the participants (Aldrin) and the "ceremonial act" he had committed on the Moon. If Knight and Lomas were right, then Aldrin's communion ceremony had no conventional Christian significance at all; it was, in fact, a direct offering by a Freemason to "the ancient Egyptian gods" that his Craft most revered.

Or, even if it was an offering to Jesus (as publicly explained ...), it was (in this unique context) an offering—by the first Freemason to set foot on *another world*—to a revered Masonic figure who himself must have taken part in rituals paying homage to the same ancient Egyptian stellar deities who "eternally resided" above the Apollo landing site in such precise locations.

Either way, this was hardly the ostensible reason given to the taxpaying American public for the $20 billion-dollar Apollo 11 Mission.

Johnston also suggested that the timing of the ceremony—*thirty-three minutes* after landing, when Sirius was at 19.5º above the landing site—might have something to do with the tradition of "the 33º" in the Scottish Rite. In the Ancient and Accepted Scottish Rite, the 33º is considered the highest level of enlightenment that a Freemason can achieve. Coupled with the tetrahedral 19.5º altitude of Sirius above Tranquility Base, this suggested that there was some significant connection between these "holy numbers."

Yet it was too easy to simply say that these Egyptian rituals were a conspiracy by the "Freemasons" inside NASA.

Over the centuries, Freemasonry has been the target of derision, persecution and suspicion that is not generally justified. The vast majority of us know of the society through their good works, like the Shriners' hospitals we see in every major city, which usually provide medical care to the poor and the young free of charge, and by seeing our grandfathers march in Fourth of July parades along with other members of their local lodges.

However, there is a significant difference between the Masonic Craft in general, and the more specific institution of which Aldrin was a member—the Ancient and Accepted Scottish Rite. The Scottish Rite is an "appendant" body of Freemasonry, meaning it is not directly connected to the Grand Lodges of the Craft. The vast majority of Freemasons throughout the world are members of the Grand Lodges only. After achieving the first three degrees of the Grand Lodges (Entered Apprentice, Fellow Craft and Master Mason), the apprentice is said to have completed the so-called "blue degrees," the base knowledge required to be a Mason. This is also where the term "getting the third degree" comes from.

If a Mason desires to continue his studies of the spiritual and ethical teachings of Freemasonry, he may elect to pursue degrees in one of several appendant bodies, of which the York Rite and Scottish Rite are the two most prominent. While the Scottish Rite is not recognized by the Grand Lodges in several countries (including England), there is no prohibition against a Master Mason joining. In the United States, the Scottish Rite is duly recognized by the Grand Lodges, and its rituals are viewed as a continuation of the base knowledge attained by the Master Mason degree. In other words, all Scottish Rite Freemasons are Master Masons, but not all Master Masons are members of the Scottish Rite.

So, when someone talks about a "33º Freemason," they are actually talking about a member of the Scottish Rite *appendant body,* as it is the Scottish Rite which confers the fourth through the thirty-third degrees. It is also not commonly known that while any Mason may elect to take any of the Scottish Rite degrees up to 32, he must be *invited* by the Supreme Council to attain the honorary rank of the 33º—it (like the U.S. Senate) is a very exclusive club.

The United States Supreme Council divides the Scottish Rite into two separate sub-bodies—the Northern and Southern Jurisdictions. As Aldrin was a member of the Southern Jurisdiction (which controlled all the Scottish Rite lodges in Washington, D.C., Houston, Florida, Alabama and every other major city that held a significant NASA manned space flight facility), we focused exclusively on this organization. It quickly became obvious that the Scottish Rite could and did have a significant influence on the Agency ... including many of its *contractors* and *employees.*

Aldrin himself made no real secret of his Masonic associations, but neither did he overtly comment on them (nowhere in *Men From Earth*, for instance, did he mention his Masonic ties). Yet he did, on occasion, engage in specific acts to indicate his support for the Craft. When he posed for the official Apollo 11 crew portrait [Fig. 5-14], he made sure that his Masonic signet ring was prominently visible, an act he repeated when Armstrong took his photo in the Lunar Module shortly after landing [Fig. 5-15]. One wonders how Mrs. Aldrin felt when she discovered that her husband chose to wear his Masonic signet ring, rather than his wedding ring, in these two historic instances.

With the help of Johnston, Hoagland discovered that Aldrin had also carried a Masonic apron with him to the Moon and a flag from the Supreme Council of the Southern Jurisdiction. Upon returning, he had delivered both personally to Luther A. Smith, then the Sovereign Grand Commander, Southern Jurisdiction, Ancient and Accepted Scottish Rite at the temple in Washington, D.C., in a very solemn ceremony [Fig. 5-16].[94] The Grand Lodge of Texas also claimed that Aldrin had staked claim to the Moon in the name of Freemasonry by a ritual he performed during one of the Apollo EVAs, and subsequently formed "Tranquility Lodge 2000," which endeavored to eventually have its meetings on the Moon itself.[95]

With Johnston's help (only a fellow Mason may inquire about the membership of another), we quickly discovered that several more astronauts were Scottish Rite Freemasons, including some of the most famous. The obvious question, then, was just how many in the NASA hierarchy (besides Aldrin) were Freemasons or, more specifically, *Scottish Rite* Freemasons—and even more specifically, members under the Southern Jurisdiction?

Beyond that, how many of *them* were in position to directly influence the selection of Apollo landing sites and landing times?

As it turned out, just about "everybody who was anybody" at NASA had some ties to the Craft. Not only that, but we soon discovered that the Freemasons were *not* the only "secret society" operating behind the scenes at

NASA in the 1960s, nor were they alone in their dedication to the Egyptian legends of the great gods Isis, Osiris and Horus.

The true history of NASA cannot be understood without appreciating, not just the influence that these gods of ancient Egypt had on Freemasonry and the other secret societies, but also the corresponding influence that the Freemasons and the other groups had *on NASA itself.*

And for that, we have to go back to the beginning ....

The Early Years: 1930–1960

"History, sir, will tell lies as usual."
– George Bernard Shaw, *The Devil's Disciple*

NASA, as we know it, actually evolved from several earlier organizations. The National Advisory Committee for Aeronautics, or NACA, was the primary source of early NASA personnel. The NACA director, Dr. Vannevar Bush, was influential in many major aerospace projects and companies. He was co-founder of Raytheon systems (still a major defense contractor) and was director of the Office of Scientific Research and Development, which oversaw the Manhattan Project. He was also President Roosevelt's scientific advisor and played a key role in bringing many of the German rocket scientists, like Wernher Von Braun, to the United States.

He was also a 33° Scottish Rite Freemason. That in itself is not entirely remarkable—but as we will show, Bush's fraternal association with the Masons ended up having a very significant impact on the American space program of the 1960s, that still echoes today.

In many ways, Dr. Bush is the lynchpin around which many spokes of the NASA wheel have turned. With the help of Bush, the Pentagon supplied a steady stream of brainpower to NASA through its secret ballistic missile programs led by Wernher Von Braun. According to author Linda Hunt (*Secret Agenda*), Bush originally played a key role in bringing Von Braun and other "ardent Nazis" to the U.S. illegally, in direct violation of President Truman's executive orders.[96]

According to Hunt, when certain Nazi rocket scientists were deemed as "too unimportant" to be brought to the U.S. by a military panel, Bush intervened with a scathing letter to the Joint Intelligence Objectives Agency (JIOA) in which he derided the military for not knowing "even elementary information on Germans whose names are as well known in scientific circles as Churchill, Stalin and Roosevelt are in political circles." He insisted that several of the scientists were "intellectual giants of Nobel Prize stature." Bush's efforts were clearly a significant factor in some of the Germans being brought to U.S.—and the reasons for his enthusiastic embrace of men with shady pasts are, on the face of it, not entirely clear.

Shortly before the war, Bush was also working closely with Dr. Donald Menzel, an astronomer at Harvard University, on the development of a "differential analyzer"—the world's first (modern) analog computer. It was this pioneering work with Menzel which eventually—critically—enabled NASA to accomplish some of its "occult" goals.

In Bush's testimony on behalf of Menzel in loyalty hearings in 1950, he stated that "I first knew DHM [Menzel] of Cambridge, MA, in either 1934 or 1935 when I was engaged in designing and building a machine known as a differential analyzer at MIT, where I was then VP and Dean of Engineering. Dr. M., who was then an assistant or associate professor in the astronomy department of Harvard University, was much interested in the possibility of applying the differential analyzer to the solution of certain astronomical and astrophysical problems. This mutual interest led to a technical association of some intimacy over a period of about a year. Thereafter, until I became associated with the Carnegie Institution in 1939, we met in connection with scientific or technical matters fairly frequently, usually in connection with the development of specialized machinery for astrophysical use."

Although not overtly stated, Bush and Menzel's objective was to develop a computer that could allow them to predict—and *centuries* ahead—*the future positions of stars and planets from any point in the solar system.* This would later allow NASA to accurately predict the appearance of the skies above a specific planetary landing site—on a specific date and time—with the amazing accuracy Hoagland would later discover. It could also, of course, allow NASA to pull off an astronomically improbable "coincidence" ... like Aldrin's fabled "communion ceremony" at Tranquility Base, staged simultaneously with Sirius at "exactly 19.5°."

Bush and Menzel were also tied together in another curious twist of "hidden history."

In 1984, microfilm documents were mailed to the home of UFO researcher

Jaime Shandera. When enlarged and printed out, the documents appeared to be a genuine top-secret briefing memo to then President-elect Dwight Eisenhower. The memo described the now famous crash of a flying saucer in the New Mexico desert in 1947 near Roswell and the recovery of bodies and a subsequent cover-up of the same events. It also listed a group of twelve members of a new organization tasked with dealing with the "alien problem." The group, called Majestic or "MJ 12," included both Bush and Menzel as founding members. At first, debunkers used the presence of Menzel on the list as "proof" the documents were forgeries, since Menzel had been a frequent UFO debunker publicly (having written three books on the subject) and was the point man for the attacks on Immanuel Velikovsky, a psychologist who had written a popular book arguing for a catastrophic model of the solar system's origin. Menzel, who served as Dr. Carl Sagan's mentor at Harvard, was essentially the Carl Sagan of his day and seemed a highly unlikely member of such a super-secret organization.

However, long-time UFO researcher (and nuclear physicist) Stanton Friedman went painstakingly through Menzel's papers as part of his research for a book on the MJ 12 documents, and discovered that Menzel had led an elaborate double life. Friedman found numerous references to Menzel's participation in a variety of intelligence projects and committees, including some top-secret weapons programs that would have made him an ideal candidate for MJ 12. Friedman also makes a compelling argument for the validity of the MJ 12 documents, and the existence of MJ 12 itself.

Whatever the reality of MJ 12, there is no question that Vannevar Bush had a significant influence on what would eventually become known as "NASA." From his position close to so many powerful men, he was able to influence a great deal of U.S. space science and rocket research.

The Caltech Rocket Programs of the 1930s

At virtually the same time as Bush and Menzel were developing their differential analyzer on the East coast, a pioneering group of chemists and self-taught rocket engineers was working in the West on the research that would eventually lead to the development of real "Moon rockets." Led by Hungarian immigrant, Theodore Von Karman [Fig. 5-17], a small group of these maverick scientists had been heavily involved in the development of rocket fuels and engine technology for the military's early jet-assisted needs for heavy aircraft in the 1930s and 40s. In fact, though based exclusively on rocket technology, the term "jet-assisted" was coined *specifically* to counteract a negative "Buck Rogers reputation" within the technical and government communities that attached to "rockets" at that time.

 Only *much* later would the disparaged "rocket" find its ultimate uses in the U.S. space and missile programs ....

What eventually became NASA's "Jet Propulsion Laboratory" (because of the aforementioned prejudice against anything with "rocket" in the name ...), actually began in the 1920s as an aerodynamics testing facility, under the fledgling "California Institute of Technology" in Pasadena. Called the "Guggenheim Aeronautical Laboratory, California Institute of Technology," or GALCIT, this early aeronautics lab was funded by the Guggenheim Foundation, but administered directly by Caltech.

In 1926, the lab was put under the direction of Von Karman. Von Karman's chief experimental "rocket scientist," formally hired in 1935, was John Whiteside Parsons [Fig. 5-18]—a brilliant chemist and engineer who ultimately made huge strides in the field of solid- rocket propulsion.

Parsons, however, led an amazing double life. Later killed under mysterious circumstances in his Pasadena lab in 1952, Parsons had a long fascination with magic and the occult, and regularly practiced ritual sex orgies in his Pasadena mansion. His compatriot in many of these bizarre rites was one L. Ron Hubbard [Fig. 5-19], who later went on to form the controversial "Church of Scientology"—which is still a major influence in Hollywood today.

Parsons took many of his most bizarre occult ideas from the equally controversial Aleister Crowley [Fig. 5-20], the self-proclaimed "wickedest man in the world." After Parsons spent several years as a member of Crowley's Pasadena lodge, Crowley ultimately appointed Parsons to *head* this Southern California center of Crowley's world-wide organization—while, *simultaneously*, Parsons was still employed by Cal Tech to develop rockets for the U.S. Army.

Parsons, perhaps not coincidentally, was born on October 2, 1914, the same day that Charles Taze Russell, the founder of the Jehovah's Witnesses, predicted that the beginning of the end of the world would commence. Parsons was named for his father, a philanderer who was divorced by his mother soon after "Marvel" (later "John" and "Jack") was born. It is safe to assume that Parsons viewed his parents' broken marriage as a consequence of the Victorian sexual mores of the time, since he spent most of his adult life raging against those very same societal values. This probably led to Parsons' hatred of traditional marriage and religion, and his embrace of the occult.

Parsons spent the other portion of his adult life developing one breakthrough after another in rocket propulsion. He eventually co-founded Aerojet Corporation with several other members of the GALCIT team (including Von Karman)—which still builds a lot of the solid-rocket boosters for NASA and the Department of Defense.

Although Parsons had not completed his formal education, Von Karman quickly came to realize he had a "rich talent for chemistry," and allowed him, his friends and associates to use the GALCIT labs. By the mid-1930s, Parsons and his pals were testing their small "home-made rockets" in an area of Pasadena called "Devil's Canyon"; today, the Jet Propulsion Laboratory sits on that very land. Parsons and Ed Forman, his compatriot in chemistry and rocketry, are also known to have had early contact with German rocket pioneers Herrmann Oberth, and Oberth's "wunderkind" successor and protégé, Wernher Von Braun. They also frequently corresponded with Robert Goddard, who by that time was working alone in the deserts outside Roswell, New Mexico.

Apparently, Parsons and his group learned little of value from these famous contacts, since Goddard and Von Braun were focusing primarily on liquid-fuel rockets. Though he and the GALCIT team did some experimentation with liquids, Parson's personal quest was to create powerful *solid-fuel* boosters, which could someday—by being far simpler, thus cheaper and more reliable—supplant the liquid fueled complexities (and spectacular accidents) of Goddard and Von Braun's painful efforts at practical rocket development.

Parson's ultimately simpler (and more cost-effective) vision is directly embodied in NASA and the Department of Defense today—a result of the chemistry breakthroughs in solid-rocket fuels that Parson's essentially single-handedly achieved. The two slim, extremely powerful solid-rocket-boosters seen flanking the Orbiter on television every time the Space Shuttle lifts off from Cape Canaveral (providing over 80% of the take-off thrust, with about *three million* pounds apiece at liftoff), is Parson's lasting legacy to today's space program.

NASA is now focusing on a successor to the Shuttle, to be called "Ares I," which will be a combination of a single, *solid-fueled* booster—with a liquid upper stage—launching a brand new spacecraft called "Orion" (much more about that later ...), in place of the aging Shuttle orbiters. Designed to replace the current Shuttle system after 2010, and ultimately take astronauts "back to the Moon and on to Mars" ... Ares is based 100% on Jack Parson's fundamental ~60-year-old breakthroughs in solid-rocket propellant mixtures and designs.

As noted, aside from Parson's obvious chemistry and engineering genius, he had a deeply mystical side. Von Karman too fancied himself as "something of a mystic," and he was known to have claimed on many occasions that one of his ancestors had actually fashioned a "golem"—an "artificial human being" in Hebrew folklore, endowed with "life." Von Karman also loved to tell people that Parsons ("a delightful screwball ...") used to "recite pagan poetry before each rocket test."

Von Karman apparently shared many of Parsons' deepest occult beliefs. He was primarily responsible for the creation of "JPL" (literally, on *Halloween* 1936, no less) at a rocket test in the Arroyo Seca area of the Devil's Canyon. JPL still refers to this test as the "birthplace" of the Laboratory, and even trots out what they call a "nativity scene" every Halloween to celebrate the event [Fig. 5-21]. The MJ-12 document researchers have also identified Von Karman as *a possible member of MJ-12*—based on his name appearing on certain new documents they have been recently unearthed.[97]

It was around December 1938, that Parsons fell in with Aleister Crowley's Ordo Templi Orientis (OTO) at their temple in Los Angeles. Initiated into the Order in 1939, Parsons made a major impression on his fellow members, including Jane Wolfe, an actress who had spent some time with Crowley. She wrote:

> "*Unknown to me, John Whiteside Parsons, a newcomer, began astral travels. This knowledge decided Regina to undertake similar work. All of which I learned after making my own decision. So the time must be propitious.*

> "*Incidentally, I take Jack Parsons to be the child who 'shall behold them all' (the mysteries hidden therein. AL, 54-5).*

> "*26 years of age, 6'2", vital, potentially bisexual at the very least, University of the State of California and Caltech, now engaged in Caltech chemical laboratories developing 'bigger and better' explosives for Uncle Sam. Travels under sealed orders from the government. Writes poetry—'sensuous only,' he says. Lover of music, which he seems to know thoroughly. I see him as the real successor of Therion (Crowley-MB). Passionate; and has made*

the vilest analyses result in a species of exaltation after the event. Has had mystical experiences which gave him a sense of equality all round, although he is hierarchical in feeling and in the established order."

By the early 1940s, Parsons was a rising star in the OTO, and he and Crowley were exchanging letters with great frequency. Crowley was almost a buffoonish character, deeply committed to his hatred of established religion, a member of a multitude of secret societies (he was a 33º Scottish Rite Freemason, for one), and a general rabble-rouser who wanted to overturn the entire religious underpinnings of Western civilization.

Yet when you read Crowley's writings, such as the *Book of the Law* (which he claimed was dictated to him by an extraterrestrial named "Aiwass"), they weave a complex but internally consistent tale of one man's search for intellectual freedom and spiritual knowledge. Whether Crowley truly "knew something" or not, there is no doubt that he acquired a substantive and perhaps even unique amount of knowledge about the world's ancient religions and occult beliefs.

He and Parsons set about expanding the membership of the OTO lodge in Los Angeles, and Parsons was apparently successful in getting some of his fellow GALCIT scientists to join. During this period, he was investigated frequently by the local police for reports of "public nudity and bizarre rituals" taking place in his Pasadena mansion. Each time, he was able to convince authorities that he was "an upstanding citizen and a 'rocket scientist'" and that "nothing of the sort was going on," indicating the wide gap between how "rocket scientists" were viewed even by the police, because of popular culture of the time, compared to his fellow scientists.

Parsons eventually left Cal Tech to start Aerojet with several of his compatriots from GALCIT, including Von Karman, and they soon had a bona fide government contract to produce Jet Assisted Take Off (JATO) devices for the Army. After the War a succession of "oddballs and bohemians" made their way through the his Pasadena mansion (and the OTO), until one very special individual showed up in 1945.

How exactly L. Ron Hubbard came to Jack Parsons' Pasadena "temple" is something of a mystery. The Church of Scientology claims that he was sent by the U.S. Navy to infiltrate and break up a "black magic cult" operating in Los Angeles. Parsons' own letters (and those of other observers at the time) tell a very different story. By their accounts, Hubbard found "a kindred spirit" in Parsons and the two quickly fell into a variety of "magikal workings," mostly sexual rituals designed to achieve one short-term goal or another. By 1946, Parsons had bigger plans.

He and Hubbard decided that they would recreate a famous series of séances performed by Dr. John Dee, the Royal Astrologer to Queen Elizabeth in the sixteenth century. In order to do this, Parsons decided he needed an "elemental"—a female magical partner with "red hair and green eyes," that could help him conjure the spirits he was looking to raise. He and Hubbard meditated intensely in the California desert for eleven days, until one night Parsons declared "it is done" to Hubbard, and returned to the mansion.

A red-haired, green-eyed woman by the name of Marjorie Cameron subsequently appeared [Fig. 5-22].

Cameron is something of a mystery woman, having apparently made a beeline directly to Parsons' Pasadena mansion—after abruptly resigning her position as a secretary to the Chairman of the Joint Chiefs of Staff, in Washington DC, in 1946. Cameron would later appear in Kenneth Anger's film *Inauguration of the Pleasure Dome*, and was an artist of some renown and a primary force in the New Age "Goddess" movement in the 1970s and 80s.

Parsons quickly became enamored of her, and in a letter dated February 23, 1946, wrote to Crowley "I have my elemental! She turned up one night after the conclusion of the Operation, and has been with me since."

Cameron was only too happy to participate in Parsons' sex magick (with both of them blithely ignoring the small inconvenience that Parsons was still, technically, married to one "Betty Northup" and also lived with her in the mansion ...). Parsons was now free, certainly in his own mind, to begin his "major magickal project"—the so-called "Babalon Working."

In magical circles, the Babalon Working is considered a masterpiece of the form, certainly on a par (at least in its level of ambition) with Dee and Edward Kelly's attempts to "communicate with angels" in the 1500s. The rituals used by Parsons and Hubbard employed the Enochian calls, or angelic language, of John Dee, and specifically required the calling of the Egyptian god Osiris.

The purpose of the Babalon Working was to give birth to a "Moonchild" or homunculus, a version of Von Karman's golem. Fundamentally, the operation was designed to open an "inter-dimensional doorway," effectively setting the stage for the appearance of the goddess Babalon in human form.

As writer Paul Rydeen points out in his extended essay *Jack Parsons and the Fall of Babalon*:

"The purpose of Parsons' operation has been underemphasized. He sought to produce a magickal child who would be a product of her environment rather than of her heredity. Crowley himself describes the Moonchild in just these terms. The Babalon Working itself was preparation for what was to come: a Thelemic messiah. To wit: Babalon incarnate as a living female, the Scarlet Woman as consort to the Antichrist, bride of the Beast 666. In effect, Parsons also claimed the mantle of Antichrist for himself, as the magickal heir of Crowley prophesied in Liber AL: 'The child of thy bowels, he shall behold them [the mysteries of the Apocalypse]. Expect him not from the East, nor from the West, for from no expected house cometh that child.'

Without the Scarlet Woman, the Antichrist cannot make his manifestation; the eschatological formula must first be complete. In whiter words, with the magickal rites of the Babalon Working, it was Parsons' goal to bring on the Apocalypse."

From the perspective of more than fifty years hence, the notion that Parsons and Cameron could give birth to a daughter (whom Parsons, as the self-proclaimed Antichrist, apparently planned to eventually impregnate) seems not only sick, but also absurd. Parsons and Crowley believed that mankind was muddled in the so-called "Aeon of Osiris"—in their view, a dark time when men were ruled by arbitrary laws that denied them their proper birthright. By giving birth to the Whore of Babylon, Parsons believed he was sowing the seeds of the destruction of the Western world and clearing the way for the illuminated, freer "Aeon of Horus."

However, when Crowley caught wind of what Hubbard and Parsons were attempting, he became alarmed. "Apparently Parsons and Hubbard or somebody is producing a Moonchild. I get fairly frantic when I contemplate the idiocy of these louts." Crowley evidently considered the actual raising of a "Moonchild" to be an incredibly risky endeavor. In his view, neither man was experienced enough in the methods (and consequences) of using Crowley's "Thelemic sex magick" in this way.

Despite Crowley's misgivings, in Parsons' *Book of the Anti-Christ* he vividly describes a dream that came to him while he attempted to call forth the Moonchild in the last "Work of the Wand." In it, he saw a startling vision of exactly the same type of 2D "tetrahedral symbology" ... imprinted on a mysterious "tower" ... that Hoagland would ultimately discover marking the Sinus Medii ruins on the Moon [Fig. 5-23].

Raising the intriguing question: what had Parsons inadvertently tapped into?

After several months of trying, Parsons and Cameron apparently failed in their attempts to produce Crowley's Moonchild. Eventually, they gave up, and hatched a scheme to make money buying yachts on the East Coast, sailing them to California, and selling them at a profit. The whole notion collapsed when Hubbard went to New York with Parsons' money (and his still legal wife, Betty), then ran off to Florida—with the boat, the girl and the *rest* of Parsons' cash.

In his frequent correspondence, Parsons wrote to Crowley and told him that "when he almost caught up to Betty and Hubbard in Florida, they escaped in one of the boats"; hearing about the couple's whereabouts too late to physically pursue them, Parsons set up a magical circle and summoned the god Bartzabel (a form of Mars) to "conjure up a storm." Indeed, Hubbard and Betty did encounter a fierce storm at sea, and were forced to return to Miami—where Parsons promptly had them arrested. In the end, Parsons recovered some of his money, and Hubbard went on to marry Betty and eventually form the Church of Scientology, which continues to claim to this day that L. Ron was operating on orders "from the Navy" to break up the "black magic cult" at JPL.

The whole episode soured Crowley on Parsons, to the point that he wanted to remove him as the head of the OTO in America. Because he had strong support among the Pasadena members, Parsons continued to lead the OTO for a few more years, but eventually quit. After that, his life spiraled downhill (Cameron married him, but left him after a few years, only to eventually come back)—until he was killed in that huge blast in his Pasadena lab, the very afternoon he and Cameron were to leave for Mexico for a vacation and a reconciliation.

Ironically, this was just the fate (his violent death) that Hubbard had "scryed" for him in the Babalon Working.

Whatever Parsons' odd beliefs, there is no question he was massively influential in rocket technology development and, along with Von Karman, responsible for the founding of the most prestigious unmanned space exploration center on the planet: "The Jet Propulsion Laboratory."

He achieved in five years what Goddard and other more famous pioneers could not in decades of work. By the time GALCIT became "JPL," in January 1945, Parsons' professional reputation had been sealed; buildings still adorn his name at Caltech, as does a crater on the "dark side" of the Moon; and, both Wernher Von Braun and Von Karman named him as one of the "three most important pioneers" of the rocket age. He was so influential that there is even an urban myth that Von Karman chose the name of the Lab so that the initials would *be* JPL—"'Jack Parsons' Laboratory."

It is also true that no "jet" engine of any kind was ever tested or produced at "JPL"—which was always strictly a *rocket* laboratory.

It is also possible, if not likely, that Parsons, Von Karman and Von Braun all met at least once. After the war, Von Karman was sent to Germany to interview the German rocket scientists and tour the Nazi facilities at Pennemunde and Mittlework. He did meet Von Braun there, and recommended he be brought to the U.S., which (as previously noted) was in violation of the directives of Project Paperclip.

Von Braun and the first batch of German rocketry experts were eventually transferred to Alabama in 1950 (after a curious stint at Ft. Bliss, and the White Sands Missile Range in southern New Mexico), where Von Karman and others continued to debrief them. Several letters from Parsons to Cameron during this period are post-marked "Alabama," indicating that Parsons was there as part of the debriefing process.

One wonders what it was they passed the time discussing.

The German Rocket Programs

The story of Wernher Von Braun, Kurt Debus and the other German rocket scientists brought to the U.S. after World War II may not be as strange as the story of JPL's founders, but it is, if anything, far more disturbing.

To most Americans who grew up during the 1960s, Dr. Wernher Von Braun is a "hero" of the American space program, largely credited as the single most important figure in the Moon rocket programs of the 1960s and 70s. Without him, there might not have been a Saturn V to carry American astronauts to the Moon. Von Braun is rarely mentioned these days, and when he is, he is usually portrayed as either a dedicated empirical scientist or a lovable buffoon, as in *The Right Stuff*.

Yet, as Shaw says, history tells lies, and the history of Von Braun and the "American" space program is no different. Von Braun was far more than just a "German rocket scientist" or even a mere member of the Nazi party, as some histories have freely admitted. Documents serendipitously obtained via an "Enterprise Mission" research foray into the National Archives in the mid-1990's, confirm that Von Braun during World War II was nothing less than a *Major* in the SS, the fearsome and fanatically loyal arm of the Nazi war machine entrusted to carry out the most inhuman acts of the regime. Linda Hunt found survivors of the Nazi missile factories at Mittlework and Peenemunde who told her that Von Braun not only "witnessed executions and abuse of prisoners at those facilities," but on at least one occasion *ordered* executions.

Recently, new documents also show that Von Braun was commissioned to come up with a plan near the end of World War II to use one of his V-2 rockets to bomb New York City with a radioactive device, a nefarious scheme which would have immediately killed thousands, to say nothing of the untold later deaths from cancer. That such a man could spend his later years walking with presidents and giving speeches on "the wonders of space exploration"—rather than rotting in prison for Crimes against Humanity, where he belonged—is clear testament to the political expediencies of the Cold War.

Born on March 23, 1912, in Wirsitz, Germany, Von Braun was the second of three children born to aristocrats Baron Magnus and Baroness Emmy Von Braun. Baron Von Braun served as Minister of Agriculture toward the end of the Weimar Republic, and brought his family to Berlin in the 1920s.

Von Braun began his interest in rocketry at a young age. He received a telescope along with a copy of rocket pioneer Herman Oberth's *A Rocket Into Interplanetary Space*, which influenced him to experiment with rocket motors. At the age of sixteen, he organized an astronomical observatory construction team, and then went on to study mechanical engineering at the Berlin Institute of Technology, where he became a member of the German Society for Space Travel. In 1932, he enrolled at Berlin University, where he assembled a team of over a hundred scientists. Von Braun's team, which included his younger brother, Magnus, performed early experiments in rocket development.

After World War I, Germany wanted to improve its artillery capabilities but they were specifically prohibited from working on "big guns" by the Treaty of Versailles. Recognizing Von Braun's early research and development skills in "alternative munitions," they offered the young scientist a grant to conduct various experiments on liquid-fueled rocket engines—which technically, were *not* prohibited by Germany's post-War Treaty restrictions.

Von Braun, who was still a student, operated at a secret laboratory on the Baltic Coast (near Peenemunde, a favorite summer vacation destination of his family), where his classified research doubled as his doctoral thesis. His design was successful, and the German rocket program led the way in rocket development in the early 1930s.

When Hitler came to power in 1933, Von Braun remained in Germany, even though he could have left and gone to France, England or America. Von Braun's team had the eyes of the Führer on them almost from the moment Hitler assumed power. Colonel Walter Dornberger was personally assigned by Hitler to oversee the research for "military applications."

In 1939, Adolf Hitler himself visited the young scientist's labs where he was treated to an impressive demonstration of the capabilities of the rockets Von Braun's team had been developing [Fig. 5-24]. The Führer came away impressed with the demonstration, but pressed Von Braun to escalate his timetable for full-scale development. Even though no new monies were immediately released to the Von Braun team, a few weeks after the meeting Von Braun received a letter offering a commission in the SS from Himmler *personally.*

The canonical history is that he was "pressured" by the Third Reich to become involved in many Nazi organizations and that he agonized over the decision, but this actually seems increasingly unlikely.

Von Braun's aristocratic parents no doubt had many high-level connections, even in Hitler's government, and many European aristocrats were members of secret societies—like the Masons or Rosicrucians. Joining the Nazi party, even the SS, might not only have seemed like "a good career move," but may have come naturally to a man who was taught his higher place in society was a birthright.

In fact, it is now clear that Von Braun was a close personal friend of German Reichsführer Heinrich Himmler, who offered him the SS degree and rank of Major. Himmler was responsible for the growth and maturation of the SS, from Hitler's initially small personal bodyguard, to a massive organization with its own unique mission, objectives, army and internal structure.

It was also a secret society.

Designed specifically by Himmler to be a German counterpart to the "Bolshevik, Jewish, Free Mason" influence in Europe, the SS led the persecution of Freemasons (and later Jews and other racial minorities) in Hitler's Germany.

According to Dr. Nicholas Goodrick-Clarke of Oxford University and author of *The Occult Roots of Nazism*, Hitler and Himmler considered the Nazi party to be a direct lineal descendant of the Teutonic Knights, and Hitler himself to be a reincarnate of Frederick Barbarossa, the founder of that offshoot of the Knights Templar. Influenced by various Aryan cults of the early 1900s, like the Thule Society, the SS created its own set of rituals and degrees modeled on the Masonic rites.

That Von Braun (along with other Nazi rocketry experts) was a *voluntary* member of this quasi-occult/military organization is not all that far-fetched; for a man of Von Braun's clear ambition, he might have seen his membership as not only desirable, but perhaps even a requirement—considering the secrecy involved in the German rocket projects; clearly, if he could show Hitler and Himmler he could "keep a secret," they would trust him that much more with money and resources (like the slave labor) needed in the secret factories.

Von Braun was certainly considered a key member of the Reich's upper echelon. A testament to this is the fact that at his induction into the SS, the only time he is known to have worn his SS uniform, Himmler himself was in attendance, and the two were photographed *together* [Fig. 5-25]. It was most certainly a rare occasion when the Reichsführer himself attended an SS induction ceremony.

Once he had secured the money from the Nazi regime, Von Braun set right to work on his task. After a number of failed attempts, Von Braun perfected the A4, the world's first true ballistic missile, in the early 1940s. Shortly thereafter, the weapon was being mass-produced at the Mittelwerk concentration camp. In compliance with Hitler's orders, the A4 flying bomb was deployed against Britain in 1944. The London town of Chiswick was devastated by the bomb, and Joseph Goebbels renamed the A4 the "Vengeance Weapon 2," or V2.

Von Braun knew well before the end that the War was lost. As the Allies neared the capture of the V2 rocket complex, Von Braun engineered the surrender of 500 of his top rocket scientists, along with plans and test vehicles, to the Americans. Von Braun and his compatriots were brought to America under Project Paperclip, despite some objections by investigators in the JOIA.

The simple fact that Von Braun, Arthur Rudolph, Kurt Debus, Humbertus Strughold and many other German scientists with elaborate Nazi pasts were brought to the U.S. at all is a testament to Cold War convenience. President Truman's executive order authorizing the program was very specific: no "ardent Nazis" or war criminals would be allowed into the U.S. under Paperclip.

Clearly, as a Major in the SS, Von Braun met the minimum criteria of "ardent Nazi." Fortunately (for Von Braun) he had "friends in high places," not just in Germany, but in the U.S.—in the personages of Drs. Bush and Von Karman. Their influence was so powerful, that doubts expressed by the field investigators about Von Braun's SS Nazi past were ultimately "conveniently" overlooked.

In its evaluation of Von Braun, written on June 25, 1947 (coincidently the same date as the "founding" of modern Freemasonry in 1717, and one possible date given for the infamous UFO crash at Roswell, New Mexico), the JOIA basically punts in its evaluation of him, clearing the way for him to emigrate to America:

"It cannot be ascertained by this office what the reasons were which caused VON BRAUN to become a member of the SS. Neither can it be determined whether his positions in the SS were honorary, required by the Nazi party, or desired by VON BRAUN. No records of his arrest have been located."

Later events would rather dramatically illuminate the answers to these crucial questions.

Once they cleared the hurdles imposed by Paperclip, Von Braun and his team were quickly sent out to White Sands Missile Range, in New Mexico. There, they conducted tests on the captured V2s and worked on developing bigger and more powerful rockets. Von Braun immediately showed his expertise and abilities as an organizer, and was eventually rewarded by being appointed technical director of the U.S. Army ballistic weapon program.

In 1950, Von Braun and his team of "former" Nazis were transferred to the Army's Redstone Arsenal near Huntsville, Alabama, where they developed the Redstone, Jupiter and Jupiter-C ballistic missiles. Even before this period, Von Braun and his cohorts were so certain of their "value" that they made no attempt to conceal either their Nazi pasts or, apparently, their love for the Third Reich; having beaten the system through Paperclip, they evidently didn't care who knew of their "ardent Nazi" histories. Most, if not all, had elaborate Nazi records, and were still heavily committed to the party's ideologies and apparently even openly practiced its "sacraments."

The group freely displayed swastikas and other Nazi symbols on their clothes and on signs in the camps where they were kept for years after their immigration to the U.S. [Fig. 5-26].

During this same period, Von Braun used various avenues to "sell"

Americans on the idea of space travel. He wrote several articles in the popular quarterly magazine Collier's, and the success of these articles attracted the notice of 33° Scottish Rite Freemason Walt Disney. Disney hired Von Braun to produce and star in three popular television films about space travel in the future. The first, *Man in Space,* aired on ABC on March 9, 1955. The second, *Man and the Moon,* aired the same year, and the final film, *Mars and Beyond,* was televised on December 4, 1957. These programs helped establish Von Braun as America's most notable space expert.

These programs also had their curious share of thinly-veiled occult and Masonic symbolism; in *Mars and Beyond,* for instance, the traveling astronauts discover evidence of an ancient, abandoned civilization on Mars —but ... that pales in comparison to the *first* program:

Man in Space.

In this seminal "space propaganda film," Von Braun discusses how a manned reconnaissance of the Moon might be handled in the near future. He tells the audience, in extensive detail, how such a mission would be designed and carried out. Using a model of his rocket concept, Von Braun then sets up a "classic Disney" dramatization/animation of man's first mission to the Moon.

In the dramatization, Von Braun's rocket takes five days to get to the Moon, with a chance for one pass around the "dark side" before returning to Earth. As the ship reaches the Moon, the crew begins to launch flares to illuminate the darkened portions of the lunar landscape far below. The film does a remarkable job of accurately depicting the Moon's surface, until a dramatic event takes place. As the ship is flying along in its single orbit taking readings, a crewman suddenly announces that he has "a high radiation reading at 33º!" A radar operator then announces that he has an "unusual formation" coming up in front of them, and a flare is launched immediately.

When it detonates tens of miles below the orbiting spacecraft, the blast of light reveals the unmistakable, *geometric* outline of an installation—starkly highlighted on the cratered surface far beneath them, on the Farside of the Moon. There is no question that this formation is completely different than anything we have seen in any of the other recreated views of the Moon to this point in the film, and no question that the formation is also, definitely, *artificial* [Fig. 5-27].

So, what is the reaction to this sight by the "astronauts" on board?

Absolutely *nothing.*

Not a word is spoken about it in the dramatization, nor at any point later in the film. Von Braun simply inserted it without comment.

One possible implication of this (in regards to the dramatization) is that the astronauts were *required* to keep quiet about what they saw, a sort of "pre-Brookings Brookings Policy" in action (but then, why wouldn't the crew at least talk about it among "themselves" ... isolated a quarter of a millions miles from Earth ... in the dramatization?). It struck us as highly unlikely that this depiction of what the first manned mission to the Moon would find—and *where* it would find it (at "33º latitude") — was, in any way, an accident.

Von Braun and Disney obviously *intended* to portray the Moon as "previously inhabited" ... and, simultaneously, pay homage to the Masonic "33º." What they seemed to be foreshadowing was that the two are, in fact, inseparable.

In 1955, Von Braun became a U.S. citizen and served as the primary army engineer for space exploration support. In 1957, after the Soviet Union successfully launched the Sputnik satellite, the Navy was tasked with matching the Soviet accomplishment with its Vanguard rocket. Von Braun boldly predicted that Vanguard would fail, but was expressly forbidden from launching a U.S. Army satellite with his separate group in response.

When Vanguard ignominiously blew up on the launch pad with the whole world watching, Von Braun's team was then, suddenly, frantically recruited "to save the day"—and get the U.S. some kind of "win" in the onrushing space race.

Seizing this opportunity, the German team launched the first successful U.S. satellite, Explorer 1, on January 31, 1958, in conjunction with JPL (which had deep Army ties—via those initial JATO contracts). Shortly after the mission's historic success, the National Aeronautics and Space Administration (NASA) was formally established—and Von Braun had finally stamped the ticket he'd been looking to cash in on for decades.

The Birth of NASA

The National Aeronautics and Space Administration (NASA), was created by an act of Congress[98] on July 29, 1958. Its ostensible purpose was to act as a "civilian science agency" for the betterment of Mankind, and simultaneously to "enhance the defense of the United States of America." We have always been taught that NASA is a public agency, beholden only to the will of the people through their representatives in Congress. The Act itself, as Hoagland noted in the Introduction, paints a very different picture.

From the beginning, NASA was under the thumb of the Department of Defense, subject to the whims of the Pentagon on any issue judged to be "necessary to make effective provision for the defense of the United States." It was required under the Act to make available "to agencies directly concerned with national defense... discoveries that have military value or significance." Such determinations were to be made solely by the President of the United States (obviously on the recommendation of DOD, the NSA, CIA, DIA, etc., etc., etc.)—and were *not* subject to Congressional oversight.

The upshot of this is that the "civilian" Space Agency was compromised from its inception. A civilian figurehead director (NASA Administrator) was trotted out for the public to consume, but he was always taking orders from the Pentagon on any question it determined was in the interests of "national defense"—and the Pentagon was accountable to no civilian branch of government on these issues (other than the president—who, in practice, defers to "military recommendations" on matters of national security 99% of the time).

Thus, NASA—as mandated in its Charter—was/is beholden to its Pentagon masters (through the White House) first and foremost; to its own interests, second; and to the general public thirdly ... if at all.

Thus, regardless of its well-cultivated public persona—as "a civilian, primarily science-gathering" Agency—NASA has always been under the thumb of the defense/intelligence establishment.

This de facto political reality, which has been quietly in place from NASA's inception a half-century ago, is finally "out in the open" for anyone paying attention; recent remarks by the current NASA Administrator, Michael Griffin, have (at last) revealed the real nature of NASA's underlying structure and allegiance ... and it's clearly not to "pure scientific investigation," or "the scientific community" ....

Griffin, a former CIA employee, was appointed the new head of NASA in 2005 by President George W. Bush, after the previous NASA Administrator, Sean O'Keefe, suddenly resigned. Amid his first actions, in keeping with the White House announcement in 2004 of the President's "Vision for Space Exploration," Griffin began severely cutting long-term science programs in the

Agency—"stealing from Peter to pay Paul"—to fund the two biggest financial drains on-going: the aging manned Space Shuttle and increasingly expensive Space Station Programs; and the new "Vision for Space exploration": the still amorphous plan "to return astronauts to the Moon" within the next 13 years.

When members of Griffin's own "NASA Advisory Council" publicly balked in 2006 at what they saw as "a serious imbalance in priorities," Griffin immediately fired two of the scientists protesting, accepted the (pressured) resignation of a third, and soundly rebuked the whole idea of "scientists protesting his policies" re NASA's priorities, in a pointed memo to the entire NASA Council:

> *"The scientific community… expects to have far too large a role in prescribing what work NASA should do," Griffin bluntly told the NAC [NASA Advisory Council] members. "By effectiveness,' what the scientific community really means is 'the extent to which we are able to get NASA to do what we want [it] to do'."*

And if NAC members disagree with Griffin's priorities for NASA?

> *"The most appropriate recourse for NAC members who believe the NASA program should be something other than what it is… is to resign [emphasis added]."*[99]

Finally—some *honesty* from NASA!

Looking at the original management team that was placed at the Agency in its beginnings, it's easy to see that finding men with strong Pentagon/intelligence ties to run the place—not "scientists"—was (and still is) the Agency's *first* priority; Griffin's comments just finally made it "official" …..

Shortly after the formation of NASA back in 1958, then-President Eisenhower surprised many in the scientific community by passing over the highly respected, apolitical Hugh L. Dryden (Director of NACA since 1949, after Dr. Bush left), naming T. Keith Glennan instead as NASA Administrator.

Glennan had been President of the Case Institute of Technology in Cleveland, was a former member of the Atomic Energy Commission (with the highest level security clearances), and was a staunch Republican to boot. Glennan also had an extensive *military* background, having served as Director of the U.S. Navy's Underwater Sound Laboratories during World War Two. As the first Administrator of NASA, Glennan immediately proceeded to set-up a compartmentalized internal structure, far more akin to an *intelligence-gathering agency* than a civilian scientific program. Dryden was appointed to the post of Deputy Administrator. Under this structure, Glennan would furnish the administrative leadership (and policy direction) for the new entity, while Dryden would function as NASA's scientific and technical "overseer"—essentially, doing what Griffin now wants *his* entire science advisory council to do: just carry out the Administrator's orders.

Nothing's really changed.

NASA absorbed more than 8,000 employees and an appropriation of over $100 million from NACA when it was first formed. Under the terms of the Space Act, accompanying White House directives and later agreements with the Defense Department, the fledgling Agency also acquired the Vanguard project from the Naval Research Laboratory; the Explorer project and other space activities at the Army Ballistic Missile Agency (but not the Von Braun rocket group); the services of the Jet Propulsion Laboratory, as noted, hitherto an Army contractor; and an Air Force study contract with North American for "a million-pound-thrust engine," plus other Air Force miscellaneous rocket engine projects and instrumented satellite studies. In addition, NASA received $117 million in appropriations for "military space ventures" from the Defense Department.

Von Braun was appointed director of the new Marshall Space Flight Center in July 1960 and given the task of developing the rockets for the new Agency.

Glennan moved quickly to establish the Agency's monopoly over space exploration. One of NASA's first acts was to commission the Brookings Report which, as we covered in Chapter Three, makes crystal clear the discovery of extraterrestrial artifacts fall under the dark blanket of "national security."

When John F. Kennedy took office in 1961, he moved swiftly to replace Glennan and restructure NASA to accomplish one of the major goals of his Presidency: placing a man on the Moon by 1970. To this end, Kennedy (on the specific recommendation of Vice President Johnson[100]) appointed James E. Webb as the new Administrator. It is under Webb—another 33° Scottish Rite Freemason—that the true influence of the various secret societies within the new Agency came into its own.

Within a few months, Webb had appointed Kenneth S. Kleinknecht as director of Project Mercury. Ken Kleinknecht was the brother of C. Fred Kleinknecht, who was the *Sovereign Grand Commander of the Supreme Council, 33°, Ancient and Accepted Scottish Rite Freemasons, Southern Jurisdiction for the United States of*

America, from 1985 to 2003. Their father, C. Fred Kleinknecht, Sr., was also a 33°
Scottish Rite Freemason and member of the Scottish Rite Supreme Council.

"Kenny" Kleinknecht had already been selected in 1959 as one of two
"single points-of-contact" between NASA and the DOD. In this dual role, he
was able to monitor information that traveled back and forth between Project
Mercury and the Pentagon. With a lengthy history as an engineer in a variety
of black military programs in the1950's, he was ideally suited for this job. He
went on to become a "technical assistant" to Mercury Program Director Robert
Gilruth in 1960, and became Project Manager for Mercury on January 15, 1962.
Kleinknecht also became *deputy* Project Manager for the Gemini Program, and
was the Apollo Program Manager for the Command and Service Modules.

If there was a plan for the Masons to place "their" men at *the highest levels*
of the space program—it could not have been more successful.

Von Braun and the German rocket team also made their moves.

Once established as the head of the Huntsville rocket development site,
Von Braun placed many of his old Nazi cohorts into key positions in the new
Space Agency. At Von Braun's behest, Kurt Debus, a former colleague of Von
Braun's in World War II, was made the first Director of the Kennedy Space
Center. Debus, like Von Braun, was also a Nazi party member and he organized
the space center at Cape Canaveral along the lines of the German rocket
programs at Mittlework and Peenemunde, minus the slave labor, of course.

Once these organizations were in place, the task of selecting the
astronauts for the manned program began. Here again, a clear preference for
Freemasons was expressed. Of the original "Mercury Seven" astronauts, John
Glenn, Wally Schirra, Gus Grissom and Gordon Cooper were all Scottish Rite
Freemasons. Of the twelve men who walked on the surface of the Moon, four
were Scottish Rite Freemasons (as were several more astronauts who orbited
the Moon). Other astronauts may also have been members, since membership
is not publicly acknowledged except by the personal choice of the member or
by request of a fellow Mason. Even then, the requestor must have a good idea
which lodge they were members of, and the archivist must do a thorough job
of checking the local lodge records. There have been persistent rumors that
Neil Armstrong and Alan Shepard were also members, but it has never been
confirmed, although Armstrong's father was certainly a Mason.

There could, of course, be perfectly mundane reasons why so many of the
astronauts were Masons; many aspects of the lunar programs were secret (because,

among other things, the propulsion and navigation technology for going to the Moon was intimately shared with the ballistic missile programs of the Department of Defense), and the potential candidates may have seen membership in the Order as a means of demonstrating their ability to "keep a secret." They may also have simply noticed that, with Webb in charge of the Agency, they might have a better chance of being selected for the "choicest missions" if they joined his Order; becoming a member of a civic or fraternal organization—certainly in the U.S.—has long been a well-trod path to "connections" for enterprising individuals, and thus greater business or career success.

However, when we look at the core belief systems of not only the Masons, but of Hitler's SS and Crowley's "magicians," a picture emerges which makes any such "prosaic" explanations for the overwhelming Masonic presence inside NASA seem pretty lame [Fig. 5-28].

Once they had established themselves in all the key positions throughout the new Space Agency—Von Braun, Webb, Von Karman, El-Baz and all the rest were then able to proceed with plans they had apparently been incubating quietly for many years "pre-NASA." With "Brookings" as a specifically-designed political excuse for keeping key future NASA discoveries secret, the elite leadership of this clandestine occult hierarchy were able to set in motion an "inner program"—carefully hidden from the general public and the "honest" side of NASA—which appears to have been no less than a massive technological effort to confirm their shared *religious visions*—of a literal "Duat"—*on the surface of the Moon, and beyond ... to which they ... and they alone ... deserved sole* access.

This, of course, now explains why—for well over thirty years, following Apollo—there are still *no* "Hiltons" hosting tourists in Earth orbit; no "Pan Am Shuttles" flying to and from Moon bases; and no cities for civilians on the Moon. And why, the "new" NASA replacement for the Shuttle—forming the foundation of the President's vanguard "Vision for Space Exploration" for the foreseeable next decades—is actually based on a *30-year-old version of* ... Apollo. Thus, there will still be no "democratic access" to Space, even if and when these "new" programs are successfully brought on-line ... certainly not based on past performance if NASA continues to design and run the Missions.

In other words, in the warped vision of these "ritual elitists"—who have, if we are right, *literally stolen the entire space program* for themselves from the rest of all Mankind—"Space" is destined to remain the *sole possession* of only those with these "proper bloodlines and perspectives" ... but *not* for any of the rest of us.

While on the surface the Freemasons, the SS and the "magicians" would appear to have little in common in pursuit of such an astonishing, arrogant, elitist vision, in fact, the exact opposite is true. As we have established, Freemasonry holds the most ancient gods of Egypt—Isis, Osiris, Horus and Set, and their complicated, incestuous relationships—as being the cornerstone of a cosmology and a religion older than "civilization" itself ... of which they see *themselves* as now playing out major, crucial roles ....

The same is true of the SS.

In *The Occult Roots of Nazism,* Dr. Nicholas Goodrick-Clarke shows that the early Aryan cults from which Hitler's Nazi party sprang, traced their lineage back to the Teutonic Knights (a Germanic offshoot of the Knights Templar), at which point their ascendancy merges with the Templars and traces back again to *ancient Egypt.* Goodrick-Clarke also shows that Hitler and Himmler believed that these Egyptian gods themselves came from "Atlantis"—which they believed was "a high civilization established on Earth by extraterrestrials."

In this view, the ancient, uninterrupted bloodline from Horus to the present was the ultimate source of the natural supremacy of the "Aryan race" itself. It was this "divine right of descent" which gave the modern Nazis, in their view, their prerogative to rule all *other* men on planet Earth.

All of these Aryan cults, especially the Theosophical Society led by Madame Helena Blavatsky (which, especially, formed this mythic backdrop for the Nazis), had this closely-related "Isis and Osiris veneration" at their core.

The same was also true of the Crowley and Parsons' JPL "magicians."

The magick being practiced by Parsons and Hubbard in Southern California in the 1940's, as JPL was born, also saw its roots in ancient Egypt, and ultimately involved the invocation of Osiris, Horus, and the rest of the *same* Egyptian pantheon. Without "Osiris"—supreme leader of these ancient, literally divine "magicians"—in Parsons' and Hubbard's eyes, there would be no Earthly "magick" to behold.

So, if all three of these secret societies had at their core a deep obsession with the gods of ancient Egypt, it was entirely plausible that all three might equally have had the motivation to manipulate the Apollo 11 landing in the ways we have described. They certainly had—given the documented deployment of key members of each of these three groups, in strategic positions throughout the Agency at its formation—the *means* ....

If that was the case, then Ken Johnston was right: there *had* to be some deep connection between the Masonic "33º" and the tetrahedral "19.5º" Hoagland discovered at Cydonia.

But *what?*

During the Aldrin communion ceremony, Sirius had not been on the meridian or the horizon, as would have been the case if its alignment had been based on a purist interpretation of ancient Egyptian cosmology; rather, Sirius had been precisely located at the "tetrahedral" elevation of "19.5º."

What could be the possible connection between *this* number—only astronomically significant in the context of "possible ruins at Cydonia" and "planetary energy upwellings"—and the Scottish Rite's canonical "33º?" Was there, in fact, another *secret* code ... of sacred "celestial power positions," known only to a select elite of even the Egyptian priesthood, and carefully preserved down through the ages ... first by the Templars ... then the Masons ... all the way to its 20th Century practitioners in NASA?

In order for this premise to be correct, then NASA must have known about the crucial concepts underlying "tetrahedral physics" *almost from the beginning of the Agency, back in 1958* ... for more than a *decade* before its ritual appearance in "Apollo"—a physics which Hoagland and Torun had only first deduced from the redundant geometry present at Cydonia, in 1989.

One of the nagging mysteries of Scottish Rite Freemasonry is just what is the true meaning of the invitational 33º? Some argue that the number "thirty-three" *has* no significance; that it is just "the next level after the 32º," after which the founders of the Craft just didn't have anything more to teach their initiates. However, given the crucial importance of each and every other Masonic symbol (recall how Aldrin carefully carried the Masonic Apron to and from the Moon—and then presented it in another ritual at the Scottish Rite Temple in Washington DC) in the day-to-day activities of the Craft, it seems preposterous that Scottish Rite founder Albert Pike simply pulled 33º out of his hat.

In fact, throughout antiquity there is a pattern of paying special homage to the number "thirty-three." Clearly, the authors of the Old Testament believed that the number itself was the key to many things, that it somehow held tremendous power. Some Biblical scholars have referred to Jeremiah 33:3 as "god's phone number," the moment of darkness for Jeremiah, where God shows him how he can be reached and how the powers he possesses can be accessed: "Call on me in prayer and I will answer you. I will show you great and mysterious things which you still do not know."

So if "thirty-three" is a key code to figuring out how to access the "power of the gods," why do we see Sirius at *19.5º* above the Apollo 11 landing site, instead of 33º? How do the two numbers connect—if at all?

Engineer and probabilities expert Mary Anne Weaver (who would later do some crucial probability work on Hoagland's developing "ritual alignment model") has studied the possible mathematical linkage between the two numbers. She first pointed out that one of the basic trigonometric functions of a circumscribed tetrahedron, the sine of 19.471—the canonical "circumscribed tetrahedral '19.5º' angle" at Cydonia—is .3333. That would be merely interesting if it were the only mathematical link between the numbers, but there is another, even more "symbolic" link.

As we discussed in the first Chapter, a tetrahedron is one of the so-called "Platonic solids," so named because the Greek mathematician Plato was one of the first to popularize them (although he actually "borrowed" them from the earlier insights of Pythagoras). Each of these Platonic solids (there are only five) are "regular" polyhedra; polyhedrons that have regular polygonal faces, or faces with a straight-sided figure with equal sides and equal angles. In other words, they will all fit neatly *in a sphere*—with no edges or angles protruding through the surface. Of these, the simplest, and therefore the "first" among them, is the by-now-familiar "tetrahedron" [Fig. 5-29].

Each of these Platonic solids is commonly identified among mathematicians with the mathematical notation {p, q}, where p is the number of sides (edges) each face and q is the number faces that meet at each vertex. This number is called the "Schläfli Symbol" in this nodal system. Thus, the tetrahedron has three sides on each face, and has three faces that meet at each vertex. As a result, its nodal designation in the Schläfli system would be {3,3}—or, obviously, "33."

By simply taking out the comma, we can see that—every bit as much as the ubiquitous "19.5"—the number "33" says "look to your tetrahedrons." It's just a little less obvious, a little harder to figure out ... a little more *coded*.

There is yet another mathematical form, called a "pentatope," which fits into this same numeric system. The Schläfli Symbol for this object is 3,3,3; or 333, obviously. The pentatope is the simplest regular figure in four dimensions, representing the four-dimensional analog of the solid tetrahedron. As such, it would hold the "key" to accessing spatial dimensions higher than our traditional three, and is essentially a 3D tetrahedron as it would be seen in four dimensions. The 2D form of this 4D shape is a pentagon—with the vertices connected by lines, which just happens to bear a striking resemblance to the D&M pyramid: Torun's "Rosetta Stone" at Cydonia [Fig. 5-30].

So, it makes exquisite sense that an ancient "mystery school," or one of the three "secret societies" we've identified in NASA—perhaps with a actual ancient knowledge of "hyperdimensional physics" at the core of its highest mysteries—would choose the "thirty-third" level as its symbolic "highest level of enlightenment." Most outside observers would never figure it out. At least, not before modern humans got a good look at Cydonia. Certainly, in that stunning example, the hyperdimensional aspects of Jeremiah 33:3 could literally unlock these "great and mysterious things" that the Lord is alluding to.

It didn't take long, once these fundamental mathematical connections were established, for Torun and Hoagland to go back and check the significance of the number "thirty-three" in NASA's ritual pattern. As they looked at key places and events in the space program, and the history of the men who ran them, a distinct "Masonic-Tetrahedral" ceremonial pattern jumped out from the background "noise" of NASA's officially published history.

When one of the current authors (Bara) "came on board," that pattern was more formally codified—revealing the now-obvious "Masonic Matrix" we're presenting here ...

First, of course, we began by looking at the Apollo landing sites again.

We noted that the Apollo 17 landing—the last one in the Apollo program—was as close as you could safely get to 19.5° latitude—where, there seemed to be a very odd "hexagonal-shaped (double tetrahedral) mountain," including a very peculiar feature called "Nansen" that the astronauts specifically investigated ... *at* 19.5 degrees.

In looking at the White Sands Missile Range, where Wernher Von Braun conducted his first V-2 tests in America, it came to light that there had, in fact, been only *one* launch pad at the range, yet it was numbered—

"Launch Complex *33*." [Fig. 5-31]

And the one and *only* landing strip at the Kennedy Space Center, at Cape Canaveral?

"Runway *33*" ... of course.

Interestingly, if you stood at the southern base of the Great Pyramid at Giza, and then set your compass heading to "333"you would eventually find yourself at the door step of ... the modern Jet Propulsion Laboratory.

This relentless, repeating "NASA ritualistic pattern" didn't restrict itself to just the numbers, either.

NASA was demonstrably enmeshed in the sorts of strange mythologies that groups like the "Masons" and the "SS" thrived on; the Apollo 11 Lunar Module was named "Eagle," while the Scottish Rite flag Aldrin had taken aboard Eagle to the Moon and then returned to Earth bore the symbol of *a double-headed Eagle*—the official crest of the Scottish Rite [Fig. 5-32].

Simultaneously, the Command and Service Module "Columbia" also had its own "mythological Egyptian connection" (as we discussed earlier)—drawing its name (ostensibly) from "St. Columba," a revered figure in the Catholic Church who had taken an "ancient stone" (said to be the very stone Jacob laid his head upon "while receiving the vision of his ladder") from ancient Egypt to Scotland. It is still used to this day as the coronation stone for all English kings and queens.

However, if you looked closely at the sky over the Apollo 11 landing site, 33 minutes after Landing, not only was Sirius ("Isis") at 19.5º ... but, so was the *constellation* of "Columba" [Fig. 5-32]—from which St. Columba and the Apollo 11 Command Module *both* ultimately drew their names. "The Dove" came with its own sets of coded "double meanings"—relating to ancient Egyptian systems for measuring the Earth [Fig. 5-33], and "previous, now destroyed, ancient civilizations" (one version of which even shows up in the Old Testament, as the familiar Noah story of "releasing a dove after the Great Flood") ....

Even the acronym "NASA" had major hidden meanings ....

Author Richard Coombes (*America, the Babylon*) points out that "NASA" is an ancient Hebrew word, meaning literally "to lift up (unto) heaven." In parallel, our friend and colleague, Jay Weidner, discovered many years ago that "NASA" also translates in ancient Egyptian *hieroglyphs* as the very names of key members of the Egyptian divine pantheon:

> "Nephtys, Ausir, Set and Aset": the sister of Isis, "Nephtys"; Osiris, "Ausir"; Osiris' brother, "Set"; and Isis herself, as "Aset."

However, of all these recurring "ritual coincidences," one truly stuck out—because it came with a completely *different* set of "ritual fingerprints" attached.

As previously mentioned, the unmanned Surveyor 3 spacecraft had touched down in Oceanus Procellarum, just a few hundred feet from where Apollo 12 would land thirty-one months later; Surveyor 3 (third in an unmanned lunar reconnaissance series designed and built by Hughes Aircraft, and managed by JPL) had landed a few minutes after midnight, April 20, 1967 [Fig. 5-34].

What struck us immediately about that date, was *not* that it had some further arcane significance in terms of ancient Egypt (it didn't seem to, initially ... though eventually Hoagland did find that Orion's belt star "Alnitak" was some "39º beneath the southeast horizon"—exactly twice 19.5º); no, the thing that stood out most about that date and time ... was that it just "happened" to be—

Adolf Hitler's birthday.

If that wasn't enough, we soon discovered that NASA—remember, an Agency riddled at all levels by "German rocket scientists" from Operation Paperclip, former Nazis and SS members like Von Braun—had landed on the Moon on Der Führer's birthday not just once ... but *twice*—

The second time was with Apollo 16, in 1972.

Curiously, the Lunar Module on Apollo 16 had "a problem" with it's critical SPS engine just before the planned "deorbit burn" ... which delayed the intending landing time for several nail-biting hours, while NASA trouble-shot various engineering possibilities from Earth. This process dragged on and on ... until the date of "April 20ᵗʰ" clicked over on the clock. Then, suddenly, the "problem" cleared itself—and the astronauts were given the go-ahead to land on "the next rev"—

Once again—on Hitler's date of birth.

Not only that, but that particular Lunar Module had a very interesting name: Orion.

Orion was also the only Lunar Module in the entire Apollo Program which was *not* deliberately crashed back onto the Moon after the astronauts had left the Moon, returned to the Command Module, and were safely on their way home—

Obviously—because *this* Lunar Module was the literal "stellar embodiment" of the sacred Egyptian god of the Dead *and Resurrection*... Osiris. The symbolic message here was definitely chilling—regarding the potential "resurrection," specifically on his *birthday,* of "something" in the name of Hitler ....

As we looked at the skies above the Apollo 16 landing site—in the lunar highlands near a major lunar crater called "Descartes"—*and,* simultaneously, over the Mission Control Center in Houston itself, we found the star gods of ancient Egypt once again in rapt attendance; "Osiris" had landed on the Moon ... with Sirius ("Isis"—his "resurrection consort") at 33° below the horizon ... rising.

At the same moment, over Mission Control back in Houston, Orion's belt star "Mintaka" was hovering over the Space Center ... at the auspicious "19.5° [Fig. 5-35]."

In other words, NASA had landed a *second* spacecraft—this time, a Lunar Module named specifically after *"Osiris"*—on the Moon on Adolf Hitler's birthday, twice, at times and places selected by an Egyptian geologist whose father was an expert in "the ancient Egyptian stellar religion."

Sirius, the stellar equivalent of Isis, the mythological wife and consort of Osiris, "just happened" to be at 33º below the horizon at that very spot, a "stellar coincidence" which would not have been possible even two or three minutes later. In addition, at that same moment, one of the three belt stars of Orion/Osiris (the same stars which are revered by the Egyptians in the layout of the pyramids at Giza) was exactly 19.5º *above* Mission Control in Houston— and all of this was made possible by a group of Nazi rocket scientists who were members of Hitler's elite SS, who had previously shown an affinity for the tetrahedral/Masonic number "33," and who had designed and built the rocket which carried the spacecraft there.

This new "ritual coincidence," a second NASA mission *deliberately* landed on the day of Hitler's birth, finally brought the astonishing set of alignments into crystal clear, sharp focus. For, by these repeated "ritual coincidences"— commemorating the infamous Leader of the Third Reich—the key players behind this entire NASA lunar ritual were now overwhelmingly identified as none other than the NASA members of the former Reich. NASA—at the highest levels—had effectively been "taken over" from the Masons by Von Braun.

Further, it was now clear from the "ritual timing" of Kennedy's critical address to that joint Congressional session (May 25, 1961), that the Nazis had carefully set up the President on what the *real* objectives of Apollo would become, including the insitu reconnaissance and return to Earth of artifacts from the ancient ruins that the Nazis somehow knew about and clearly viewed as being left—

By their own ancestors!

No wonder Kennedy was murdered, immediately after his repeated offer to share this priceless "Nazi heritage" with their worst enemies, the Russians, was finally accepted.

*   *   *

Of course, all of this was just "one big, not-so-improbable coincidence" ... according to critics of our new ritual alignment model.

However, there was far more to come...

As we went further back in time, more and more hits just fell out of the ritual system. What we were trying to do was "reverse engineer" somebody else's religion—without access to any of the "catechisms" needed to unlock the codes; something akin to writing about the beliefs of the Christian Church without benefit of having a copy of the New Testament.

Yet still, this did not satisfy critics like Van Flandern—who insisted that we needed to make a *future* prediction based on the model in order for it to have any validity. As it turned out, we were about to have exactly that opportunity ... only, not exactly in the manner that we thought.

It was now July of 1997, and NASA was going *back* to Mars.

Mars Pathfinder

> *Still, we cannot deny that the act of placing a tetrahedral object on Mars at latitude 19.5 contains all the necessary numbers and symbolism to qualify as a "message received" signal in response to the geometry of Cydonia. Moreover, such a game of mathematics and symbolism is precisely what we would expect if NASA were being influenced by the type of occult conspiracy that Hoagland, for one, is always trying to espouse. —Graham Hancock,* The Mars Mystery

When Mars Pathfinder bounced to a halt on its innovative airbags after its unprecedented meteor-like decent to the Martian surface on July 4, 1997, most of us were convinced—watching its initial panoramas—that we were simply looking at another boring Martian desert filled with rocks—just like the scenes relayed from the previous Viking Lander missions some twenty years before. Overshadowed by Hoagland's highly advertised public prediction— that the Pathfinder Lander would be diverted to Cydonia, instead of arriving at its announced touchdown point a thousand miles away—many of us in the "anomalist community" were initially distracted from viewing those first Pathfinder images in any great detail. In the months intervening since the initial announcement of his suspicions around the stellar alignments, he'd found even more connections that convinced him the "ritual NASA model" was real. Viking 1, for instance, originally scheduled for July 4, 1976 (like Pathfinder), had been redirected (after encountering "landing site problems") to set down at a new site ... and on the *twentieth*—the same date as the Apollo 11 landing and Aldrin's "Osiris/Isis ceremony." So, Hoagland had a lot of confidence in analyzing Pathfinder that he'd "figured out the system."

One of the things that "threw" Hoagland off was the sky above the landing site: viewed before the actual July 4 arrival, it did not correspond *at all* to the stellar pattern he had found—multiple times—throughout NASA's previous manned and unmanned missions. At the announced time of Pathfinder's projected landing, the key "ritual star," Sirius, and the critical belt stars of Orion, would *not* be in their expected ritual positions above or below the horizon—or even *on* the horizon or meridian. By sharp contrast, the sky at Cydonia *did* fit—not for July 4, but for the "perfect" ritual date, July 20.

Because of this, Hoagland had gone on *Coast to Coast AM* and boldly predicted that Pathfinder would not, *could* not, land as scheduled, but would indeed end up at Cydonia on July 20th.

In hindsight, it turned out that NASA definitely threw in a "ritual change-up" for this Mission: Hoagland was simply looking in the wrong direction, based on prior NASA actions; instead of looking at the stars *above* the site, he should have been looking at the announced landing site *itself*—and then have focused on those clues. For its choice could not ultimately have been more "symbolic," or more embedded in NASA's (by now) amazingly consistent pattern; the initial "landing ellipse" was centered precisely on "19.5° N x 33.3° W"—and the spacecraft settled well inside those key "ritual" parameters on landing.

In reality, the Pathfinder mission and its landing site could hardly have been more clearly "tetrahedral"; the spacecraft itself was shaped like one big tetrahedron [and, when unfolded after landing on its airbags, resembled an even larger, solar-cell-studded *2D version* of the same 3D geometric figure—an *equilateral triangle*]; the redundant, "tetrahedral motif" of the 19.5° N x 33.3° W target of the landing was therefore blatant *reinforcement* of "the message."

In fact, looking back at the launch of the Pathfinder probe itself, there were even more alignments and important, tell-tale clues: Mars was rising—*at 19.5°*—when the Pathfinder spacecraft was launched toward the Red Planet; and Earth was hovering at 19.5º above the landing site on Mars when Pathfinder successfully touched down. So, even though the traditional "Isis/Osiris/Sirius/Orion alignments" hadn't been associated with this Mission, it did have a plethora of collateral ritual "tetrahedral messages" overwhelmingly, redundantly, embedded in the flight plan [Fig. 5-36].

Pathfinder's unique tetrahedral spacecraft design geometry, coupled with the totally "recursive" tetrahedral geometry of the landing site, and the placement of the Earth and Mars above its launch *and* landing, were obviously all specifically intended by NASA "ritualists," those shadowy figures setting hidden mission priorities somewhere "behind the scenes," to celebrate the two key *hyperdimensional numbers* underlying *all* these NASA rituals—"19.5" and "33."

Within hours after its successful touchdown, Pathfinder—according to a pre-loaded on-board computer program—began to transmit image after image of the scene around the landing site, far more (it inexplicably turned out) than originally advertised or programmed. Unlike JPL's highly conservative pre-announcements, the images were *not* just "looking down at the edges of the spacecraft solar panels and the airbags"—but were of *the entire Martian panorama* ... from the edge of the spacecraft to the distant red horizon.

It was in those first, unplanned and totally uncensored images, that several startling features—clearly *not* belonging on "a lifeless desert world"—could unquestionably be seen:

Enterprise Mission consulting geologist Ron Nicks also began looking, and started independently seeing major "inconsistencies" about the landing site as the unexpected panoramas swept across the screen: multiple, blatantly geometric "anomalies" around the Lander, showing up on CNN's live television broadcasts [Fig. 5-37].

Worldwide viewers, watching the NASA unmanned Mission "live on CNN," began calling in about "the strangely geometric rocks" that they were seeing in the sudden flood of images. The number of calls was so significant that CNN's science anchor, the late John Holliman, eventually felt compelled to ask one of the Pathfinder scientists about it on the air.[101]

Meanwhile, Nicks and Hoagland, taking careful notes and taping everything from two separate locations, quickly realized that what they were seeing was *not* the expected "wind or water eroded rocks" of Ares Vallis—but a debris field *filled* with a variety of apparently *corroded metal objects*; "canisters" with opposable handles, shimmering "glass-like geometric structures"—and even a couple of deformed but recognizable "*tracked vehicles.*"

A totally separate research effort—the Near Pathfinder Anomaly Analysis Group (NPAA)—also sprang up, triggered by the same "anomalous images" appearing live on television [Fig. 5-38].

Other objects seen close-up looked unmistakably like miniature *pyramids*— or (as Hoagland would later realize ...) *the exposed tops of massive, buried obelisks* [Fig. 5-39].

However, after the initial—and still totally unexplained—windfall of those first new Mars surface images, seen by tens of millions *live around the world* (but, unfortunately, at television's limited resolution), the crucial *digital* post-landing analysis of these remarkable anomalies was immediately frustrated by the maddening (and also inexplicable) lack of clear Pathfinder *web* images— promised immediately across the Internet from JPL.

Over twenty years had elapsed between Viking's mechanical facsimile-type surface cameras in 1976, and Pathfinder's state-of-the-art CCD imaging technology of 1997—but judging by the digital Pathfinder photographs publicly available from JPL ... *you'd never know it!* Unusual (and still unresolved) discrepancies were found repeatedly, between the original (paradoxically, much clearer) initial television transmissions from CNN's live Pathfinder coverage that July 4 afternoon, and the later electronic versions of those same images released across the web by JPL.[102]

The simple truth was that the processed internet digital files should have been *much* sharper than the original raw video feed coming from Pathfinder that first afternoon, especially after being rebroadcast over television's limited "525 scan lines." Instead, JPL's web versions were full of astonishingly amateurish compression artifacts, assembly errors and blatant color registration problems. JPL even adjusted the color of the Martian sky to appear the traditional "Viking pink" instead of the forecast (from concurrent Hubble imagery)—natural Earth-like blue.

The implications of this blatant image tampering and crude manipulation were as abhorrent as they were sickeningly obvious (to everyone, but the mainstream press that is): JPL was clearly trying to hide (and not too well, for some reason ...) "something" in these images.

It soon became obvious exactly "what."

Besides all the near field Pathfinder anomalies [Fig. 5-40], Nicks and Hoagland began studying the so-called "super-resolution" images of the two distant (almost one kilometer) "mountains" imaged on the horizon of the landing site: the celebrated "Twin Peaks." Nicks, in particular, soon realized that these features showed definite signs of *engineering,* as opposed to natural erosional processes. And, although he recognized that the area (as NASA had previously advertised) obviously had been subjected to some kind of catastrophic flood, he could not explain some of the strikingly geometric features he was seeing on the Peaks as "simple geology"; there were obvious, repeating block-like structures (particularly on South Peak), and some very unusual orthogonal 3D layering on the exposed ("downstream") surfaces [Fig. 5-41].

The Twin Peaks seemed to have the definite geometric shapes of *pyramids—* but highly eroded pyramids (especially North Peak), which had apparently had some of their casing literally ripped off in that same massive flood that had devastated the entire area countless millennia before.

The debris field in the foreground, which had initially captured Hoagland and Nicks' attention, consisted of a myriad objects possessing multiple sharp points and edges. They could *not* therefore be merely water-eroded (or water-tumbled) rocks. If they were, the sharp edges would have been smoothed out eons before. They obviously had to be made of much harder materials— potentially artificial objects, and even metallic *machinery,* ripped from the exposed interiors of the not-so-distant, exposed arcologies, including, probably, remnants of the technology that actually built them.

The flood that had happened at this site had been as short-lived as it was catastrophic.

All of which was quite fascinating, and ultimately irresolvable—for without better resolution images from future rover missions, or cleaned-up versions of the Pathfinder originals (as opposed to the degraded and obviously sanitized versions deliberately placed on the web by JPL), there wasn't much more that could be done to test the artificial nature of the region.

That is, until more than five years later, when NASA introduced a new technique called "super-resolution surface modeling."

In the several years following Pathfinder, NASA scientists had been able to take multiple, overlapping images of the Pathfinder landing site and enhance them (via a technique called Beysian interpolation") well beyond their original resolution; Giuseppe Pezzella, a reader of the *Enterprise Mission* website from Naples, Italy, pointed the authors to the official NASA image archive for these images. After downloading a few, it quickly became apparent that something—completely unnoticed at the time, and even more interesting than the potential nearby artifacts originally analyzed by Nicks and Hoagland—might have been imaged on the Martian landscape, *between* "South Peak" and Pathfinder itself [Fig. 5-42].

What had been merely a huge, dark "blobby" shape on the original images, suddenly emerged as recognizable on the new, highly processed images. What it seemed to be—at least to our eyes—was *a sphinx.*

Like some ancient, strange sculpture, this large object (or, set of objects) stood guard at the base of South Peak. The clear, geometric, rectilinear relief on the "Peak" was strikingly evident in the new image, confirming Nicks' earlier assessment—that it could, indeed, be another "shattered arcology on Mars." The "sphinx" lay some distance from the Pathfinder Lander, at the edge of the previously mapped debris field near the base of South Peak and, like its counterpart on Earth, also faced due east—directly toward the Martian equinoctial sunrise. There were even a couple of what appeared to be "vertically-faced buildings" to the left of this potential "Martian sphinx"—they could easily be viewed as "a temple" (in the Giza model), or a distant entrance to the background Pyramid arcology itself.

In ancient Egypt, "sphinxes"—later-Kingdom, much smaller versions of the ancient "Great Sphinx" on the Giza Plateau—were routinely used to guard temples, tombs and monuments. The grandest example of these and probably the earliest on Earth *is,* of course, the Great Sphinx itself—forever guarding the three Great Pyramids on the Plateau. As a sphinx carries out this task, it is always in the same repose: lying flat on its stomach, forepaws extended outward, ready to "pounce" into action at a moment's notice. Sphinxes invariably have the head of a man (or woman?) to go with their lion's body. The head, in turn, is framed by the characteristic banded "nemyss" headdress (which is meant, of course, to signify the lion's mane ...).

Even from this compressed narrative perspective, you can see in these NASA-enhanced close-ups [Fig. 5-43] that this "Martian sphinx" has all the classic earmarks of its Egyptian counterpart: there are two attached and extended "forepaws," a body (complete with what looks to be a feline hind leg) and a very clear rounded face, encompassed by a symmetrical nemyss-like Pharaohic "headdress." The headdress even has two opposing angles just below the "chin"—extending outward at about a 45° angle. These characteristics alone—the extended symmetrical paws and the rounded face framed by that familiar headdress—coupled with its context, lying to the east of a strikingly pyramidal-looking structure, would normally be enough to call this formation's natural genesis into serious question, but when you view it side-by-side (above) with its counterpart on Earth—the Giza Sphinx itself—the resemblance is uncanny.

And remember—*this* Martian sphinx is guarding an obvious pyramid on Mars ... at 19.5° N by 33° W.

After all this, the authors were forced to conclude that the Pathfinder landing site presented an enigma at least as important, and potentially as rich in artifacts, as Cydonia itself. Its location, the Lander's tetrahedral shape, the games NASA had played with the images and even the revelation of the possible Pathfinder sphinx, all pointed to a "sacred site"—selected by NASA precisely because they suspected they would find exactly the treasure trove we were seeing. We considered it additional confirmation not only of the ritual alignment model we were constantly refining—but also of the notion that NASA was, in fact, closely following Hoagland's earlier work on geometric relationships at Cydonia.

The fact that the landing site, "Sagan Memorial Station" as it came to be known, after "The Carl" himself, was at this critical lat./long. location was also interesting on another level to us. The Lander had no official name, but the little Rover, which spent several months trolling the near field to analyze various objects, did receive a moniker; "Sojourner." The canonic NASA line is that the name was selected in an essay contest about famous American heroines. A 12-year-old girl submitted an essay about Sojourner Truth, an African-American abolitionist who lived during the Civil War era.

We of course, always suspected a double meaning. In Freemasonry, a "sojourner" is a special term applied to a Mason who has also joined a military service. This obviously implies that the Rover not only served a scientific purpose, but may have also had a separate, military purpose as well. How likely is it that this is also just another "lucky coincidence?"

The critics were, of course, quick to accuse us of using the artifacts and "sphinx" just to divert attention from the fact that Hoagland's *Coast to Coast AM* prediction on the landing date had been obviously wrong. But, given the political and religious complexities of trying to figure out a hidden symbolic system *from the outside*, and the fact that—ultimately—the Pathfinder Mission was a "tour de force" of *predicted* tetrahedral symbology *in every way* ... missing the exact date when this mythology was carried out was far less significant than predicting its obvious existence in the Mission in the first place.

"Dates" can easily be changed; the underlying, obsessive, tetrahedral "Cydonia symbology" that NASA demonstrated once again overwhelmingly on Pathfinder—just as Hoagland called it—just could not.

On the artifacts discussion: given that a trained geologist (Nicks), and thousands of viewers of the initial data in real time on CNN, had come to Hoagland first—regarding "seeing artifacts"—and not the other way around, we feel this criticism also is unwarranted. The anomalous images and data—of bona fide Martian artifacts—also stand on their own.

It was now time to nail down the ritual system once and for all, and decide just what would and would not constitute a "hit." On Pathfinder, NASA had indeed followed "the ritual"—but had done so in a *way* that had indeed surprised us.

What we had to do now was figure out exactly "what they would do next."

The Ritual Alignment System

After the notable misfire by Hoagland—on using his newly-coined "ritual alignment model" in an attempt to make accurate detailed predictions, concerning specific rituals NASA would carry out during future missions— the authors met, and began to have extensive discussions regarding just what would constitute a "hit.". As stated earlier, this task was daunting, as we were faced with essentially "reverse-engineering" somebody else's most private metaphysics and religion—without access to any of the liturgy or texts.

Extending our previous metaphor with a well-known historical example:

it was akin to trying to figure out the beliefs of the early Christian Church, without knowledge of what was on the Dead Sea scrolls ....

In order for our model to obtain widespread acceptance, we would have to use our knowledge and speculations about the meaning of the alignments to make a prediction—not just about a past event, but also about a future one. While we both rejected Dr. Van Flandern's argument that an untested past event was scientifically worthless, we realized his arguments would hold sway in the anomalist community and with the public at large until we could deliver a successful *a priori* prediction to his standards.

We agreed that the first task was to set down specific and tight criteria regarding a "hit." Fortunately, we had NASA's past behavior to build on, so it was easy to sort out our initial stellar candidates.

We agreed that the star Sirius, in the constellation Canis Major which, to the ancient Egyptians, represented the goddess "Isis"—literally the "mother goddess of the world" and the source of all life—would be a prime candidate.

Next, the three belt stars of Orion were an obvious fit; Orion/Osiris was the god of "death, judgment and resurrection," and the conveyor of science, astronomy, religion and agriculture to the Egyptian people. The three prominent "belt stars" of Orion are the only members of this constellation which figure into the equation of the ritual pattern. This is for not only simplicity's sake—for the simpler a model of such stellar motions the better—but also because the Egyptians held them in such a high regard, literally recreating the belt pattern on the ground at Giza in the arrangement of the three great Pyramids.[103]

The third major figure in this pattern is the constellation Leo—evocative of the Sphinx of Giza and representative to the Egyptians of the god "Horus," son of Isis and Osiris. Horus is also frequently associated with the planet Mars, and in fact the Sphinx was at one time painted red, and they shared a name, "Hor-Dshr," literally "Horus the Red."[104] The major star in the Leonine constellation is Regulus, the blue-white "heart of the lion." It is the only star in this constellation that we focus on for the purposes of determining our pattern.

We'd also decided, after the Pathfinder incident, to include some of the planets as well. If a mission to Mars was launched at a time when Mars itself was at 19.5º above the launch site, for instance, then it made sense that this ritual positioning would be significant. This also made the positioning of Earth at 19.5° over the Pathfinder landing site a "hit." However, an alignment of the Moon in a ritual position over the launch site would not be a hit, unless the mission was to the Moon itself. There had to be appropriate context for a specific mythological or physical association.

There were clear historical examples of this particular pattern being used by NASA.

In July, 1964, Ranger 7 became the first successful U.S. mission to travel to the Moon. The Ranger series launched from Cape Canaveral on a direct impact trajectory into the lunar surface—with their TV cameras activated and running only for the last few minutes before impact.

As it launched, the Moon itself was exactly 33° below the horizon of the launch site, and almost due West, in the realm of Osiris. At that same moment, at the impact site of Ranger 7 *in three days*, Orion's belt star Alnitak was exactly 19.5° above the impact site in Mare Cognitum. Given what we know about the NASA ritual pattern and the Agency's capabilities with mission planning, this hardly seems a simply coincidental series of alignments.

We chose to exclude the constellation Taurus, which is associated with the evil god Set, brother of Osiris, since it had not shown up in any of the alignment checks done to that point. Even though Set is intertwined in the story of Isis, Osiris and Horus, he seemed to have little significance in the NASA system, at least as far as we had decoded it to this point.

So that's it.

Except for the occasional addition of Mars or the Moon, or even the Earth (if we are observing from the surface of another world), these five celestial bodies are the sum total of the ritual objects that we would search the sky for.

This left us with the subject of "temples."

While we agreed that the ideal fit would be the stars above a specific event site itself (i.e. a landing site), we also had some evidence that alignments at various other ceremonial locations –simultaneous with (and, in restricted cases, substitutions for) the Primary Event—were significant to NASA as well.

For instance, Apollo 8's lunar orbit insertion "match" had been over the Apollo 11 landing site, since fundamental celestial mechanics determined "where" on the Farside of the Moon the SPS engine on Apollo 8 *had* to fire, to successfully decelerate the spacecraft into lunar orbit; thus, the location of the stars above that point would not be "significant."

But just what would constitute a "temple?"

One obvious candidate was the Apollo 12 landing site.

NASA had originally landed Surveyor 3 there on April 20, 1967 (remember, Hitler's birthday). As it soft-landed on the Moon in Oceanus Procellarum at four minutes after midnight GMT, the Moon itself was rising through 33° directly over the Jet Propulsion Laboratory back in Pasadena—which was, of course, controlling the entire Mission.

With the discovery that this Apollo 12/Surveyor 3 site was, somehow, seen as "sacred" long *before* any spacecraft actually touched down—as gauged by the stunningly exact alignment of Orion on its lunar horizon, precisely as President Kennedy called for "sending a man to the Moon" in May of 1961—it became increasingly obvious that a lot of thought and planning had gone into selecting these "sacred temple sites" ... and far, far earlier than we had hitherto imagined.

After considerable discussion, we agreed on our own "reconstruction" of the following "ritual temple" locations:

The Giza Plateau on Earth (for obvious reasons); the Apollo 11 (because it was the location of the first manned landing) and Apollo 12 (because of the landing on Hitler's birthday and "the Kennedy Connection") lunar landing sites; Viking 1 (because it was the first NASA landing on the planet Mars); the Pathfinder landing sites (because of the 19.5 x 33º location); and, of course, Cydonia (because of the Face and City Complex).

Next, we had to decide just what constituted a "ritual position."

Since the NASA rituals seemed to be based in the ancient Egyptian system, we knew we had to include significant celestial positions from that venerable religion. In the end, our model would include only *five* key positions in the sky relative to elevation—i.e. stellar altitude above or below the horizon.

They are the horizon (the zero point); 19.5° above and below the horizon; and 33° above and below the horizon. Five simple positions.

We included star positions below the horizon because we know that the ancient Egyptians were good enough astronomers to track the stellar motions even when they were out of sight below the horizon, and NASA with its astronomy software certainly could track these same motions. In terms of Right Ascension—essentially stellar longitude—the only significant location is the meridian.

So effectively, in terms of stellar elevation, 95%—or 355 out of 360º—is off limits for a "hit" at any one moment. In terms of lateral movements across the sky, fully 99% percent of the sky (358 out of 360º of Right Ascension) is also out of play. Yet, still we are told the alignments that we cite are "common."

It is perhaps even more important to understand the symbolic significance of these reference locations in the sky.

In the ancient Egyptian stellar religions, the horizon and the meridian were the most significant; as noted previously, the horizon represented a sort of netherworld *between literal dimensions* to the Egyptians—the "world of men" ... and the "world of the gods." The meridian in Egyptian cosmology marked an object's traverse from east to west in a nocturnal rising and setting, symbolizing the daily birth and death of the sun.

The 19.5º and 33º positions are representative of the "hyperdimensional message" of Cydonia.

With the Pathfinder mission behind us, we now had to wait for an opportunity to test our modified system in a real world, "live fire exercise." Fortunately for us, NASA was about to provide us with the perfect opening ....

Chapter Five Images

Fig. 5-1 – Original Apollo "Orion" logo.

Fig. 5-2 – Project Mercury logo representing the element Mercury, rather than the "god's messenger."

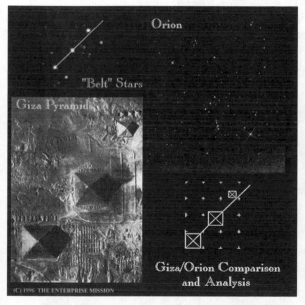

Fig. 5-3 – The Giza\Orion connection, as identified by Bauval and Hancock.

Fig. 5-4 – The Egyptian concept of "as above, so below," exemplified by the connection between Orion's Belt stars and pyramids at Giza.

Fig. 5-5 – The "Helical Rising" of Sirius ("Isis" – consort of Osiris) before sunrise over the Giza Plateau. Isis (brightest star, left) plays a crucial role in the "resurrection of Osiris" in Egyptian mythology, and the origin of their son "Horus" (Hoagland).

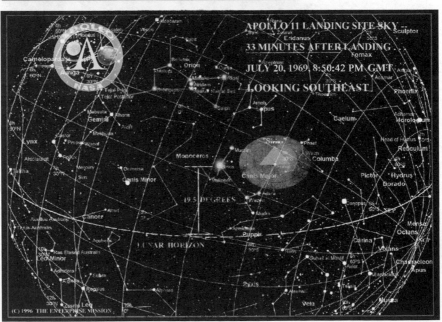

Fig. 5-6 - Lunar horizon, Tranquility Base – thirty-three minutes after touchdown. Sirius/Isis is precisely 19.5º above lunar horizon.

Color Fig. 1 – False-color enhancement of the Mary Moorman photograph showing Gordon Arnold, the "badge man" and "railroad worker."

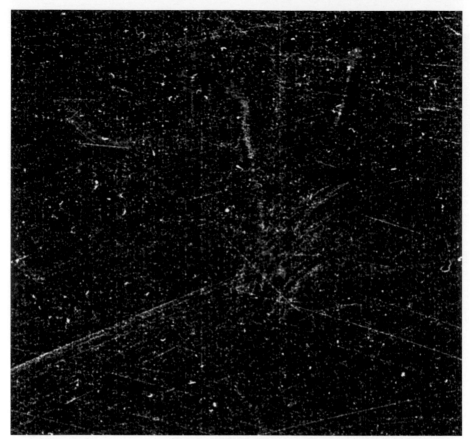

Color Fig. 2 – High contrast, false-color enhancement of "scaffolding on the Moon" over Sinus Medii. Note recurrent cross members, three-dimensional supports and connecting nodes at termination points.

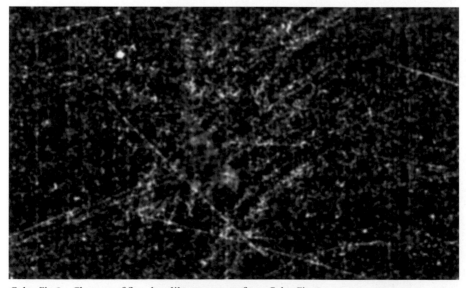

Color Fig 3 – Close-up of fine, box-like structures from Color Fig. 2.

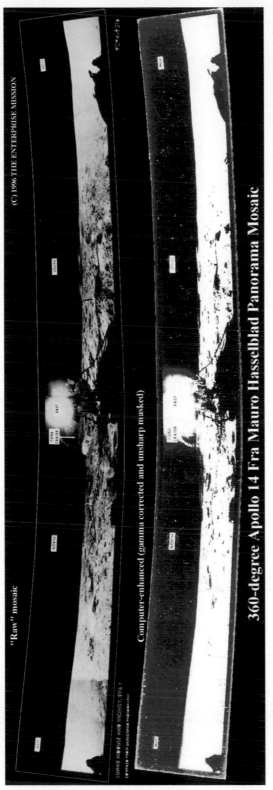

"Raw" mosaic

Computer-enhanced (gamma corrected and unsharp masked)

360-degree Apollo 14 Fra Mauro Hasselblad Panorama Mosaic

Color Fig. 4 – Apollo 14 landing site panorama from Ken Johnston collection taken by Alan Shepard. Note that the brightest portions of enhanced, geometric lunar structures above horizon are 180 ° opposite the sun—a phenomenon known as "backscattering." How would "image artifacts" or "lens flares" understand and behave according to this crucial optical physics?

Color Fig. 5 – The now-famous true color "Mitchell Under Glass" Apollo 14 panorama frame AS14-66-9301
Note brilliant specular reflection, supporting three-dimensional cross-beams and slanted buttresses (inset).

Color Fig. 6 – (Left) Apollo 17 astronaut Gene Cernan erects American flag under a half-Earth, wearing a gold-visored helmet. According to NASA statements, the gold film was designed to "protect the astronauts on the Moon from unfiltered ultraviolet light." (Below) Measured special transmission curve of gold film reveals the visor actually enhanced blue-violet light over all other wavelengths. The helmets in fact allowed the astronauts to specifically see the blue rayleigh scattered glass ruins towering over the landing site. (Spectral data: The Gold Bulletin)

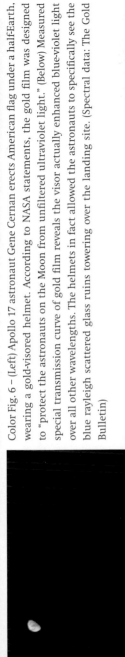

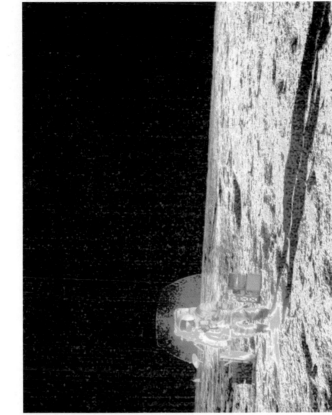

Color Fig. 7 – Apollo 12 frame AS12-46-8607 shows astronaut Alan Bean, wearing specially tuned gold visor, carrying ALSEP experiment package. Bean is surrounded by vertical "stringers" supporting studded remains of glass-like lunar ruins. Hexagonal halo around Bean is Hasselblad camera internal lens flare.

Color Fig. 9 – Color-corrected AS17-137-20990. Note colors in background including blue and purple rocks and pink hills due to refracted sunlight passing through "prisms" of overhead lunar ruins. Color "gnomon" was placed to ensure proper color calibration of orange soil found at Shorty Crater.

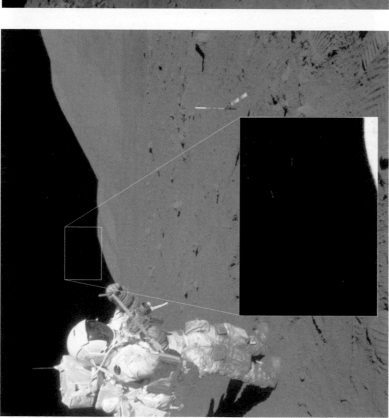

Color Fig. 8 – Gold-helmeted astronaut Harrison Schmitt holds collecting rake while photographed by astronaut Gene Cerman. Behind Schmitt is an "impossible" lunar phenomenon—a backscattered lunar sunrise. Lunar sunrise behind Cerman, refracting prismatically through the fragments of glass structure arching overhead, is projected against surviving glass structure behind Schmitt. Inset enlargement reveals one shard of glass reflecting a miniature rainbow. Color chart confirms accuracy of all colors. (Color enhancement: Hoagland)

Color Fig. 10 – (Left) AS17-134-20442. Apollo 17 astronaut Gene Cernan photographs Lunar Module Challenger looking West toward South Massif during EVA-1. Further West, the prismatic display of lunar sunrise, from a sun only 13 degrees above horizon behind Cernan, presents an impossible display of color against what should be an absolute black background. (Right) Sunset over San Francisco Bay looking East eerily duplicates Cernan's lunar scene. Colors in Earth's atmosphere are caused by normal atmospheric refraction. Colors in what NASA says is an airless lunar scene? Answer: an ancient, refractive lunar dome.

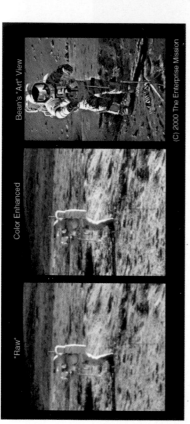

Color Fig. 11 – Comparisons of Apollo 14 original AS14-66-9301 (left), color-stretched version (center), and astronaut Alan Bean's color study titled "Colors of the Moon."

Note pink lunar regolith and "structured" background reflecting slanted, inclined buttresses seen in Apollo 14 frame AS14-66-9301 from Ken Johnston collection (right).

Color Fig. 12 – "Rock and Roll on the Ocean of Storms" by astronaut Alan Bean (left).

Color Fig. 12 – "Rock and Roll on the Ocean of Storms" by astronaut Alan Bean (left). Note pink lunar regolith and "structured" background reflecting slanted, inclined buttresses seen in Apollo 14 frame AS14-66-9301 from Ken Johnston collection (right).

Color Fig. 13 – Two prisms refracting incident sunlight with overlapping beams produce different colors than a single prism alone. Look carefully at the laboratory example above. Note the off-pink slice of color created by the overlay of two prismatic beams. Now, imagine thousands, or millions of such prisms spraying rainbow colors over the surface of the Moon from shattered lunar domes. What would you see?

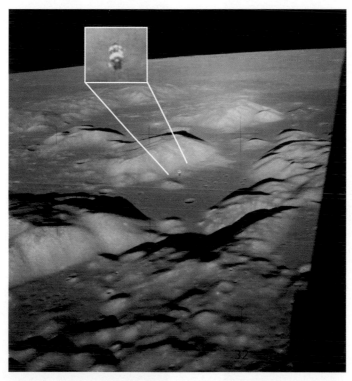

Color Fig. 14 – Color-corrected version of AS17-147-22465. Note pink, purple and blue mountains, just as seen in Apollo 17 surface views (Color Fig. 9) and prismatic examples in Fig. 13 (above). CSM America is visible in center of image just above prominent crater (inset).

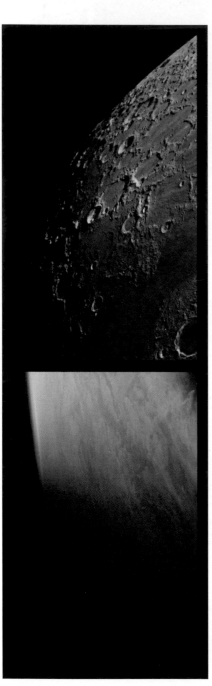

Color Fig. 15 – Light scattering sunrise on two worlds. (Left) Prismatic sunrise colors on Earth projected on atmospheric clouds. (Right) Similar phenomenon on the airless Moon is difficult to explain by any theory except the shattered lunar dome model.

Color Fig. 16 – AS15-88-12013 Post TEI view looking back toward the Moon. Color enhancement shows blue rayleigh scattered light encircling the Moon, identical to atmospheric "airglow limb" seen on Earth. Such scattering is flatly impossible on an airless body like the Moon, leaving the shattered transparent lunar dome model as a viable explanation for this photographic phenomenon.

Color Fig 17 – THEMIS IR enhancement of Cydonia. The "Fort" is to the top left of the image. (Laney/Hoagland/ASU)

Color Fig 18 – Tube with cross members running to the "train station" in Cydonia. Note open archways in "train station." Again, note eerie resemblance to aerial views of terrestrial cities.

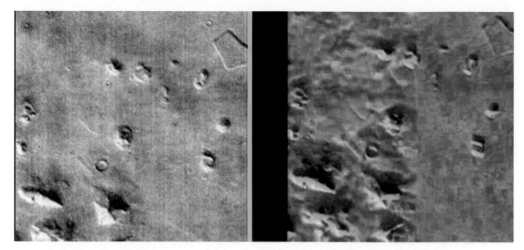

Color Fig. 19 – Thermal IR enhancements of Cydonia region by Keith Laney from the official data (left) and the "real" data (right). Note the amount of noise and banding in the official version. Identical computer processes were used on both images.

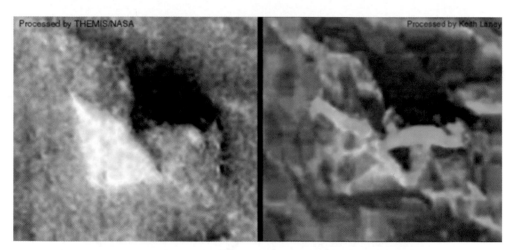

Color Fig. 20 – The D&M Pyramid enlarged from Fig. 19, processed using the official version of THEMIS IR Cydonia image (left) and the "real" version (right). Note the similarity in color distribution on both images, and the unmistakable structural details of the D&M itself in the "real" version. Both images composited, color ratioed and decorrelation-stretched using ENVI 3.5 image-enhancement software. (Laney)

Color Fig. 21 – Early Viking 1 image of the surface of Mars. Note blue sky and "Arizona-like" landscape (left). Altered version of the same image, as modified by NASA within hours (right).

Color Fig. 22 – Ludicrous "Technicolor red" skies and landscape from Mars Pathfinder (NASA/JPL)

Color Fig 23 – Three views of the "true colors" of Mars. Official NASA rendering (left) and Lane (center and right). Note the red tint of what should be a blue NASA logo in the official version (see inset).

Color Fig. 24 – Airbag colors: imagined (top) and real (bottom)

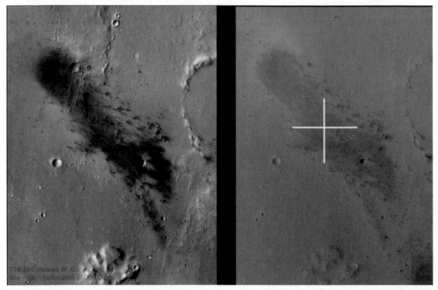

Color Fig. 25 – First Mars Express color image of Gusev Crater (ESA): original (right) and altered (left).

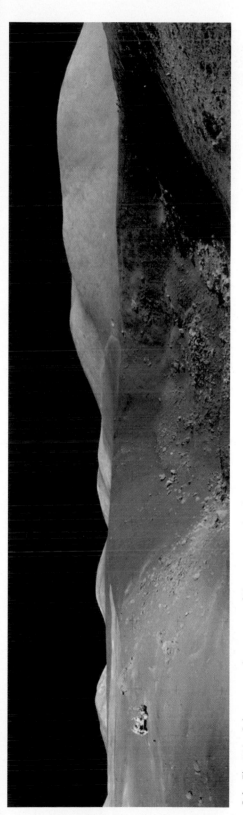

Color Fig. 26A – Color-corrected panorama of Shorty crater taken by astronaut Gene Cernan. Note orange soil (near rover), blue rocks and pink mountains in background. Interplay of color not seen in official NASA versions testifies to the level of suppression of the extraordinary, mysterious and complex environment of the Moon.

Color Fig. 26B – (far right) Long-suppressed, first Real "prismatic" color image of lunar surface – from Surveyor 1, June 1966. Note accurate color, confirmed by calibration chart on end of Surveyor 1 boom. Middle image: Surveyor 1 close-up of own footpad, with second color calibration. For left: Bean painting of his "fantasized recollections" of personal experiences on lunar surface. Note eerie color similarity (NASA / Enterprise Mission).

Alan Bean Painting

Surveyor I Footpad

Surveyor I Lunar Landscape

Color Fig. 27 – Interior of Shorty crater showing numerous multi-spectral bits of smashed machinery, including a stunning surprise — "Data's Head."

Color Fig. 28 – Enhanced close-up of "Data's Head." Red stripe is not an artifact of image processing.

Fig. 5-7 – Apollo 12 astronaut Alan Bean inspects the Surveyor 3 spacecraft, which had soft landed in virtually the same location as Apollo 12 (note Lunar Module on Surveyor crater rim), but more than two years earlier – on *Adolf Hitler's birthday* (NASA).

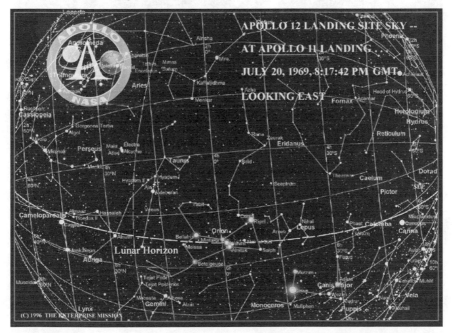

Fig. 5-8 – Apollo 12 landing site sky pattern at Apollo 11 touchdown, ~850 miles due east. Orion's belt on the eastern horizon is an extremely significant alvignment in the ancient Egyptian stellar religion – unmistakably signifying "resurrection." But … of *what*?

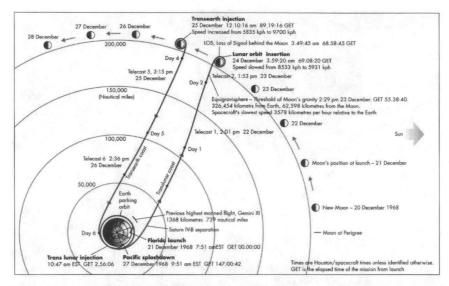

Fig. 5-9 – The celestial mechanics of landing on the Moon and returning safely to the Earth.

Fig. 5-10 – Apollo 8 Lunar Orbit Insertion, December 24, 1968 – exactly as Orion's Belt rose on the eastern lunar horizon over the *future* Apollo 11 landing site. Illustrates NASA's fanatical, relentless, redundantly symbolic message of "resurrection" (Hoagland)

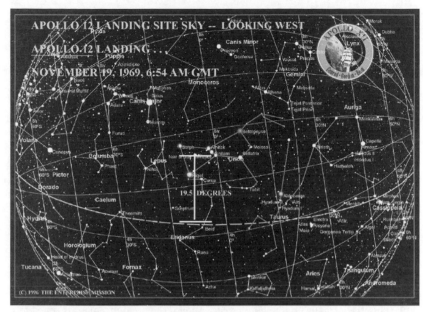

Fig. 5-11 – Apollo 12 landing site sky pattern at Apollo 12 touchdown, November 19, 1969. Orion's Belt stands 19.5 degrees above the Western lunar horizon – a mathematical coding for "dimensional transition." (Hoagland)

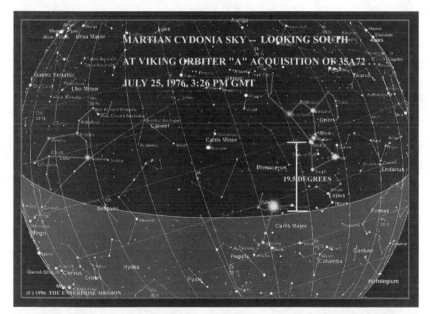

Fig. 5-12 – Sky pattern over Cydonia (Face on Mars, 41° N x 9° W) at Viking 1 Orbiter acquisition of first "Face on Mars" image, 35A72 – July 25, 1976. Orion's Belt stands at 19.5 degrees about southwest horizon, a redundant mathematical code for "dimensional transition" (Hoagland)

Fig. 5-13 - President John F. Kennedy addresses a Joint Session of Congress, May 25, 1961 (inset) – formally announcing Project Apollo. The announcement is timed as Orion's (Osiris') Belt stars rise precisely on the eastern lunar horizon, both at the future Surveyor 3 landing site (April 20, 1967 – Hitler's birthday) and the *future* site of Apollo 12's revisit (November 19, 1969). Disturbingly – in a Space Agency filled with newly-imported Nazi scientists and engineers, the symbolism is again one of "resurrection." Foreshadowing... what? (Hoagland)

Fig. 5-14 - Official Apollo 11 crew portrait. Most sources crop this image to disguise Aldrin's prominent display of his Masonic signet ring. (NASA)

Fig. 5-15 – Aldrin in the Lunar Module just a few moments after the landing of "Eagle." Note that he is once again wearing his Masonic ring on this occasion.

Fig. 5-16 - Aldrin with Luther A. Smith, the Sovereign Grand Commander, and the Masonic flag he took with him to the Moon (The New Age Magazine).

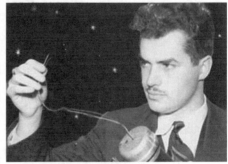

Fig. 5-18 – JPL co-founder and early solid rocket fuel pioneer Jack Parsons.

Fig. 5-17 – JPL co-founder Theodore von Karmen

Fig. 5-19 – L. Ron Hubbard, Parsons "magical" partner in the so-called "Babalon Working," during his service in naval intelligence in World War II.

Fig. 5-20 – Aleister Crowley in full Masonic garb, circa 1914.

Fig. 5-21 –Liquid fuel rocket test in Arroyo Seca, October 31, 1936 (L), with key members of the GALCIT team—including Parsons (lower right frame); (R) the "nativity scene" re-creation of the Arroyo Seca test, every Halloween at JPL —marked as "the birth of JPL."

Fig. 5-22 - Jack Parsons and Marjorie Cameron in the late 1940s.

At the time he received *Liber 49*, Parsons could not have been aware of certain letters written by Frater Achad,[25] which came to light after Parsons' death. In them, Achad expressed doubt as to Crowley's having uttered the Word of the Aeon of Horus because, having identified himself with the Beast.he was speechless; and in some inexplicable manner not yet clearly understood the Aeon of Horus itself has become merged with the Wordless Aeon. In the *Book of Anti-Christ* Parsons wrote, describing the Black Pilgrimage:

> "...Babalon called on me again, and I began the last work, that was the Work of the Wand. And I worked for 17 days, until Babalon called me in a dream, in an astral working. Then I reconstructed the temple, and began the Black Pilgrimage, as She instructed.
>
> And I went into the sunset with Her sign, and into the night past accursed and desolate places and cyclopean ruins, and so came at last to the City of Chorazin. And there a great tower of Black Basalt was raised, that was part of a castle whose further battlements reeled over the gulf of stars. And upon the tower was this sign ⊽ "

-- Hecate's Fountain (Grant, Kenneth, Skoob Books Publishing Ltd, London, 1992)

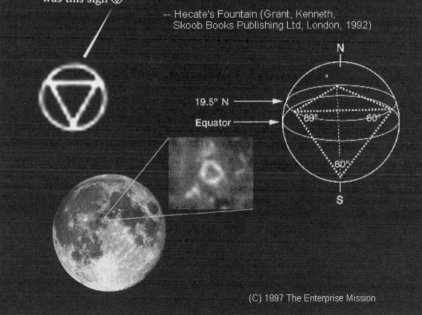

19.5° N

Equator

N

S

(C) 1997 The Enterprise Mission

Fig. 5-23 - During one of his lengthy "magikal workings," Parsons had a dream – in which he saw the same "hyperdimensional symbology" that Hoagland would later discover on the Moon and Mars (Enterprise Mission)

Fig. 5-24 -Official photograph of the 1939 briefing and rocket demonstration for Adolf Hitler (front row, center) by Von Braun's (inset) "amateur" rocket team – which led directly to Von Braun being personally offered a commission in the S.S. by Reichsführer Heinrich Himmler, and the V-2 Program at Peenemunde (National Archive)

Fig. 5-25 - The only known photograph of SS Major Wernher Von Braun in his black SS uniform, directly behind Reichsführer Heinrich Himmler. Taken at Von Braun's induction ceremony into the SS. (Source: National Archives)

Fig. 5-26 - Wernher Von Braun and fellow "Project Paperclip" scientists spend off-duty time in El Paso, Texas – from nearby "White Sands Missile Range," in 1945. Note symbol on the "Billy the Kid" sign above them ... in an "American" town, just after WWII.

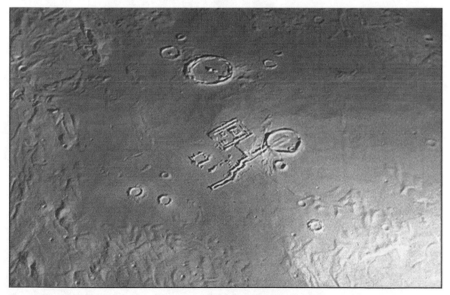

Fig. 5 27 Screen capture from Von Braun\Disney production "Man in Space," showing an *alien base* on the farside of the Moon "at 33 degrees ...." The fact that Walt Disney was also a 33° Scottish Rite Freemason was, of course, purely coincidental. (Disney)

Fig. 5-28 - Organization chart showing Masons, SS members and "magicians" in key positions of power all throughout NASA in the 1960s. Everybody who was anybody at the Agency, from the director on down, was a member of one of these three secret cults.

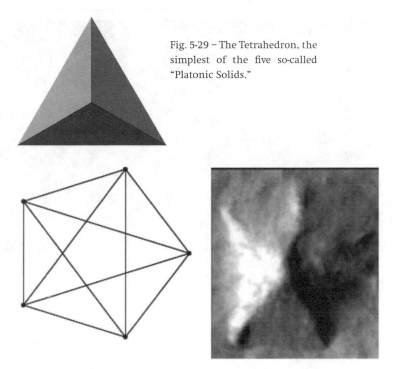

Fig. 5-29 – The Tetrahedron, the simplest of the five so-called "Platonic Solids."

Fig. 5-30 – The Pentatope, a 4D representation of the 3D tetrahedron as compared to the pentagonal D&M Pyramid on Mars.

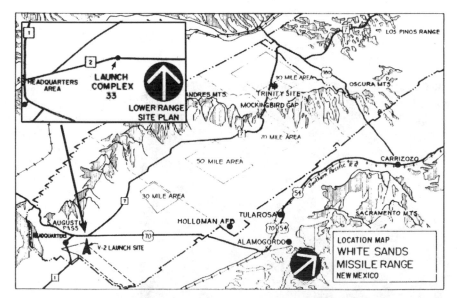

Fig. 5-31 – Map of White Sands Missile Range, where the only launch pad was numbered "Launch Pad 33," just as the singular airstrip at Cape Canaveral was named "Runway 33."

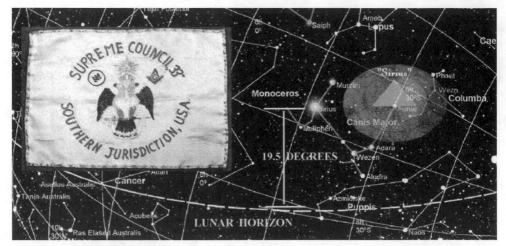

Fig. 5-32 – Official flag of the Supreme Council 33º, Southern Jurisdiction – taken by Buzz Aldrin to the Moon on Eagle, bearing the official double-headed Eagle of the Scottish Rite (Hoagland).

Fig. 5-33 - Ancient Egyptian iconography links doves with ancient geodetic markers called "omphalos stones." The "doves" were actually homing pigeons, used to create survey maps in Ancient Egypt, eventually linking "doves" and "navigation," as in the tale of Noah's Ark (Temple)

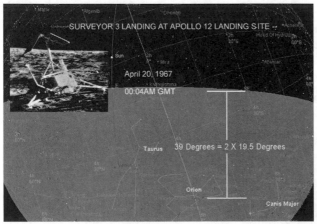

Fig. 5-34 – Sky over unmanned Surveyor 3 (inset) lunar landing site, at touchdown, April 20, 1967, Hitler's birthday. Note Belt stars of Orion (Osiris), "Ancient Egyptian god of resurrection," 39 degrees below SE (rising) horizon. 39 degrees is TWICE 19.5. Symbology consistent with ritual "resurrection" of Hitler, or "the Third Reich" (Hoagland).

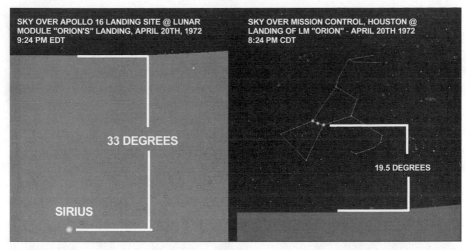

Fig. 5-35 – Alignments at the landing of Apollo 16 lunar module "Orion" on April 20th, 1972 (Adolf Hitler's birthday).

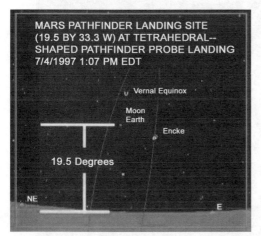

Fig. 5-37 – Screen cap of CNN coverage of first Mars Pathfinder images.

Fig. 5-36 – Tetrahedral-shaped Mars Pathfinder spacecraft lands at 19.5 degrees on Mars (inscribed tetrahedral latitude), with Earth at 19.5 degrees above eastern Martian horizon. (Bara)

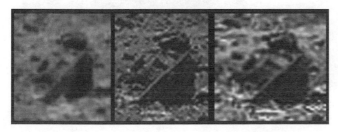

Fig. 5-38 - Three views of the "hydrant," a bizarre angular mechanism some distance from the Lander (NPAA)

Fig. 5-39 - Pyramid-shaped object seen from Sojourner Rover – days after landing.

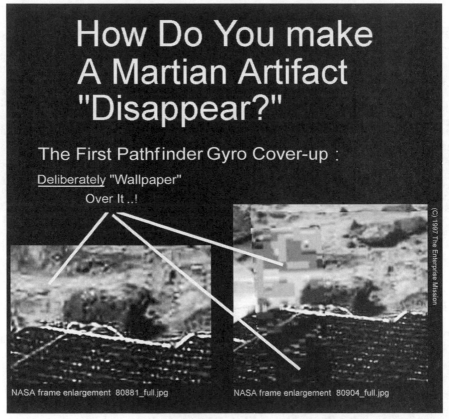

How Do You make A Martian Artifact "Disappear?"

The First Pathfinder Gyro Cover-up :

Deliberately "Wallpaper" Over It ..!

(C) 1997 The Enterprise Mission

NASA frame enlargement 80881_full.jpg

NASA frame enlargement 80904_full.jpg

Fig. 5-40 - Evidence of *deliberate* cover-up of possible Martian artifacts—on official Mars Pathfinder NASA website . (JPL)

Fig. 5-41 – Enhancement of South "Twin Peak" at Pathfinder landing site, showing anomalous "terracing" on the face of the Peak.

Fig. 5-42 – Super resolution enhancement confirming "terracing" on South Peak and strange "Sphinx" at edge of debris field.

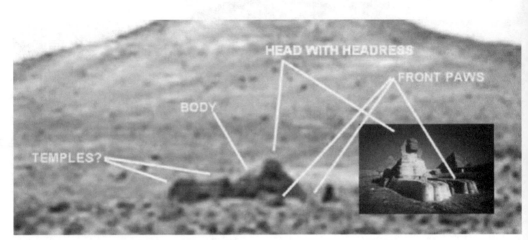

Fig. 5-43 – "Sphinx" comparison.

Chapter Six

New Mars Global Surveyor Images of Cydonia

O n November 7, 1996, well before Mars Pathfinder had even landed at its hyperdimensional site of 19.5º x 33º, NASA had already launched its next mission to the Red Planet. Mars Global Surveyor was the successor to the ill-fated Mars Observer, to be fitted with a slightly upgraded version of that probe's "Malin camera," designed by Dr. Michael Malin.

Malin Space Science Systems had won a new contract for the camera on MGS in a less-than-open competition. Attempts by the military team that ran the Clementine mission to the Moon to bid on the contract were rejected in a manner that one witness described as "frothing," in spite of the fact that the Clementine team's camera was superior in all respects to Malin's. So NASA asked Dr. John Brandenberg, an old colleague of Hoagland's from the early independent Mars investigations, to present the camera a second time to JPL's open bidding process. The decision was then abruptly assigned to a "selection committee" run by JPL, who selected Malin and his camera, basically the same 1993 technology that was on Mars Observer, over the more flexible Clementine instrument, which, among other things, could be gimbaled to point at an off-nadir target, while the entire spacecraft had to be maneuvered to take such an image in the case of the Malin camera. JPL, it seemed, *really* wanted Malin to be the camera man for the next Mars mission.

This time however, we had reason to believe that we might finally get the images we had waited so long for. The launch of Mars Global Surveyor had followed the ritual pattern we had come to expect. As MGS pulled away from the cape, Sirius was hovering at 33° below the horizon. We hoped that this meant that MGS would be successful where Mars Observer had failed.

In fact, there was even a "double alignment" of sorts at the launch, with Orion's belt star Alnitak also at 33° below the Cydonia horizon. So we had a "hit" on the occasion of the launch at both the launch site and the anticipated target zone.

With Malin's well-demonstrated hostility toward the Cydonia issue clouding everything around the mission, Stan McDaniel's SPSR organization arranged a clandestine meeting with NASA's Dr. Carl Pilcher, the Acting Director of Solar System Studies, in November of 1997. At the meeting, which was attended by McDaniel, Carlotto and Brandenberg among others, Pilcher feigned interest and promised that Cydonia would be imaged at every opportunity during the "science mapping phase" of the mission. He later dismissed the meeting, saying he "just took the meeting to get SPSR to stop bothering us." However, in a report on the meeting made by Carlotto at a Brazilian UFO expo, there was this one intriguing nugget:

"Moreover, they said that they are in fact very interested in these objects, for two reasons. There are two groups in NASA. One group believes that [UFO researchers] are all wrong, and they want to prove it. So they want to take these pictures to prove that you're wrong, so you'll go away. The other group—and our sense is that it's a small but growing group within NASA—believes that we have some interesting data, and they want to take a closer look at it."[105]

Unfortunately, despite Pilcher's verbal promise that re-photographing Cydonia was now "official NASA policy," SPSR didn't get anything in writing. As the Spring 1998 orbital insertion of MGS approached, Hoagland used the power of his appearances on *Coast to Coast AM* to ratchet up the pressure on NASA to formally commit to re-photographing Cydonia. He argued that Malin should not have the god-like power to decide what would be photographed, that the data stream from the Orbiter should be live, as opposed to the (up to) six months embargo period Malin was allowed under his private contract, and that Malin should be put on record as to whether he had any knowledge about Mars Observer's disappearance or possible secret resurrection.

As the public pressure mounted, Malin took to the airwaves himself in an effort to diffuse the situation and retain exclusive control over "his" instrument—which the American public had paid for.

Malin chose to give an interview to Linda Moulton Howe, a regular contributor to *Coast to Coast AM*. In the interview, he expressed indignance that anyone could view him as being responsible for what happened to the Mars Observer. He also went to great lengths to claim that getting an image of an object as "small" as the Face (which is about 2.5 x 2 km) was an iffy proposition at best, comparing it to winning the lottery. When Howe asked him what he would say to those that had waited for almost twenty years for new images of the region, Malin said "... all I can say is, jeez, I'm sorry, that's the reality of the thing."[106]

We, of course, knew this to be baloney. The targeting capability of the MGS camera was exceptional, with very little error built into the system. Malin's team had devised an excellent targeting software suite that enabled them to pre-select a Face-sized target with ease.

There are two factors affecting the targeting of a specific object with the MGS camera: downtrack and crosstrack. "Downtrack" is the path back along the spacecraft orbit. "Crosstrack" is the lateral scanning normal to the spacecraft's vertical axis. Of the two, crosstrack is by far the most difficult to account for when targeting a specific object because of uncertainties in the mapping grid developed by the Rand Corporation.

However, even in the worst case scenario, the maximum crosstrack error is about 0.15 miles, or *one eighth* of the width of the Face itself. The downtrack error margin is significantly less than even that small distance. In the words of Stanley McDaniel's "McDaniel Report," hitting a specific target the size of the Face is "About as difficult... as hitting a door with a baseball from a distance of about one foot."

Malin's attempts to deceive the listeners with his "lottery" comments were obviously very worrisome to Hoagland and the other Cydonia researchers. In a page on his website, Malin had made his feelings plain as recently as 1995, when he stated that "no one in the planetary science community (at least to my knowledge) would waste their time doing 'a scientific study' of the nature advocated by those who believe that the 'Face on Mars' [is] artificial."

This was of course a gross misstatement of the argument. No one, not even Hoagland, had expressed a specific belief that the Face or any of the other objects at Cydonia were artificial. Although we all strongly suspected they might be, we'd merely argued that they deserved further study and should be imaged at every opportunity. But Dr. Malin had a long history of distortion and obfuscation when it came to the Face and Cydonia.

According to Vincent Dipietro, during the "Case for Mars" conference in Boulder, Colorado in 1981, Dipietro and his research partner Gregory Molenaar were audibly assaulted by Malin. Malin had set up a display table near to the one hosted by Dipietro and Molenaar, who were trying to generate interest in the Face and their image enhancement process. Every time anyone approached the table and started to look at the images of the Face, Malin picked up a *megaphone* and began yelling into it, creating so much noise that he was able to drive away many a curious conference-goer.

It seemed obvious that he and JPL remained extremely hostile to the idea of even trying to target the region. Further, Hoagland noted that Pilcher's promise was to target Cydonia only during the "science mapping" phase of the mission. This was key, because MGS was about to enter a pre-science mapping orbit that would have it pass over Cydonia every nine days.

In March 1998, after receiving literally tens of thousands of faxes from Art Bell's *Coast-to-Coast AM* audience demanding that NASA take new photos at the first opportunity, NASA finally caved. In a prominent agreement, later posted on the official JPL website, the Mars Surveyor Project agreed to "announce these [Cydonia] imaging opportunities *in advance*..." and then "[shortly after receipt] to post the resultant images on the internet." According to the specific terms of this agreement, "we [the Mars Surveyor Project] expect that there will be widespread scientific and public interest in the new results from Mars. As such, there is a strong commitment by NASA and the MGS scientists to release data to the public on a *timely basis*. The project will be releasing data *shortly after receipt* on the internet in a manner similar to that seen on the Clementine, Near Earth Asteroid Rendezvous and Mars Pathfinder missions.

"The Mars Global Surveyor spacecraft is only able to photograph features on the surface of Mars that are directly below it as it makes each orbital pass. The spacecraft will fly directly over the Cydonia region, where enigmatic features were observed in the Viking mission, a few times during its mapping mission.

"The Mars Global Surveyor project will announce these imaging opportunities in advance and will post the resultant images on the internet." [Emphasis added.]

The fax and e-mail campaign generated by Hoagland and Bell paid off a few weeks later when NASA announced its first Cydonia imaging opportunity for April 5, 1998, during the science phasing orbit period. At last it seemed we were going to get the pictures that might settle the argument once and for all.

Playing in the Catbox

In the anxious days leading up to the April 1998 imaging opportunity, NASA released even more details on the targets planned for the science phasing orbit images. They released a list including Cydonia, the Viking 1 and 2 landing sites, and the Pathfinder landing site. If it was designed to be a run through of all of the possible "temple sites" on Mars, it could not have been more obvious. The announcement emphasized that these sites were chosen because they could provide scientists with crucial "ground truth" measurements to compare with what they had seen from the landing sites themselves, even though they acknowledged that the Landers would not be visible. Exactly how these images could be of any use in making such comparisons when the Landers weren't even visible escaped us, but it was certainly true that, visible Landers or not, this ritual of re-photographing the most "sacred sites" on Mars established by man could serve a powerful ceremonial function. In keeping with Malin's previous statements, the announcement maintained the idea that actually imaging a specific target would be problematic:

"The probability that the targets of interest will be within the camera's field of view varies between thirty and fifty percent."

A further update announced that the Cydonia image would be taken on orbit 220, at 08:33 UTC (Universal Time). This gave us an excellent chance to look for alignments for that image. The first place we looked would logically be Cydonia, where the image was being taken. There were no alignments there. However, when we checked the sky over JPL, from which the command to take the image was sent, we did find a significant alignment—Sirius/Isis hovered below the JPL horizon at precisely the tetrahedral 19.5° altitude when the image was taken. So the ritual pattern continued.

Heading into the April 4 weekend, NASA announced that the Cydonia image targeting the Face on Mars would be released on Monday morning, April 6 at 10:30 a.m. PDT. Getting such a precise date and time aroused our curiosity, and as we waited over the weekend for the image to be taken, we occupied ourselves by checking the alignments for the image release.

Although there was no alignment over JPL at that time, there was a significant alignment of Orion's belt star Mintaka at 33° below the horizon at Cydonia. In addition, Mintaka was almost dead on the Meridian, making this an extraordinary "double hit" in the ritual system. We concluded after finding this alignment that we would indeed get something very significant that Monday morning.

On the morning of the sixth, we all huddled around our computers, anxiously awaiting the image release. Precisely at 10:30 a.m., in accordance with the alignments that had been established, the link appeared on our screens. When we clicked on it, we saw what we had been waiting for over two decades to see—sort of.

The image that was released was a black, grainy, essentially blank image. Was this a joke? [Fig. 6-1]

Unfortunately, it was not a joke (that was to come a few hours later). It was a raw image, supposedly of the data that had been uploaded to the spacecraft. However, even though the image was taken in the full light of mid-day, it was virtually black. This was not what anyone, even the media, had expected. Hoagland's phone began to ring off the hook, asking him if he still "believed" in the Face on Mars and if he had any comment. He simply responded that he was disappointed with the quality of the image, since we had all been led to believe that it would be much better than Viking's raw data, when in fact it was worse. He urged caution until we saw if NASA would release a processed version later in the day.

That "processed version" came a few hours later, from JPL's Mission Image Processing Laboratory (MIPL). At last we could see enough detail to tell where in the image swath the Face actually fell.

Contrary to his previous claims about "hitting the lottery," Malin was able to nail the Face dead center in the image on his first official attempt. Rotating the spacecraft "off-nadir" as it approached Cydonia, Malin and the JPL navigators positioned the probe and its camera to capture the Face nearly perfectly. It was dead center on both the crosstrack and downtrack of the image swath. There was only one problem; not only didn't the Face look much like a face, it didn't look like much of *anything* [Fig. 6-2].

Instead of a high resolution overhead view of the Face, we got a low-contrast, noisy and washed-out image that was apparently taken well after MGS had passed the Face mesa. This resulted in a view that was looking up from below and dramatically to the left side. Details of the right side, previously shadowed in the Viking images, were visually compressed by perspective and hidden behind the "nose." The image swath had extended far beyond the

Face to the north, showing a featureless plain, and south all the way to the D&M Pyramid, capturing almost one full corner of that enigmatic object. The image was so lacking in contrast and detail that it gave the impression of a flat, blank desert landscape, with virtually no elevation at all.

This was not what the researchers or the public had expected to see. But it was clearly what JPL and MSSS wanted the media to show them. Within an hour of the release of an image that they surely knew was far below the quality of what *could* be obtained from the raw data, JPL spin doctors had spread out to the various news media pronouncing the Face to be natural. Obviously working from pre-arranged talking points, these spin doctors— employees of JPL mostly—insisted that even though they were NASA scientists (and by implication smarter than most of us) they were not speaking for NASA or JPL, but only for themselves.

NASA and JPL pronounced that neither would take an official position on the image, thereby draping both organizations in a fallacious robe of objectivity. Surely, they knew what their employees were doing on their lunch hours, since it was all over the television.

The end result of this was to insulate NASA and JPL from direct criticism on the matter. Any of their employees subsequently found to have made false statements or unscientific arguments over this issue could be dismissed as "loose cannons" that acted outside the purview of their responsibilities at the agency. This meant that there could never be a second "McDaniel Report," proving NASA's complicity in a campaign of misinformation and ridicule of a scientifically testable hypothesis. At the same time NASA could then claim that it acted openly and honestly by releasing data quickly and allowing its scientists to comment on the matter.

Hoagland and the other independent researchers were caught off guard by this well-coordinated media assault against them. Still trying to process the raw image themselves and hopefully get a better version of it than had been provided by JPL, they were ill equipped to deal with the media circus around the image release. Facing deadlines, the major media couldn't wait around for Hoagland or anyone else to process a better version. When the six-o'clock news rolled around, they went with the MIPL image.

They were almost uniformly hostile. None other than brilliant planetary scientist Dan Rather pronounced it "a pile of rocks." NBC's Tom Brokaw called the image "proof of what we already knew." Only CNN's John Holliman, who had been friendly to the independent investigation over the years, was somewhat sympathetic, saying that the independent researchers needed more time to properly evaluate the image. He concluded his report by saying "NASA has always said the Face is merely a trick of light and shadow. Some trick." JPL's spin team had done the job.

Then, within three minutes of the last national six-o'clock newscast sign off, a second image suddenly appeared—again without comment—on the various NASA, MSSS and JPL web and mirror sites. The "TJP" (Timothy J. Parker) enhancement was a significant improvement over the earlier MIPL image. It contained far more contrast and detail, and less noise, than the image that had dominated the newscasts [Fig. 6-3].

Parker, a JPL geologist, had produced this second, vastly superior version of the "raw" data using mostly standard Photoshop tools, and posted his steps on the web. His version had detail that was far more visible and confirmed many Face-like features—including clearly unmistakable nostrils, of all things—but it came too late. Only after the major news organizations had broadcast their reports and made their pronouncements did this considerably improved, much more obviously Face-like image miraculously come to light. Even so, it was still improperly ortho-rectified and gave a less-than-ideal perspective on the object.

That night on Bell's show, Hoagland tried to explain all this to an obviously disappointed public. He pointed out that, given the distance the spacecraft's vertical track was from the Face, such a side-view was the best that could be hoped for. However, it wasn't necessary to have waited as long as they had to take the image, resulting in a view looking back at the Face from "below." While this angle did show some new secondary facial characteristics, such as the "nostrils," it could not be used for a realistic symmetry study that a true overhead shot might have provided.

Bell, however, was indignant at the political aspects of the day. He considered it a joke that the horrible "MIPL" version had been the only one available as the TV news had gone to air, and asked Hoagland why it might take another seven hours for Malin's team to release their "TJP" version. Hoagland admitted that the TJP version should have only taken about thirty minutes to produce, and lamented the fact that the MIPL version made it appear that there was no "Face" at all. "Well, looking at that image, Richard, I'd have to conclude as well that there is no Face on Mars," Bell said, "and my question now is, where the *hell* did it go?"

Bell summed up the MIPL image by saying that it reminded him of a pattern his kitty might scratch up in her litter box. It was from that moment forward that the MIPL image would forever be known as the "Catbox" version of the Face on Mars.

Honey, I Shrunk the Face

Within a day of the release of the new MGS image of the Face, Hoagland had become suspicious of its quality, and then reasoned that there might be a way to discern if the image had been altered or degraded.

The Mars Orbiter camera, in its "narrow-angle mode," is composed of a single line of detectors—a 2048 Element Line Scan CCD Array. The camera produces images by electronically "cross-track sampling" the array, while the physical motion of the spacecraft around the planet moves the entire line of detectors over the Martian surface at right angles to that scan (normally straight down at the planet's surface).

Inevitably, each individual CCD element in such an array possesses slightly varying sensitivity compared to its neighboring elements, across the width of the detector. Thus, any image produced by the "line-scan CCD array" will inevitably display a series of irregularly-spaced, vertical bright and dark lines—like scratches on an old print of *Casablanca*, stretching the length of the entire image at right angles to the scan.

Normally, these vertical irregularities are removed from the final image by appropriate computer processing; however, in "raw" or incompletely processed images, these scan lines can serve as unique detector fingerprints of that particular CCD array. No two line-scan cameras will imprint the same spacing, intensity or number of such lines on any of its images. Thus, like matching bullet markings in a murder investigation through a ballistics test, comparing lines on various CCD line-scan camera images can uniquely determine crucial aspects of those images—including which camera took which image.

The next day, a listener named Fred Hoddick, acting on Hoagland's conversation about these CCD idiosyncrasies, discovered that the MOC indeed imprinted a unique line-scan fingerprint on every Mars Surveyor photograph.

One such image he investigated was a close-up section of the spectacular Vallis Marineris, the "Grand Canyon of Mars." In comparing the line-scan signature visible in narrow-angle image with the pattern of faint lines seen in the raw version of the MGS Cydonia image, Hoddick indeed made a major, startling discovery—the spatial dimensions of the Mars Surveyor image of Cydonia released by JPL were only half of what should have been acquired [Fig. 6-4].

When the bright, clear, full resolution image was reduced in size by 50%, its scan-line signature precisely matched that of the Cydonia raw image. Thus, the raw frame displayed on all the NASA websites only presented half the spatial data apparently originally imaged by the camera. This radically reduced the ability of image processors to detect (if not unambiguously identify) any artificial sub-structures present in the image.

When this blatant spatial image tampering is added to the extremely limited grey scale presented in the same MGS raw image (only forty-two out of a possible 256 shades of grey were present in the raw data) the result is an extremely noisy image enhancement. Because of the morning light aspect of this MGS Cydonia photograph, this reduced number of grey levels further distorted the raw Mars Surveyor Cydonia image, effectively eliminating meaningful comparisons with the previous Viking data. This comparison is further hampered by NASA's choice of the spacecraft imaging angle—oblique, as opposed to Viking's overhead frontal view.

Finally, NASA's choice of imaging enhancement tools for this bland image—high-pass filtering—further reduced the MIPL version of the Face from Mars Surveyor to a black and white "cartoon," what Art Bell termed whimsically "the Catbox image."

Michael Malin quickly responded by saying that the image was reduced in size by 50% over what could have been acquired in order to avoid a downtrack error from occurring and missing the Face. Of course, as we have already discussed, the major navigation problem with MGS was not down track, but *cross track*, and that error range was so slight as to be insignificant in targeting an object as large as the Face—and to extend the downrange "footprint" of the Surveyor's imaging, from slightly under seven miles to twenty-six miles, in an ostensible effort to guarantee successfully re-imaging the Face (but at the cost of cutting the surface resolution in the camera in half) simply doesn't make sense. And the spectacular success of the JPL navigation team confirms this—the actual location of the Face in the image JPL released was almost dead center of the "downrange" footprint, and just left of the east/west "cross track." (In the words of one of Malin's own associates on that Monday afternoon: "We nailed it!")

Other nagging inconsistencies also remained. If "trading off" imaging resolution for a larger photographic footprint was a deliberate pre-Cydonia strategy reached by the entire project, why didn't anyone at NASA (including Michael Malin) *say so* before the Cydonia attempt? Why did they wait to offer an explanation for this "surprise" only *after* Hoagland caught them at it?

And, finally, there is the little matter of the "corrected caption" that then appeared on JPL's own website:

> *"CYDONIA PHOTO CAPTION*
>
> *"as stated on: Mon 04/06/98 10:30 AM PDT Image dimensions: 1024 x 19200 pixels, 4.42 km x 82.94 km*
> *"This was a typographical error for which we appologize [sic].*
> *"Actual image dimensions: 1024 x 9600 pixels, 4.42 km x 41.5 km."*

Somehow, it's hard to imagine anyone typing "19200" in place of "9600," 82.94 km for 41.5 km—even in a government contracted typing pool at JPL— but if the original transmitted imaging resolution from Dr. Malin's camera was 2048 pixels across, subsequently downsized on Earth to 1024, then the corresponding down track dimension would have been precisely 19200 pixels—exactly what the original NASA caption read, exactly consistent with Hoagland's discovery that the image was somehow missing 400 percent of its expected resolution.

In other words; they did it.

Responding to the other major criticism that greeted the first raw Cydonia image—that it was simply too dark, resulting in a lack of grayscale contrast range—Malin posted data claiming that the MGS raw image in fact "wasn't all that dark..." He attempted to compare the new MOC data of Cydonia with the twenty-two-year-old Viking image histograms, insisting that in truth "the MOC data actually have more grey levels than the Viking images..." There is only one small problem with Dr. Malin's analysis—he's wrong.

The major difference is that both Viking Cydonia frames were taken in the late afternoon, with long shadows obvious in even the raw data. The MGS image was taken at 10:00 a.m. local Martian time—yet the image histograms showed more levels of gray in the Viking images than the MGS image, and all this with a camera that was on the order of ten times better than Viking's

In the end, we were forced to conclude that the supposedly raw image was a second-generation copy filled with noise and devoid of crucial detail—and an SPSR member was about to show us that the "Catbox" image was an even bigger ruse...

Unmasking the Catbox

Even as the controversy raged over the question of the raw data, other independent researchers remained focused on the MIPL, or Catbox, enhancement. However much detail and contrast had been removed from the raw data, it paled in comparison to the hatchet job done on the Face by the unnamed creator of the Catbox image. One researcher, Lan Fleming of SPSR, was also a NASA contractor by day. He spent weeks trying to recreate the Catbox image with standard software processing tools, to no avail. No matter how hard he tried, he was unable to reproduce the flat, featureless look of the "enhancement."

Then he decided to try a new combination of techniques. By first applying a high-pass filter (which removes high frequency data from an image) and then a low-pass filter (which removes low frequency data), he got very close to the Catbox look. This was also probably how so many of the gray levels were removed from the raw image presented by JPL. He then applied a noise filter, which introduced more noise into the image, to reproduce the "graininess" that was so prominent in the Catbox enhancement.

But he was still lacking a crucial "something." Fleming had noted that a boulder near the Face in the Catbox image was producing a shadow that pointed almost due north, essentially the twelve o'clock position in the image, implying that the light was coming from the six o'clock position. He knew this had to be a false shadow, since the light was coming from below the Face, in the four o'clock position, when the image was taken. He was now stuck trying to find a filter that could reproduce this effect [Fig. 6-5].

Eventually, he tried an emboss filter, a software tool that works by turning lines and edges into a relief. These edges then become illusory ridges and depressions, depending on the direction that is chosen for the false lighting. This has the effect of creating false visual cues for elevation, effectively scrambling an image to make it less visually coherent.

By adding these two additional filters, the Catbox image was revealed as a simple fraud. As Fleming put it:

"After JPL removed most of the tonal variation in the original image that gives the observer the visual cues to the real three-dimensional shape of the object, they added false visual cues to give the object its rough, jumbled appearance, inadvertently falsifying the appearance of the surrounding terrain as well... the Catbox is not a 'poor' enhancement, as it is often called; it is a crude but very effective fraud perpetrated by employees or contractors to the United States government. Even if the Face is proven to be completely natural, this is inexcusable misconduct and a gross abuse of power. If the Face ultimately is proven to be artificial, the Catbox will certainly come to be regarded as the greatest, most malicious and most destructive scientific hoax since the Piltdown Man, and perhaps of all time."

In other words, in order to get from the original raw MOC 22003 image to the eventual Catbox enhancement, which defined the Face to the majority of the public and academia for several years afterward, NASA/JPL/MSSS had gone to the following trouble:

1. Reduced the resolution of the original 2048 x 19200 image strip to 1024 x 9600, some time after acquisition of the image;

2. Removed almost 85% of the tonal variations by using high-pass and low-pass filters on the "raw" data;

3. After initial processing, applied another high-pass filter;

4. Applied a noise filter to induce more noise into the image than had already been created by the previous processes;

5. Used an emboss filter to delete visual elevation cues and induce false visual cues into the image.

And all of this, just to discredit an investigation that "no one" at NASA or JPL supposedly took seriously.

Just what was it on that original raw data that was so threatening that it would require this degree of suppression? We may never know.

Reaction

Over the following weeks and months, Carlotto and a whole host of amateurs performed enhancements on the new Face image. Although hampered by the degradation of the "raw" data and the poor light and angle of the image, some remarkable work was eventually accomplished. Carlotto produced a better ortho-rectification than the NASA version, and still others produced even better versions. Eventually, the "Mark Kelly enhancement" came to be viewed as the best that could be gleaned from the limited source data.

Although hardly an ideal rendering, the new Face image at least confirmed many of the assumptions and predictions of the early independent investigations. There was indeed a "brow ridge," apparently on both sides and roughly symmetrical. The beveled "platform" upon which the Face rested could also be confirmed as being close to 98% symmetrical, a condition that was almost unheard of in any natural formation. Beyond that, there seemed to be a curled lower lip, and fairly unmistakable "nostrils" in the nose, right where they should be if they were indeed intended to represent nostrils. There was also a hint of a pupil in the eye socket.

To Dr. Tom Van Flandern, these obvious secondary facial characteristics were compelling. He argued that such features were inherently predicted by artificiality hypothesis, and that their existence represented strong enough evidence to conclude that the Face was artificial.

"The artificiality hypothesis predicts that an image intended to portray a humanoid face should have more than the primary facial features (eyes, nose, mouth) seen in the Viking images," he wrote on his website. "At higher resolution, we ought to see secondary facial features such as eyebrows, pupils, nostrils and lips, for which the resolution of the original Viking images was insufficient. The presence of such features in the MGS images would be significant new indicators of artificiality. Their existence by chance is highly improbable. And the prediction of their existence by the artificiality hypothesis is completely *a priori*.

"By contrast, the natural-origin hypothesis predicts that the 'Face' will look more fractal (e.g., more natural) at higher resolution. Any feature that resembled secondary facial features could do so only by chance, and would be expected to have poor correspondence with the expected size, shape, location and orientation of real secondary facial features. Any such chance feature might also be expected to be part of a background containing many similar chance features."[107]

He finished by saying, "In my considered opinion, there is no longer room for reasonable doubt of the artificial origin of the face mesa, and I've never concluded 'no room for reasonable doubt' about *anything* in my thirty-five-year scientific career."

If Van Flandern was more than satisfied, others were not quite as enthusiastic. Carlotto contended that the Face might be artificial, but that if it was, it was in a "highly eroded" state. Graham Hancock, a sometime Face proponent who later authored a book about Mars and the Cydonia controversy (*The Mars Mystery*), appeared on *Coast to Coast AM* a few days after the Catbox image was released and stated "I have to say, I believe that the advocates of the Cydonia hypothesis have been dealt a blow." The authors' own conclusion was that with the source data so hopelessly compromised, there was little that could be decided about the new image. Our focus had now turned forward, to the next two Cydonia imaging opportunities that were coming up. NASA, regardless of Van Flandern's protestations, had won this round. They had successfully suppressed interest in Cydonia to the point that no major media would touch the subject for the time being. What we wanted now was to get more out of the next imaging opportunities instead of arguing the last one.

On April 10, 1998, a few days after the release of the Catbox image, NASA/JPL released a document on their website announcing the second set of "targeted imaging opportunities." It contained a map of Cydonia with a predicted image swath through the city, with the so-called "city square" as the primary target. The document explained that NASA had decided to ignore the Face on this second pass since it had already been "successful" in capturing the Face on the first Cydonia over-flight. While we all contested the accuracy of that statement, the question of artificiality at Cydonia had always rested on far more than the Face, and it would be good to get images of some of the other pyramidal structures and the odd, Giza Pyramid-sized mounds scattered around the Cydonia plain [Fig. 6-6].

The document also contained one specific statement, which caught our attention:

"Results of the Cydonia imaging will be posted on the internet, in the same manner as in the first observation attempt, at approximately mid-evening Pacific Time on Tuesday, April 14."

The authors immediately recognized this as an opportunity to run a true, indisputable *a priori* test of our ritual alignment model. Even though we had been successful in pointing out the ritual alignments on the first image release, on this occasion we did not even have a specific time for the release, just a general "mid-evening" timeframe. It was up to us to construct a falsifiable experiment that would verify our hypothesis.

Fortunately, the stars cooperated. We decided to first look at the sky over JPL, since that was where the image release was being controlled and where the servers providing the images were almost certain to be located. It didn't take us long, looking at the stars through Red Shift 2, to pin-point the timing for the image release. At 6:55 p.m. (PDT) on the anointed date, Sirius would be passing through 33º, right over JPL. This moment in turn would open up a "window" in which the three belt stars of Orion would all pass through that 33º altitude over the next ten minutes.

So we had an unmistakable opportunity to forever silence the naysayers. But we also realized that a public prediction could alter the conditions of the test, by providing a heads up to JPL that we were tracking them. We decided against a published prediction, since that would make it far too easy to delay the image release by a few minutes and scuttle our test. Instead, we decided to e-mail our prediction to specific researchers and members of the press, including Tom Van Flandern and Art Bell. We predicted an image release at 6:55 p.m. PDT, when Sirius was at 33º above the JPL horizon. Although we had the 10 minute window, we knew that we would have to pick a specific moment for the test to have complete validity. JPL did not disappoint us.

As we were watching the JPL websites, with Bara on the main JPL site and Hoagland watching the primary mirror site, we continued to refresh our browsers to insure we got the image at precisely the moment it was posted. Then at 6:55 p.m., exactly as we had predicted, the new image link appeared on the JPL main web page.

Even though we'd scored a "hit," we continued to monitor the primary JPL mirror site, which had not been updated with the new image. Then, exactly ten minutes after the image had appeared on the JPL main page, a link to the new image appeared at 7:05 p.m. PDT on the JPL mirror site. So they had released the image twice, in effect, once at the opening of the "alignment window" and once at the close. But unquestionably, undeniably, they had followed the ritual pattern.

As a side note, some months later, when reorganizing the Cydonia images under a single web page, NASA/JPL changed the release date and time stamp to show that the second Cydonia image was posted at 6:30 p.m., PDT, taking the release outside the alignment window. The authors can categorically state that this was not the case. The images were released in just the manner described above, at 6:55 and 7:05 p.m.

Even more interesting was what the image contained. Although it had missed the targeted "city square" by some two miles, the image swath managed to capture a sizable chunk of the so-called "Western Pyramid" in the City, along with some of the (remember, *tetrahedrally* arranged and shaped) mounds and a sizable portion of the landscape beyond the City. There were some surprises.

One of the objects in the city, the Western Pyramid, had been noted and named by architect Robert Fiertek while looking at the Viking data. At its base, just above an enclosed "courtyard," was what appeared to be a small knob that seemed possibly to be faceted, but in the original data it was not clear enough to tell for sure. In scanning the new image (which appeared on the web in both poor [MIPL] and good [TJP] versions) this "knob" stood out dramatically.

It was a pyramid [Fig. 6-7].

Indeed, not only was this knob distinctly pyramidal, it was overtly layered and faceted on all four sides, like a cross between a Mesoamerican pyramid and the structures at Giza, although it was nearly five times the scale of those pyramids. Although collapsing on the northern side, it maintained a rigid four-sided structure that gave the impression of monumental architecture. From its southwest corner, a wall extended that terminated in a multi-tiered structure Hoagland dubbed the "Castle of Barsoom."

A fuzzy fog oddly obscured the southern and western faces, though "fog" had to be a false description, since the MGS camera went well into the infrared and should have effectively cut any local haze. Instead, the fog seemed to be areas of the image that had unaccountably lost all contrast and detail. This is generally a characteristic of a blended, modified section of a digital image. An examination of the histogram showed major compression of grayscales in these areas.

A geometric reconstruction of the shape revealed that the best visual fit for the object was a two-tiered pyramid shape, with a 45° slope angle for the first layer and a 60° slope angle for the cap layer. Critics argued that the "Giza Pyramid" was simply a product of "mass wasting," a known geological process that results in debris piling up at the base of a mountain. However, the chances of such a pile of rocks forming themselves into a pyramidal shape with at least two obvious profile edges and two tiers sloping upward at 45° and 60° respectively, as Hoagland put it, were "pretty remote."

Then came the issue of the tetrahedrally arranged mounds of Cydonia. Two of them were captured in the new image, mound "P" and mound "O." Of the two, mound P was the most extraordinary. Not only was it a neat little wedge shape, as the image from Viking had predicted, but right next to it was something those images hadn't revealed—a hexagon [Fig. 6-8].

In fact, there were two distinct hexagons on the image swath, one next to mound P and another at the bottom of the image, in a rugged patterned area of the frame called the suburbs. Both were the size of baseball stadiums and both were unnaturally regular. Mound O was unfortunately just at the edge of the image swath, and although it too looked faintly hexagonal, the resolution just wasn't there to reach any definitive conclusions.

Hoagland considered the presence of hexagons on the ground at Cydonia to be especially significant. In the works of Maxwell, upon which Hoagland based his hyperdimensional model of the message of Cydonia, he postulated about the physical properties of multi-spatial dimensions and their interactions in our familiar 3D world of energy and matter. The numbers predicted that the behavior of a spinning sphere, such as a planet, would outwell higher dimensional energies at the key tetrahedral latitude—the now-ubiquitous 19.5° connection. A lesser-known aspect of this model was the prediction that there would be inwelling points in the system as well, and that they would be hexagonal. There had been some confirmation of this idea in images taken of Saturn and the sun. Both sets of images showed hexagonal rings of clouds around the northern poles of both bodies, making the turns at high velocity. No known physical phenomena could account for this behavior.

Further, if Hoagland and Torun had been right all along in their declaration that the arrangements of the monuments at Cydonia were intended to inculcate knowledge of tetrahedral physics to observers, then placing hexagons all around the area would be a dramatic reinforcement of that message.

The Final Image

On April 20, 1998 (yes, Hitler's birthday), NASA again posted a document announcing the third set of targeted observations of the Viking 1 and 2 landing sites, the Pathfinder landing site, and Cydonia. Again, this process seemed to be almost ritualistic, as if they were cycling through the Martian temples one by one. This time, the release even took the care to tell us exactly what date and time each image would be taken, almost daring us to look ahead to what they were planning. Of course, we took them up on it, and although there were no significant alignments at any of the landing sites, once again the Cydonia image followed the ritual pattern with Sirius dead on the horizon from the vantage point of JPL when the image was taken.

The image itself contained even more anomalies, ranging from a flat triangular-shaped ruin that had a complex latticework of supporting struts, to rectilinear room-sized cells on the main pyramid. The triangular ruin was especially controversial, since some members of Stanley McDaniel's SPSR tried to dismiss it as an old-fashioned trick of light and shadow, while others pointed out that it pointed directly due north, quite a coincidence for a trick of shading.

The long-sought City Square was finally captured in the image. Initially, it appeared to be a set of four fairly unremarkable mounds, albeit mounds with highly unusual reflective properties. After some proper enhancement work, however, the four mounds took on a distinctly more geometric quality. In the end, Van Flandern's assessment of the third image was perhaps the most descriptive: "Triangles and hexagons are rarely found in satellite imagery, except at Cydonia, where they seem to be common."

Admittedly, we ended the first round of new Cydonia imaging with some very substantive evidence to support not only the artificiality hypothesis, but our ritual alignment model as well. Still, NASA had scored a big win in the political battle with the Catbox fraud. By deliberately manipulating the data, they had managed to relieve the pressure from the mainstream media. With the Catbox as cover, they had the perfect opportunity to take their Cydonia studies back under the dark blanket of Malin's "exclusive rights" contract. For the next couple of years, that's exactly what they did.

Chapter Six Images

Fig. 6-1 – original "Raw" data image of the Face on Mars (Mars Global Surveyor\JPL\MSSS)

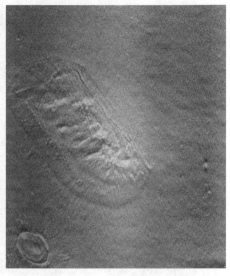

Fig. 6-2 – The infamous "Catbox" version of the Face on Mars, from NASA image 22003. (NASA\JPL\MSSS)

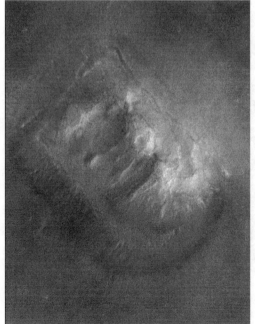

Fig. 6-3 – The "TJP" enhancement of Mars Global Surveyor image 22003 -- released 3 minutes after the major television news networks in 1998 completed their 6:00 PM reports, based on the previously-released "catbox" image (JPL).

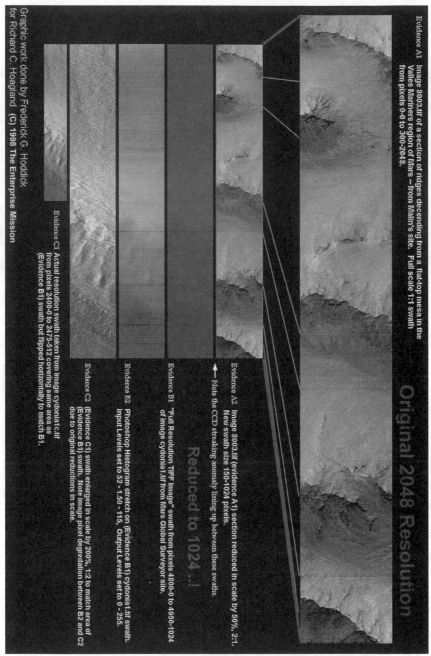

Fig. 6-4 - Comparison of previously released Mars Orbiter Camera image with "raw" Catbox image data. The earlier image has been reduced in resolution by 50%, resulting in an exact scan-line match with the Catbox image. The inescapable conclusion: The Catbox image was "missing" 50% of the data which should have been present.

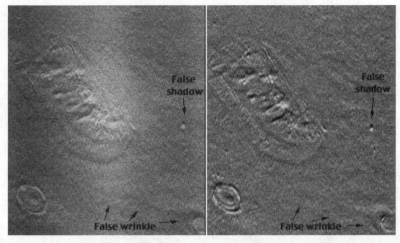

Fig. 6-5 - The "Catbox" enhancement (L) and Lan Fleming's recreation (R). Fleming's analysis proved that the Catbox image was deliberately degraded and modified to make it appear like a natural "pile of rocks."

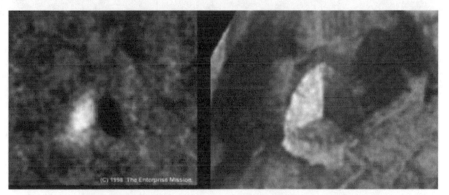

Fig. 6-6 - Giza Scale tetrahedral pyramids from original Viking data (L) and MGS (R).

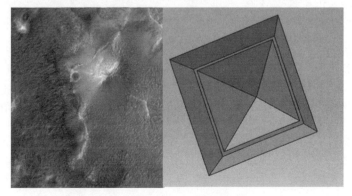

Fig. 6-7 – Mesoamerican style pyramid at the base of the "Western Pyramid" in the City, and a 3D reconstruction of its original shape.

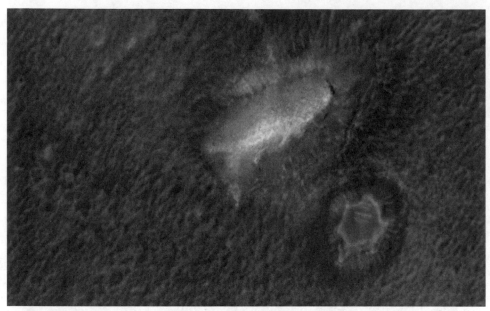

Fig. 6-8 - The "mound P" hexagon, from Hoagland's original analysis. Hexagonal object is about the size of a baseball stadium.

Chapter Seven

An Eye for an Eye

On July 24, 1998, with no prior announcement, the Society for Planetary SETI Research (SPSR) quietly delivered to NASA a report on the new Mars Global Surveyor images of the Cydonia region. One day later, the summary of this report was placed on the web at the home site of SPSR member Dr. Stanley McDaniel. No mention was made of which NASA officials received the report, and the actual contents of the report were listed but not released to the public.

The summary, as written by Dr. Horace Crater, was a timid document that barely scratched the surface of the anomalies present in the three new MGS images. It focused almost exclusively on the Face, mentioning the other plainly anomalous features of the region only in passing. Some obviously geometric areas supportive of Hoagland's arcology model of the Cydonian monuments and the hyperdimensional physics hypothesis derived from them were completely ignored. In addition, Crater misrepresented Hoagland's descriptions by cryptically referring to "speculative individuals" who claimed the city objects were pyramids and that the City Square was "a major object of concern."

Hoagland had proposed the arcology (architectural ecology) model back in the early 1990s, and the City Square was but one of numerous objects cited by him in the geometric relationship model of Cydonia. Dr. Crater also ignored the fact that Hoagland—on nationwide radio—had asked for images not of the City Square, but of the Fortress, D&M and the Cliff. These were hardly the acts of someone obsessed with the City Square. As for the pyramids reference, the major features of the city were given identifiers (Main Pyramid, Western Pyramid) by architect Robert Fiertek, not by Hoagland.

The SPSR report seemed to (falsely) link Hoagland with features or theories not well borne out in the new images (the central mound in the city square is somewhat indistinct, and the so-called pyramids are obviously not Giza-style colossi). In reality, the new images were littered with symbolic and overt structures reinforcing his stated concepts. If such "speculative individuals" were really so speculative (read—not thorough or careful), why the need to lie about their positions?

The summary also asserted that "as far as we know, the work done by SPSR scientists constitutes the only careful study of the images in relation to the Viking data..." This flatly specious statement hinged its mendacity on the word "careful." By inserting this single word along with the opening equivocation, Crater swept aside assessments and enhancements done by Hoagland and others immediately after the release of the images. All that was required was an evaluation (by SPSR of course) that these studies were not "careful" enough.

The assessments of the Face and other objects that actually made it in the report were remarkably banal. SPSR mentioned only the most indisputable anomalies of the landforms (like the symmetry of the Face platform), and did so in a conservative, almost bashful manner. Mark Carlotto continued to refer to the Face as "highly eroded" when it is in fact quite remarkably well preserved (far more so than our own earthly Sphinx, for instance)—and he was even unwilling to acknowledge the spherical pupil in the western eye socket, despite its presence in three confirming images over three separate missions. Crater then cited a discovery of "considerable importance"—a crater in the area with ice in it. Now, to be fair, the presence of water ice on the surface of Mars has some geologic significance, but compared to the various triangles, hexagons and buildings in the images, it's pretty tame. SPSR seemed to feel a need to justify further imaging of the region with a legitimate geological anomaly. Hardly the bold sort of leadership required to bring such a controversial subject into the mainstream.

All this added up to a markedly meek and naïve report. Little was made of the poor marksmanship of Malin's camera in the new report, for instance, and there was no mention of the data reduction that Hoagland has found on the original Face image. SPSR seemed to be more concerned with winning favor with NASA hierarchy than with providing a clear picture to the American people of just what their newest Mars probe had found.

The implication of these observations was that SPSR, far from being an independent oversight group, was playing an inside game with NASA brass. Dr. McDaniel had flip-flopped on NASA's motives and conduct regarding the Cydonia question, contradicting his own earlier report and proclaiming the process and principals "honest." While deploring the conduct of individual JPL employees, the new report failed to point out the obvious, that the PR spin fest after the release of the new Face image was a well-planned and executed political operation, and was entirely consistent with NASA's previous behavior on this issue (as documented by McDaniel himself). After declaring NASA's earlier conduct to be highly suspicious and even "contrary to [its]... stated policy" in favor of full disclosure of any discoveries of artificial structures in the solar system, McDaniel now seemed ready to ascribe the whole thing to misunderstandings and honest mistakes.

With this latest political move, SPSR had started to take on the foul air of a sell out. Far from pushing NASA to keep public commitments or questioning its behavior, they seemed to be spending more energy dragging their feet and attacking the agency's harshest critics. Nowhere in their report did they push NASA director Dan Goldin to keep his promise to "continue taking images until everyone is satisfied."[108]

The whole issue had taken on added importance because of additional research Hoagland had been doing on the original three MGS Cydonia images. In going back over the original Cydonia Face strip, MOC 22003, he'd taken a closer look at an oddly symmetrical mesa just south of the Face [Fig. 7-1].

It is highly unusual for any kind of natural formation—let alone a mesa hundreds of feet across—to erode symmetrically. However, this sort of fractal degradation is exactly what one would expect to see from an artificial object of the type proposed for Cydonia. Even more interesting, just beneath the southern edge of this odd mesa was something else—a seemingly collapsed, eroded tetrahedral pyramid.

The only part left of this fractally-eroded pyramid was the lower portion of the southern Face, a bit of the left sidewall, and a bright apex node pointing roughly north. Still, it was fairly easy to reconstruct the original shape from the surviving faces and equidistant corner nodes. Hoagland also noted that his proposed reconstructed apex lay on a line 19.5º off of the central "Face-D&M" line in the geometric relationship model. Another interesting coincidence?

All of this paled however, to what he found next. As he was studying the lower portion of MOC 22003, the part that covered about one fifth of the northern portion of the D&M, he noticed something strange. As he worked to enhance it, it became more obvious what it was [Fig. 7-2].

It was writing. Hebrew-Arabic script. On the D&M Pyramid.

Letters From Mars?

Hoagland's mind raced. He'd been working for months, hoping to find more proof of what he'd suspected from the beginning; that the Catbox image was a second generation fake, or at best a degraded copy. Here, he felt, he finally had what he was looking for. Specific words were difficult to make out, but one of them might have been "Barsoom," the name for Mars in Edgar Rice Burroughs' novels, and also the name of Michael Malin's MSSS website.

Before he could act on this new data, he got an excited call from SPSR's Tom Van Flandern. He'd found the letters too—before Hoagland had.

This complicated matters, because Hoagland wanted to go public with the information, while Van Flandern wanted to hold the letters for future use. In the course of the discussions, Hoagland discovered that the official position inside SPSR was that these symbols were features embedded on the D&M Pyramid itself. They steadfastly refused to consider the alternative—that someone placed these symbols on the image at Malin's lab, or even JPL. As a result, they were forced into a totally absurd scientific position.

Let's consider this for a minute. If these are really letters on the D&M, then that has all sorts of implications about who put them there. The first assumption must be that Martians—three million years ago, at least—used Hebrew-Arabic lettering in their ancient communications. And not only that, but they scrawled these modern English language letters on the D&M, like some sort of cosmic graffiti, just for *us* to find, literally millions of years later. And that these "symbols" just happened to be oriented on the pyramid in such a way that they could be read without even having to rotate the image to make them appear right side up.

A far more plausible explanation is that somebody at MSSS or JPL simply *put* the letters there. These letters, which are obviously on the original image on the official NASA site, were put on the D&M, an object of obvious interest to us, as a message. They are clearly meant to confirm that the image has been altered.

Whether this was done as a whistle blowing-move, or simply to rub in the impunity with which MSSS and JPL felt they could manipulate the data, this is clearly a far more logical explanation than SPSR's anonymous "Martian graffiti artists." Yet, SPSR (through Van Flandern) refused to budge from their ridiculous stance that these symbols were genuinely on the D&M. Beyond that, since they had found them first, they requested that Hoagland not make mention of them. Out of respect for their priority, he agreed.

So Hoagland literally had the smoking gun he had sought to prove that JPL and Malin had been tampering with the Cydonia images, and he was unable to use it. It was small consolation that SPSR claimed they were making headway in an effort to get their papers on Cydonia published in either *Science* or *Nature*.

As a result of this fiasco, he decided he could no longer trust the members of SPSR. They were now so entrenched in their "honest but stupid" model of NASA that they couldn't see proof of NASA's duplicity, even when it stared them in the face. However, that stance would be sorely tested by the events of the next few years.

Oh My God! They Killed MARCI!

On September 23, 1999 at 9:06 a.m. UTC, JPL lost contact with the $125 million Mars Climate Orbiter (MCO) as it passed behind Mars on its final orbit insertion burn. The spacecraft was carrying the Mars Color Imager (MARCI) camera under the control of Dr. Michael Malin. JPL scientists were unable to re-establish contact with the spacecraft when it should have re-emerged from behind the Martian disk, and immediately began a series of emergency measures to communicate with the spacecraft. As with the Mars Observer in 1993, these efforts failed when it was discovered that a navigation error had likely plunged the spacecraft into a terminal crash dive into the Martian atmosphere.

Mars Climate Orbiter was part of NASA's "faster, better, cheaper" Mars Surveyor 1998 program. Teamed with the Mars Global Surveyor and the upcoming Mars Polar Lander, it was to give NASA an unprecedented opportunity to study the geology and environment of Mars, not only from orbit but also from MPL's 195 (19.5?) longitude southern pole landing site.

According to a terse press release,[109] the spacecraft was thrown off course when one navigational team in Colorado and the other at JPL used two separate measurement systems (metric and imperial) in key navigational calculations. The Lockheed/Martin team transmitted acceleration data in the English system, and controllers at JPL assumed that it was metric. According to press reports, this had been ongoing since the December 1998 launch.

This highly implausible and bizarre set of circumstances caused some minor upset on Capitol Hill, but otherwise created nowhere near the firestorm that surrounded the loss of Mars Observer. Still, this very suspicious and convenient "error" had all the usual earmarks of another NASA ritual killing.

The notion that this error could have been induced from the beginning of the mission and gone unnoticed is ridiculous. The considerations of such an undertaking, the relative positions of the two planets at launch, their relative speeds as they orbit the sun, their rotational speeds and the gravitational effects of not only the two major bodies but also of most of the other objects in the solar system—all must be accounted for in the elegant dance called "celestial mechanics." While exceedingly complex, these factors are also exceedingly well-defined and predictable.

This is why all missions are set up on the concept of waypoints. As the spacecraft travels the millions of miles from one world to the next, it passes a series of check points at which its position, speed and trajectory are checked against the predictions. Any deviation is immediately noted and a course correction burn is initiated as necessary—and the farther away and faster it goes, the more pronounced an error would become. According to AP:

"The bad numbers had been used ever since the spacecraft's launch last December, but the effect was so small that it went unnoticed. The difference added up over the months as the spacecraft journeyed toward Mars."

To anyone with the slightest understanding of measurement systems and orbital mechanics, this statement, apparently sourced from inside NASA, is ludicrous. The conversion factor from pounds of thrust (imperial) to Newtons (metric) is 4.44 Newtons to the pound. This means that from the beginning of the mission, the velocity calculations should have been off by some 75%! And nobody noticed this "minute" error?

In fact, by the time of the MOI burn, the spacecraft should not have been anywhere near Mars. That is why NASA's explanation (excuse) is so unbelievable. If there were any error introduced and not noticed by the "honest folks" at JPL, it would almost certainly have had to come at the MOI burn itself.

Even Van Flandern, SPSR's resident anti-conspiracist and defender of the system, was forced to admit the obvious:

"I wouldn't take the cover story literally—it's just an excuse the public might be able to understand and sympathize with, but with little connection to reality. We'll find out the truth when the outside investigation releases its report."[110]

But there are, as always, other reasons to be suspicious of JPL's motives and explanations. In watching the stars above JPL at the moment that the fatal burn was initiated, we see the expected Masonic/Egyptian ritual stellar alignments. Alnilam, the center belt star of Orion/Osiris, the Egyptian god of death, resurrection and the afterlife was positioned at the ubiquitous 19.5° tetrahedral altitude as the engine was fired. Five minutes later, as MCO slipped from view behind the occultation of Mars itself, Mintaka had assumed the 19.5° ritual position.

The mission patch reinforces this tetrahedral pattern for Mars Polar Lander. It depicts a triangular shape, with Earth, MCO and Polar Lander in the vertices, and Mars in the center. It even shows a hint of Mars' rich watery past, with an ocean and dense atmosphere covering half the planet. And of course the Polar Lander was scheduled to touch down at 195 W longitude when it landed.

It is the authors' position that this implies a deliberate, ritual "tetrahedral" act on the part of Malin/JPL, either to destroy MCO or to take it "black" for their own purposes. It could be, however, that the alignments were originally intended to commemorate another successful Mars arrival to the in-crowd. This would be a further indication that whatever was done to the spacecraft was a last minute decision, not an error induced months before. However, the feeble and obviously hastily conceived "metric" excuse speaks to a desperate need to take the spacecraft "dark."

So just what could be fueling this desperate need? Malin has often feigned indignance at the notion that he had anything to do with the disappearance of Mars Observer, and has claimed it cost him money and prestige. Yet, despite his complaints, as we have seen, he was subsequently given the exclusive contract for the camera on the Mars Global Surveyor spacecraft and then the Mars Color Imager. He remains NASA's one and only "Mars boy" when it comes to orbital images of the Red Planet.

This may have been the problem. MARCI was in many ways a better camera than the one on the MGS. Its wide angle camera was capable of images of the Martian surface at an average of 7.2 km per pixel. However, under optimal conditions it could get resolutions of one km per pixel, good enough to significantly illuminate the arguments *vis-à-vis* Cydonia—but the medium resolution camera was even better. According to *Aviation Week & Space Technology* (September 27, 1999):

> "MCO carried a Mars Color Imager (MARCI), designed to observe atmospheric
> processes on a global scale and study the interaction between the atmosphere and
> the surface of the planet. Medium and wide-angle coverage were to be provided
> in ultraviolet, visible and near-infrared wavelengths. The medium-angle portion
> of the system was to have a resolution of forty meters (130 ft.) in eight [sic—10]
> colors to characterize surface properties and changes in surface dust cover."

In other words, this "medium" resolution camera was capable of providing images some 20% better than Viking. The addition of multi-spectral color, infrared and UV bands would have placed the instrument on a par with the highly degraded Face image released by MSSS in April 1998. Its near infrared capability would have also given the instrument a significant degree of "ground penetration" in the images sent back.

So, just as with the disappearance of Mars Observer (remember, under intense political pressure), taking MCO and its medium resolution, full multi-spectral color camera "black" would have given Malin and co. plenty of opportunity to review just what would be revealed by MARCI or a comparable instrument (like the upcoming THEMIS imager on the Mars Odyssey 2001 mission). Was history repeating itself?

A few months later, the second half of the Mars Surveyor 1998 mission also went missing when the Mars Polar Lander simply *disappeared* during its de-orbit burn. As with Mars Climate Orbiter, there was a significant alignment over JPL that fit the ritual system. This time, we saw Regulus, the heart of the lion in Leo, dead on the horizon when the command was sent to MCO to begin the de-orbit burn. While Leo represents Horus, this was the first time such an alignment had appeared in the system. We took note of it, but did not consider it necessarily a major "hit," despite the fortuitous timing.

Following this second failure in as many missions, NASA director Dan Goldin appointed a commission headed by Thomas Young, formerly of Lockheed Martin, to investigate not only the MPL disappearance but also the entire Mars program at JPL. Then, just a few days before Young issued his report, former NASA chief debunker James Oberg published a story on UPI that accused JPL employees of knowing full well that the MPL was doomed (due to software problems related to the spacecraft's landing legs) from very early on in the mission. JPL employees rabidly denied the report, using words like "bunk," "complete nonsense" and "wacko" to describe their reactions to Oberg's charge.111

Young's scathing report was subsequently delivered to Congress, the White House and the press—in addition to attempting to identify the immediate cause of the Mars Polar Lander failure, the Report went on at length to probe an insidious series of more fundamental "management shortcomings" that had taken place at JPL. The Report underscored that the underlying reason for these shortfalls was Goldin's own decision to mandate the, "faster, better, cheaper" space management philosophy.

In response to the report's grave findings, Goldin immediately moved to institute major changes at JPL. He not only appointed a new "Mars Czar" at NASA Headquarters to oversee all future Mars exploration programs (thus taking such management away from JPL), but the man he picked—Dr. Scott Hubbard—came from one of JPL's long-term rivals within the "NASA family": the NASA-Ames Research Center in Northern California.

This was apparently only the first step in a series of new moves, designed ultimately to win back agency management authority over Mars from JPL. Based on this, we began to wonder if we were seeing the playing out of the scenario that Mark Carlotto had referred to after the initial SPSR meeting with Carl Pilcher in November 1998. Was there a growing friction between those at the agency that wanted the ground truth of Mars revealed (NASA headquarters) and those that still wanted to hold onto the idea of a cold, dead Mars (JPL)?

If it was the goal of NASA Headquarters to seize control of Mars from JPL, then deliberately allowing the lab to gradually overextend itself with "faster, better, cheaper" would, in effect, push JPL into the inevitable spacecraft failures it was now experiencing—but that left us with a fairly difficult question: if NASA headquarters wanted to usurp control over the unmanned Mars programs coming out of JPL, why didn't Goldin simply *order* it to be done? As NASA Administrator, he (theoretically) had authority over every division of his agency.

Yet following NASA's damaging previous two weeks, stretching from Oberg's initial bizarre UPI accusation to the release of the Young Report, Dan Goldin flew on March 29, 2000 to JPL itself. His mission: to address the beleaguered personnel, scientists and engineers of the laboratory, and to advise them of the new political and engineering realities, while simultaneously exhorting them to continue to new heights under more stringent NASA management.

Goldin's speech was provocatively titled "When the Best Must Do Even Better." It was in the second paragraph of his prepared text that Goldin gave the game away:

I'd also like to acknowledge Admiral Inman, head of the JPL Oversight Committee at Caltech. He couldn't be here today, but I talked to him by phone. His commitment to the team here is also unwavering. And I thank him for that...[112]

The "Admiral Inman" he was referring to was Admiral Bobby Inman, former director of the National Security Agency, deputy director of the Central Intelligence Agency, vice director of the Defense Intelligence Agency and former Director of Naval Intelligence. He was once memorably referred to by *Newsweek* as "a superstar in the intelligence community." A White House press release,[113] issued on the occasion of President Clinton's 1993 recommendation that Inman be confirmed as Secretary of Defense, noted: "As he rose through these posts, Inman won the Distinguished Service Medal, the Navy's highest non-combatant award, and the DIA's Defense Superior Service Medal for "achievements unparalleled in the history of intelligence."

So what was the nation's most celebrated spook doing heading an oversight committee at one of its leading private universities and, specifically, a committee overseeing all civilian unmanned exploration of the planet Mars? Could it have anything to do with Viking's discovery, a quarter of a century before, of a set of artificial ruins at Cydonia and the potential national security consequences of that discovery?

The blatant flagging by Administrator Goldin of Admiral Inman's name was obviously not accidental. The staggering implications of the most accomplished alumnae of the current intelligence community having legal oversight responsibilities of JPL's supposedly civilian space activities drove home all the "alternative scenarios" that we'd discussed for many years, certainly going back to the missing Mars Observer. In light of Inman's presence, the idea that these latest Mars missions also weren't truly lost at all seemed far more likely. It was reinforced even more strongly when a Caltech spokesman revealed he'd been on the committee for "at least eight years"—in other words, since just *before* Mars Observer disappeared.

This also meant, obviously, that he served on the Caltech committee while in his active capacity as Secretary of Defense of the United States. At this point, we could smell blood in the water.

Hoagland and Bell (and subsequently Mike Siegel, who replaced Art as host of *Coast to Coast AM* when Art abruptly retired from the air for family reasons) took the opportunity to use the radio program to inundate Washington, and specifically Senator John McCain of Arizona, with a fax and e-mail campaign. It had been nearly two years since the first three MGS images of Cydonia had been released, and no one had held Goldin to his promise to keep taking pictures of the region. Over that same time period, there had been at least a dozen Cydonia imaging opportunities during the primary science mapping period of the mission. McCain was on the senatorial committee that oversaw NASA, and he obviously got the message. McCain called Senate hearings on the contents of the Young report, and lambasted Goldin in front of the committee:

"If the media reports are true—that NASA *withheld critical information* from the public and elected officials—then the trust that is vital between this government and its citizens has been violated and warrants *a very serious examination of how this agency operates.*" [Emphasis added.]

And then, after a few more days of the fax and e-mail campaign:

"This report is an embarrassment to the agency. I believe it's important that this committee exercises more rigorous oversight of NASA from this point forward."

This latter comment must have sent shockwaves through the NASA hierarchy, because just two days later, on the second anniversary of the Catbox fiasco, everything came to a head. Without notice, and in violation of NASA's stated policy on Cydonia, Dr. Malin released nine previously unseen images of the region to the internet, some of which he had held on to for more than a *year*. Although Malin's "exclusive rights" contract with NASA/JPL gave him the right to withhold images for up to a one-year period, Administrator Goldin had specifically exempted Cydonia from that constriction.

Clearly, this new image release was cynically designed to curry favor with congressional critics and reinforce the idea that NASA was open and honest. But, by proving what the authors had asserted all along—that Malin had far more images of the region than he had released—it in fact showed that the system of public accountability had all but broken down at JPL.

Reconstruction of the orbital parameters revealed fifteen science mapping imaging opportunities for Cydonia in the previous two years, not to mention the Science Phasing Orbit opportunities between October 1997 and May 1998, when MGS was passing over Cydonia every nine days, creating several dozen more chances to take images of the area. So the question hung there: If there had been fifteen imaging opportunities and Malin had released only nine, what had happened to the other six?

Malin, in various scientific meetings over the previous two years as well as in a published interview in Smithsonian Magazine (September 1999), had gone to great lengths to describe the "terrible arm twisting by NASA Headquarters" that ultimately forced him against his better scientific instincts to acquire the three original MGS Cydonia images in April 1998. In these public protestations, he also vowed (because it was "simply awful science") to "never do it again"— despite what NASA had agreed to:

"His least favorites (images) are the ones NASA ordered him to take of the so-called Face on Mars... According to Malin, it cost $400,000 to take the new pictures. There were other targets that could have been viewed on that same orbit, including volcanoes on Elysium that would not likely come into view again. 'Does the government spend money on ghost research?' Malin asks. 'Or the Loch Ness monster? Or the lost continent of Atlantis? I think the Face was a kind of stupid thing to spend money on.'"

Yet here we had found, when push came to shove, that he had been privately imaging the hell out Cydonia, despite the fact there was virtually no public pressure to do so.

Indeed, several of his nine new images had required Malin to point the MGS camera "off nadir"—essentially taking an image at an angle other than straight down. Since MGS is a "nadir pointing" spacecraft and Malin's camera cannot be gimbaled, the entire spacecraft must be rotated in order to take an image of anything on the Martian surface that is not directly below the spacecraft.

He cannot do this alone. While Malin has total control over the imaging sequences of the MGS camera, he has no say in how the spacecraft is oriented to acquire those images. In order for him to have even attempted to obtain many of the new images, he would have had to submit his request through channels at JPL and have the approval of project managers to spend the money to have the commands worked up and then transmitted to the spacecraft. An off-nadir image of this sort involves a fairly convoluted dance between Malin, mission planners at JPL and the uplink dishes required to send the orders to control the "momentum wheels" which actually reorient the spacecraft. So, JPL would have been aware of several attempts by Malin to re-image Cydonia.

After all this, were we really to believe Dr. Malin—who was obviously a bald-faced liar when he feigned no interest in Cydonia—if he claimed that he just didn't take the other six pictures? How plausible was it, really, that he just skipped six opportunities while in the midst of his Cydonia photography binge?

Anecdotally, both Vincent Dipietro and "Communion" author Whitley Strieber stated on *Coast to Coast AM* that JPL scientists told them that Malin had been taking so many pictures of the Face in the last two years that they could not get time on the MGS for their own research. If this was true, and Malin's obviously overwhelming interest in Cydonia makes it likely that it was, where were these images? And even more, if these images clearly showed the Face to be natural, why would he not release them immediately? After all, if they supported his public contention that "this is all nonsense," he could have ended the debate long before by simply putting out the images of these supposedly "natural" features—but he hadn't.

The nine images themselves pretty well blanketed the Cydonia complex. Malin, despite his earlier claims of targeting difficulties, had no problem nailing several of the key features of the region including the Tholus, the Fort and the majority of the City—but he seemed to have completely missed the Face itself. According to the image maps, he did make an attempt to image a portion of the Face in mid-February 2000, but a sequencing error caused the loss of most of that day's data.

What he did get was more than a little interesting. He managed to get several more of the tetrahedrally arranged mounds, and a direct, high-resolution hit on the Tholus, a rounded object completely at odds geologically with the rest of Cydonia. Earlier shape-from-shading computer enhancements had shown that the Tholus had some sort of peak at the top. The new MGS image revealed for the first time what that peak actually was—a fractally eroding tetrahedron [Fig. 7-3].

This still-discernable structure once again exquisitely reinforced our previous geometric relationship model for the entire Cydonia Complex. One of the edges pointed due north, straight through the Cliff, while another was aligned with the D&M apex.

In addition to the Tholus itself, there was also a cluster of partially buried dome-like objects, just north of the Tholus on the image strip, with regularly spaced, archway-like entrances at their bases.

Also of great interest was the Fort, imaged for the first time under better, overhead lighting. What had seemed to be an angled, interior foundation with walls was now revealed to be apparently a very strange looking "mesa."

Having extracted these nine images from Malin's cold storage box, we weren't about to let up on the gas pedal. Hoagland quickly arranged a public lecture in Senator McCain's back yard, Scottsdale, Arizona, and invited the senator, his wife and staff to attend. Ironically, just as Hoagland would take the stage at the Scottsdale Center for the Arts around 2:30, May 7, 2000, MGS would be passing over Cydonia with an ideal chance to take a direct, almost perfect overhead view of the Face under ideal lighting conditions.

We don't know if Dr. Malin ever took an image of Cydonia that day, because one has never been released. Perhaps he was too busy.

As it turned out, what we did get, two weeks later, were *twenty thousand* previously embargoed images of Mars. A quick scan of the ancillary data showed that a great many of the images were actually prepped for the web on Sunday, May 7, 2000, literally as Hoagland was making his presentation to the public. Although no new views of Cydonia were among them, this massive image release (and a subsequent release of thirty thousand more images later that year) provided a treasure trove of new data to scour—and Mars did not disappoint us.

The Glass Tunnels of Barsoom

Almost immediately, the authors, as well as numerous amateur researchers, began to find extraordinary anomalies all over Mars. Ranging from what appeared to be pools of standing water (with waves) to long tubular constructs with supporting cross-members, to meandering streams and rivers, to entire towns full of block like buildings and symmetrical installations, there was seemingly no limit to the oddities of Mars. Near the South Pole, an image strip was found which seemed to show lush, growing vegetation [Fig. 7-4].

"Arthur's Bushes," as they came to be known, bore a striking resemblance to terrestrial banyan trees, and images showed them growing and receding as the Martian summer waxed and waned. They were named for English visionary Sir Arthur C. Clarke, probably best known to followers of our investigation for taking Richard C. Hoagland's ideas about life in the oceans of Europa and using them to create his novel *2010*, the long awaited sequel to *2001*.

There was another image, even more extraordinary, which had caught Clarke's eye even before the "banyan trees"—found by Hoagland, the image is located in an ancient ocean bed which has rifted apart due to some sort of cataclysmic stress, and appears to be nothing less than a series of interlocking, reinforced and still intact translucent, glass-like tubes. [Fig. 7-5]

Reinforced by regularly-spaced, cylindrical arches, this clearly defined translucent structure seems to be running along a hollowed out section of the former ocean floor. The clear "glass" tube can be easily seen running the length of the rift, and there is a distinct edge where the clear tube wraps around the arches. The composition of the tube is given away even more directly by a brilliant specular reflection. This reflection is not associated with any kind of geologic feature (it seems to be simply hanging in space), effectively demolishing the argument that "wind polished rocks" are responsible for the many brilliant features of the Martian surface.

Critics in the past attempted to pass off similar arches as "sand dunes." To be sure, there are some superficial resemblances between these "arches" (and similar structures near the base of some pyramids at Cydonia) and real sand dunes—but on any sort of close examination, the "sand dunes" argument quickly falls apart.

We quickly found many images of real sand dunes on the surface of Mars. They are irregularly spaced, vary in length, have diffuse edges and are the same color and texture as the surrounding terrain. They are also restricted mostly to flat, wide plains, and are not parallel to each other, even when the topography does not interfere with wind patterns. By contrast, the arches on the glass tube are regularly spaced, nearly identical in length and breadth, and wrap around the surrounding features. They have completely different albedo properties than the surrounding terrain (indicating they are made from different material), and are restricted to the specific area of the glass tunnel. Note also that they are sharp-edged and tubular, suggesting that they are individual structural features rather than drifting mounds of piled-up sand.

After seeing the images on Hoagland's website, Clarke made a fuss about them at an event in December 2000 hosted by Carl Sagan's Planetary Society. Clarke's flat statement, that he believed the images constituted evidence of life on Mars (he was thinking of the "glass tubes" as potentially fossil remains), made Society chairman Louis Friedman acutely uncomfortable—so much so, in fact, that he abruptly ended the video conference with Clarke. Friedman is remembered by

Monuments readers as the guy who refused to look at images of Cydonia while attending a meeting with Carl Sagan, John Brandenburg and Dr. David Webb of the Mars Investigation Group, at one point hiding his own face behind a stack of books to prevent himself from seeing the images of the Face. Sagan eventually gave up trying to get Friedman to look, and then told Brandenburg and Webb that he would deny the meeting took place if he were ever asked about it.

Clarke upped the ante a few months later, during a visit to his Sri Lanka home by Buzz Aldrin, the Apollo 11 astronaut. Clarke used the occasion, covered by Space.com, to reinforce his earlier statements, saying, "I'm fairly convinced that we have discovered life on Mars. There are some incredible photographs from [the Jet Propulsion Laboratory], which to me are pretty convincing proof of the existence of large forms of life on Mars! Have a look at them. I don't see any other interpretation."

NASA's staff geologists were swift to respond. First, they attacked Clarke personally; second, they attacked the images themselves; and thirdly, Dr. David Pieri and others attacked the authors, implying we had altered the images. Their counterclaims were quickly disproven, however.[114] The sole remaining issue to be addressed was the question of light direction. Using photoclinometry (shape-from-shading), a pair of independent researchers ran some experiments with a ribbed plastic bottle to simulate the "tubes." The resultant experiments ended with a very close match to the tube in the visual image, strongly implying that the glass tube was just that, a convex, 3D cylindrical tube. Amateurs soon found many other examples similar to the original glass tubes in various MGS images. However, just as the debate over their reality reached a crescendo, Dr. Malin surprised us again...

An Eye for an Eye

On the last day of January 2001, with no warning to anyone in the planetary science community or the independent investigators, Malin Space Science Systems principal investigator Michael Malin released a close-up image of the western half of the Face on Mars.[115] In the same image batch, Malin (also for the first time) released an image of the Cliff, another anomalous feature of the Cydonia region [Fig. 7-9].[116]

Initially it was very difficult to determine when the image was actually taken, since the normal ancillary data was not linked to the page. It was not until several months later that the ancillary data was actually posted to the page, and it revealed that Malin had taken the Face image back in early March 2000. It was during this time that he and NASA were under intense pressure from Administrator Goldin and Senator McCain, but somehow he had neglected to include this image in his April 2000 Cydonia data dump.

By not issuing a notice that the image would be taken and then withholding it for almost a year, Malin was once again in violation of NASA's stated policy on Cydonia. In fact, it could be argued that he was in violation on seven counts, since he released six other Cydonia images taken between March 2000 and January 2001 at the same time.

Despite the fact that this was undeniably the best (though partial) view of the Face yet, there were problems. The stated resolution of the image (1.7 meters per pixel) was not exactly the whole story. As with the previous MGS view of the Face, there is a large amount of noise in the image, suggesting that the full range of contrast was not made available to the MOC. Since actual image resolution is a function of both spatial resolution *and* contrast range, the actual image resolution is more like five to six meters per pixel. This same problem on the previous 1998 MGS Face image had the effect of reducing the *actual resolution* to around fourteen meters per pixel, as opposed to the stated resolution of around five meters per pixel.

What all this induced noise does is make it more difficult to discern the fine structure of a given feature. And the basically overhead sun angle also has the effect of washing out details. That said, this new image was still remarkably revealing.

More than nine years before, former NASA imaging specialist Vince Dipietro proposed that his new analysis of the Viking Face images showed the presence of not only what appeared to be an "eye socket," but also evidence of a "pupil" of the right size and shape to be a representation of such human features [Fig. 7-7]. Despite the fact that other researchers using different imaging techniques found the same feature, his prediction was ridiculed at that time by individuals both inside and outside of NASA, and his "bit-slice" imaging technique was roundly criticized.

Now, it seemed, Dipietro would have the last laugh.

The most noticeable thing about the new image (a narrow swath cutting across the center forehead region of the Face and down across the right eye socket to the corner of the mouth) is that what appeared to be an actual "eye socket" and "pupil" in the earlier Viking and MGS images are, in fact, just that. The "eye socket" was perfectly shaped and positioned to represent a human eye (even including a tear duct) and even though the outlines of the socket are somewhat faded from the sun angle and lack of contrast, it took very little imagination (or enhancement) to determine just what the original shape truly was, and it was easy to artistically enhance the clearly present and obviously genuine contours of the socket to reflect them as they might have appeared in their heyday [Fig. 7-8].

Normally, these kinds of interpretations are dismissed as just that, an interpretation, but in this case absolutely no "fudging" was required. Simply emphasizing the lines that are flatly present on the structure produces the extraordinary result—and the presence of the "pupil," so controversial previously, could now be placed alongside most of the other predictions of the various independent Cydonia researchers, as proven beyond a reasonable doubt.

It's an eyeball.

In addition, we could also see (in spite of the noise) some incredible details in the fine structure of the Face as well [Fig. 7-9]. Around the eye socket was a set of very regular, geometric shapes that appeared to be a sort of honeycomb cellular structure on the Face itself. This very anomalous and decidedly artificial pattern is exactly what Hoagland had predicted (in *Monuments*) we would find on the Face when we eventually got a good enough look. He had argued that the Face was not just a Mt. Rushmore-type re-carving of an ancient Martian mesa, but a 3D, architectural, "high-tech" construct. That, with high enough resolution, it would begin to reveal precisely those necessary (though now badly eroded) architectural details of which it truly is composed.

So once again, the higher resolution images had confirmed earlier predictions of the artificiality model. However, after nearly three years with MGS in orbit around Mars, we had grown tired of Malin dancing around the subject at hand. It was time for a full-on high resolution image of the Face, and we were ready to push for it.

Chapter Seven Images

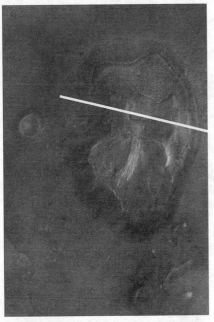

Fig. 7-1 - Symmetrical mesa and tetrahedral ruin, from MGS image 22003. Line of symmetry added.

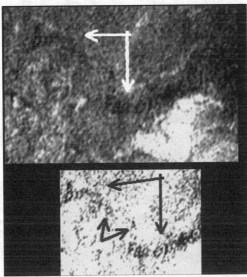

Fig. 7-2 Two versions of the "writing" on the D&M, from NASA\MSSS image 22003.

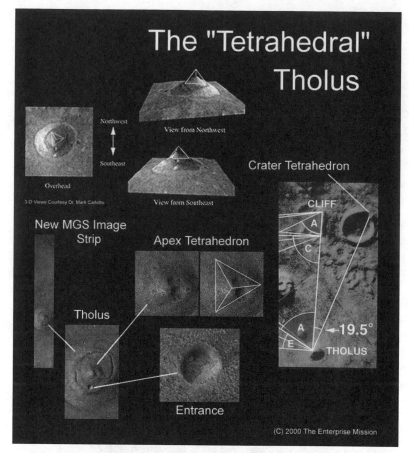

Fig. 7-3 – EnterpriseMission.com poster, showing high- resolution images of an ancient, heavily-eroded tetrahedron atop the "Tholus" in Cydonia. The structure is also located at the "inscribed tetrahedral angle" – 19.5 degrees – in relation to another tetrahedral Cydonia structure (on crater rim – right) and a linear feature ("A") called "the Cliff." (JPL/Enterprise Mission)

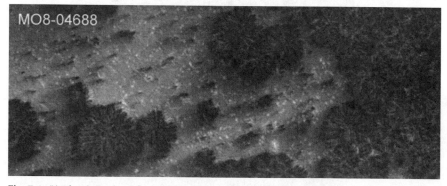

Fig. 7-4 - "Arthur's Bushes" from MOC—MO8-04688. (NASA\MSSS)

Fig. 7-5 – "The Glass Tunnels of Barsoom," from NASA\MSSS image M04-00291.

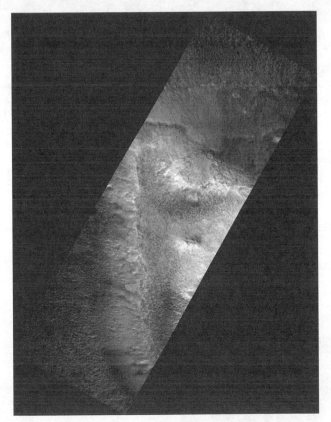

Fig. 7-6 – Full swath of image M16-00184.

Fig. 7-7 - Bit-slice enhancement of Viking frame 35A72 by Vince Dipietro. Dr. Michael Malin and NASA argued that the "pupil" was an artifact of the enhancement process.

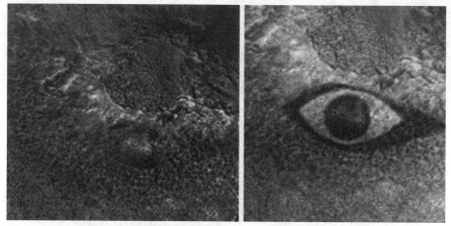

Fig. 7-8 – Enhancement of M16-00184, showing the right side "eye socket" of the Face in high resolution (L) Artistic enhancement to emphasize obvious features (R).

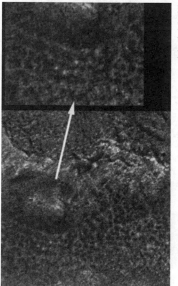

Fig. 7-9 – Close-ups showing regular patterned "honeycomb structure" around the "eye" itself – exactly as predicted by Hoagland in "Monuments."

Chapter Eight
FACETS and the Face

"All government agencies lie part of the time, but NASA is the only one I've ever encountered that does so routinely." – George A. Keyworth, Science Advisor to President Reagan and Director of the Office of Science and Technology Policy, in testimony before Congress, March 14, 1985

Just as we prepared to jump back into the political game with NASA, SPSR beat us to it. We had heard through the grapevine that Stan McDaniel, the erstwhile leader of the organization, was contemplating "retirement" from the Cydonia problem. Word was that he felt he'd been snookered—and publicly humiliated—when NASA had released the Catbox image after he had pronounced their November 1997 meeting a "breakthrough in communications" on his website. Evidently faced with either reversing his position on NASA again, or simply withdrawing from the game, he'd chosen the latter option. This void left Dr. Van Flandern as the de facto leader of the self-described "serious Cydonia researchers."

His first act in that capacity was to call a press conference at the Washington Press Club on April 2001, the third anniversary of the Catbox fiasco, after the journals *Science* and *Nature* had rejected all of their Cydonia papers—and informed them that the question of extraterrestrial artifacts was on a short list along with UFOs, Bigfoot and the Loch Ness monster as subjects that would never be published in their pages.

He spent nearly 90 minutes—after paying for the room for two hours at a cost of around $10,000—droning on about various aspects of the Cydonia research that anyone who had heard him on the Art Bell show had heard many times before. Of course, since Van Flandern himself had only about three years of active participation in the issues of Cydonia, the presentation was built solidly on the work of other researchers (mostly Hoagland)—which, incredibly, for the most part he consistently failed to properly attribute.

That was insulting enough, but he then went on to discuss the writing on the D&M. Once again, he reiterated the absurd notion that these were actually Arabic letters on the structure itself. It would have been bad enough if Van Flandern and his SPSR colleagues had stopped right there—but, as if to further erode their own remaining credibility, the press kit presentation Van Flandern handed out (and actually mailed across the country) then devolved into a ridiculous series of "pictographs" supposedly present on the Martian surface. Van Flandern at least had the presence of mind to not bring these images up at the actual Press Club briefing, but their presence as graphics on the press conference web link, and in the hard copy press kit, was an abominable political move.

These supposed "pictographs" represented what SPSR's best minds *imagined* that they saw, because no rational person could convincingly argue that they are actually *on* the Martian sands. Everything from a "scorpion" (actually a collapsed structure found by Hoagland in 1998)[117] to a "child" to an "antlered animal" to a "dolphin" to (we're not kidding here) "Nefertiti"— was found on Mars, according to SPSR's rejected "scientific paper." There was seemingly no limit to what these guys imagined that they saw without *any* collateral mathematical context or substantiation (unlike Cydonia)—and as anyone with any experience with media can easily attest, there is no quicker way to bury yourself with the press than to wildly speculate along the lines of this demonstrable projection.

No wonder *Nature* and *Science* refused to even *consider* SPSR's "scientific" paper on Cydonia. SPSR had become so intellectually trapped by their insistence that there were no NASA conspiracies that they had now publicly embarrassed themselves. Luckily, thanks to the Catbox image three years before, virtually no major press showed up at the "event." That didn't stop Van Flandern from revisiting the Catbox in order to assign the most benign motive to Malin's duplicitous behavior around its acquisition, and JPL's participation in its creation. SPSR's semi-official position was that yes, the Catbox image *was* deliberately degraded, but it was only done by JPL to "protect their funding." This dubious (if not ridiculous) position is based on the idea that admitting the truth about Cydonia would inevitably result in a manned mission to

Mars, and in such a venture JPL would be out of the funding loop. They cite the Apollo program (and the cessation of the unmanned lunar probes a few years before) as a stark example of how this would come to pass, then point to the lack of unmanned lunar missions *after* Apollo as the clincher—except, of course, their notion of history is demonstrably wrong, and their reasoning intellectually vacuous.

The simple fact of the matter is that the Ranger, Surveyor and Lunar Orbiter programs were not run independently of the manned program at NASA, but to support it. Their sole function (after Apollo was announced) was to map and examine the lunar surface to prepare for the manned landings a few years later. A Mars-manned program would presumably follow the same pattern.

In reality, a commitment to a manned Mars program would be the best thing that ever happened to JPL. They'd have more work than they could handle, sending probe after probe to map the Martian surface (as much land area as all the continents of Earth combined) to pave the way for the manned landings, just as they did in the heady days of Apollo.

It was now obvious to us that the SPSR crowd would do *anything* to avoid admitting the truth—that JPL was deliberately covering up the evidence of artifacts on the surface of Mars. To do that would be to admit that ours—and not theirs—was the correct model for the motives behind NASA's twenty years of aberrant behavior *vis-à-vis* Cydonia. And that, apparently, was politically impossible.

This absurd refusal to acknowledge that JPL's resistance to getting good images of the Face is rooted in a deep imperative to maintain political control over public reaction to the "unthinkable reality" of artificial structures on Mars led the authors to sever all ties to the group. We printed a scathing review of the press briefing on the Enterprise Mission website, and while Hoagland would continue to support Van Flandern's work on the Exploded Planet Hypothesis and other areas of agreement, we decided to go our own way on Cydonia.

FACETS and the Face

Coincident to SPSR's disastrous press conference, Dr. Malin marked the three-year anniversary of the Catbox by releasing another batch of 10,230 hi-resolution Mars images to the internet. Included in this release were three more images taken in and around the portion of Cydonia that included the Face and other artifacts. Unfortunately, Malin had managed only to get another partial image of the Face, but just miss (again) the still largely unseen eastern half of the monument. Somehow, the man who seemed able to target objects like the Cliff (which is *narrower* than the Face) with pinpoint accuracy just kept missing this most crucial piece of Cydonian real estate—the Face itself.

A clear shot of the eastern half was crucial to settle another area of disagreement with members of SPSR. In various publications they had taken the staunch position that the Face was a symmetrical human visage, while Hoagland had speculated (and predicted) as far back as 1992 (at the UN) that the eastern half had a *feline* aspect. In fact, a JPL source had recently confirmed to Hoagland that the eastern side of the Face *did* possess this puzzling "feline" aspect.

By the early 1990s, Hoagland had come to the conclusion that the Face was significantly asymmetric. While broad features, like the platform and the two visible "eye sockets" were generally aligned, Hoagland decided upon close examination of the original Viking data that the overall features would be significantly asymmetric when new imagery of the entire feature was obtained. Various possibilities for this apparent divergence were bandied about among the other Cydonia researchers at the time (including that it was not a face at all, or that the right side was "significantly" eroded), while Hoagland began seriously thinking that such asymmetry was actually *planned*. Kynthia Lynne, the Enterprise Mission art director, was in the process of sculpting successive 3D models of the Face in this time period. She saw—and even modeled—the same asymmetry, but was uncertain of the cause. It was only years later, after the acquisition of the 1998 Catbox image, that Kynthia—working to bring her 3D analog Face sculpture into conformity with the new data—became a convert to Hoagland's specific asymmetry ideal, that the right-hand (Cliff) side was specifically intended to represent a "lion."

Even afterwards, however, a few of the other scientists working the

Cydonia problem continued to argue that the Face *had* to be symmetrical, and attempted to persuade Kynthia to re-sculpt her model to conform, as a "valid reconstruction of the original design." As previously noted, Hoagland hadn't bought that the "original" shape was anything like a symmetric form—and, more importantly, not necessarily even *human*. One key reason was an experiment that he'd conducted; Hoagland had taken a series of cutouts of large photographic blow-ups of the Face from the Viking data and made himself two faces—one mirrored from the western or "City side," and one mirrored from the eastern or "Cliff side." What he found astounded him [Fig. 8-1].

When the two "city halves" were put together, they created a distinct (if primitive) proto-human form—a clear "hominid" appearance. When the Cliff side halves were placed together, they created the markedly feline image on the right. Hoagland later made a major point of this during his 1992 UN presentation, and included the feline side prediction in all subsequent versions of his book *The Monuments of Mars*. When the first MGS image was released in April 1998, Hoagland again went on nationwide radio and television, reiterating his position that the Face was *two distinct* Faces, and that one was feline. He even posted the old and new images pointing out feline characteristics of the Face on the *Enterprise Mission* website at the time.[118]

So while a full daylight overhead view of the Face might not exactly be the holy grail of Cydonia research, it could certainly go a long way toward deciding yet another crucial aspect of the argument. Was the Face a symmetrical human visage, or did it have some other, deeper and even more mysterious message to send us?

The question now was how to extract an overhead image from Malin and the lab when they seemed so intent on preventing us from having it.

Into this void stepped Peter Gersten and David Jinks. Gersten, who as the lead counsel for Citizens Against UFO Secrecy (CAUS) had fought and won previous actions against the U.S. government under the Freedom of Information Act, was designated as the lawyer for a new group, FACETS, the Formal Action Committee on Extraterrestrial Studies. Jinks, an anomalist and author, put forward a substantial amount of money to contract Gersten's services. Along with the authors, Gersten and Jinks formed FACETS as a new public interest lobby for those of us that wanted more from NASA and JPL than we had been given to date. It was thought that such an organization (with open membership) might have more pull with the reluctant space agency.

FACETS' first act as an entity unto itself was to compose a letter to NASA, specifically to administrator Goldin. The letter contained three specific requests, which were to be acted upon within thirty days:

1. Post on the internet any and all previously obtained, but still unreleased, images of the Cydonia area of Mars;

2. If not already obtained, vertically image the entire structure known as the "Face" at high resolution with reasonable high sun lighting, publicly releasing the results immediately; and

3. Consent to re-image five additional areas of Mars from a list submitted by FACETS.

As a last bit of motivation, copies of the letter were sent to Senator McCain, the *New York Times*, the *Washington Post*, and other major media outlets. Since we had no idea how this new initiative would be received, all we could do is sit back and wait.

Well, perhaps "no idea" is less than accurate. For months ahead of the March 16, 2001 letter, we'd been receiving hints from NASA through a variety of public and private statements that they might be amenable to such a formal request. A private source, with connections to the Bush White House, had informed us that things had changed behind the scenes at NASA, and that there were forces inside that wanted a more open policy on Cydonia. So, in reality, we were putting this alleged "new tone" to the test with our FACETS letter.

Still, first days, then weeks, then more than a month went by with no response. Finally, on May 15, 2001, Gersten received a reply. In a letter dated May 11, 2001, from NASA deputy director Dr. Edward Weiler (essentially Goldin's second-in-command), NASA formally responded to our requests—and also shocked the hell out of us.

First, Weiler defended NASA's conduct on the Face. He denied that NASA had ever withheld any images of Cydonia (which was laughable, but it may have been what he'd been told). But then he made a stunning revelation:

> "*None of the images acquired to date by the MGS/MOC system have been withheld and indeed, several recently (April 8, 2001) acquired images,* including stereoscopic coverage of the Cydonia feature under question, have been released via multiple public websites. *In this case,*

NASA responded to the request by FACETS... by initiating a complex set of MGS spacecraft operations to ensure that the highest possible resolution images of the Cydonia 'Face' feature were acquired. These spacecraft operations require special care and only a few can be performed each day. In addition to 1.5 m per pixel... resolution images of the Cydonia feature, NASA released a stereo 'anaglyph' of the feature that allows a viewer with colored 3D glasses to view the feature in 3D. This is the first release of a 3D image of any features on Mars acquired in this resolution. Furthermore, NASA has assembled public website access of ALL MGS images acquired of the Cydonia Face feature since the start of MGS scientific observations. Given the challenges of imaging any feature on Mars (i.e., NASA has yet to find the second Viking Lander specifically), this has involved considerable effort." [Emphasis added.]

So Weiler was claiming that not only had NASA responded to our letter by targeting the Face specifically on April 8, 2001, but he was also claiming that they had already released the image (along with a stereoscopic 3D version) some time prior to his response.

Obviously, this put us back on our heels. Had we somehow missed something? Quickly, we went back to scour all the public NASA, JPL and MSSS websites, but could find no such Face image. Satisfied that it was not in the public domain, Gersten composed another letter, this time directly to Weiler, dated May 21, stating bluntly:

"My client requests that you provide it with the specific URL(s) where these new images can be found. Your statement that 'NASA has fully and openly distributed by means of public web-sites all images obtained of the Cydonia Face feature under question' seems somewhat disingenuous in light of our inability to find the new images on the internet."

Before we received a response, rumors began to circulate of a hubbub inside NASA. Our Bush administration source (let's just call him "Deep Space" from here on out) told us that a new Face image had indeed been taken, and it was sending ripples, if not shockwaves, through the agency. Hoagland went on *Coast to Coast AM* on the evening of the 23rd to inform the audience that the latest report from Deep Space was that high NASA officials (including Dr. Weiler) were meeting "late into the night" to try to decide what to do about the Face question. There were even rumblings of a press conference being scheduled for the next day. Instead, all we got was the picture.

In the late morning of the 24th of May, NASA abruptly released the first MGS high resolution, full and mostly overhead image of the Face on Mars [Fig. 8-2]. While it was still substantially off-nadir, taken at an angle off the vertical of 24.8º, as opposed to 45º for the Catbox image, it was a significantly better representation of what the Face would look like from directly overhead.

Very quickly, it was also obvious that there were a number of issues with this Face image, as there had been with the Catbox. While the image was the full resolution 2048 pixels wide, it was only 6528 long, implying it had been cropped by about two-thirds along the downtrack. While it had 175 different tonal variations (compared to only forty-two for the Catbox) this still left about 30% of the grayscale information missing. A two meter-per-pixel spatial resolution was declared for the image by MSSS, which meant that an object as small as a jetliner could be discerned from the data available. Further, it seemed to have been improperly ortho-rectified, because features that were seen to be along the centerline in Viking data and the Catbox image were now skewed to the western side. This had the effect of enhancing the asymmetry of the two sides of the object by stretching the eastern half in proportion to the western side.

Overall, however, it was a dramatic improvement over the Catbox image. What was clear from the new image was that while the Face had a substantial general symmetry, it was not (just as we predicted) a clearly symmetrical human face. Preliminary symmetries confirmed it to be exactly what the authors had predicted, a half-and-half, human-feline hybrid.

Unfortunately, our hopes for the supposed "new tone" that Deep Space had told us about quickly evaporated in the light of day. NASA released the new image amid a flurry of extremely negative public comments simultaneously posted on several official NASA websites. Specially prepared "hit pieces" were posted coincident with release of the new image. Titled "Unmasking the Face on Mars"[119] and authored by NASA (there was no byline), the article series resorted to gross distortions and outright fallacies in their attack on the image. Obviously, these were prepared days or weeks before the image release, and it was now obvious that the late night strategy sessions were *political* strategy sessions, not scientific. A scientific approach would have been to simply release the data the day it was acquired, and allow the scientific debate to take its course. Instead, we were once again treated to a calculated smear campaign obviously aimed directly at the national media.

While we were disappointed that NASA had chosen to continue the disinformation campaign they began when the initial "Catbox" Face image was released, we were hardly surprised. What did surprise us were the rather desperate lengths NASA was forced to go to debunk the new Face image.

Making a Mountain Out of a MOL(A) Hill

In "Unmasking the Face on Mars," NASA used all the standard debunking and propaganda techniques they had honed over the previous twenty years of debate on the Cydonia issue. They described the Face as a "pop icon," never mentioned the existence of any of the other anomalies in the Cydonia complex, and used a cartoon to ridicule the idea that the Face was anything other than a common Martian mesa. Jim Garvin, chief scientist for NASA's Mars Exploration Program, was quoted as saying that the Face reminded him of Middle Butte Mesa in Idaho. Of course, the article didn't contain an image of Middle Butte, making it impossible for anyone to assess NASA's integrity when making such comparisons.

Fortunately, SPSR's Lan Fleming contacted the U.S. Geological Survey and obtained an overhead view of Middle Butte, and published a comparison in an online rebuttal titled "Unmasking Middle Butte."

Any reasonably observant person could easily conclude that Middle Butte bore little or no resemblance to the Face [Fig. 8-3]. For one thing, the Face had two parallel straight edges on either side of the base that ran straight for hundreds of meters on either side. Middle Butte was just a common cinder cone. Fleming concluded his own evaluation with the comment "I think the time has come for science to start searching for a real explanation for the Face on Mars. The public's patience with the sophistry from JPL's public relations office may eventually wear thin."

However, there was still more to NASA's hit piece that needed to be addressed. In the article, NASA used a vertically compressed, grossly distorted and upside-down version of the Face, supposedly generated by a shape-from-shading algorithm. It was so badly distorted that parallel features clearly visible in the Viking overhead shots from 1976 ended up highly (and impossibly) divergent [Fig. 8-4]).

Later in the story, they linked this image to a very impressive-looking 3D color version of separate data from the MGS MOLA (Mars Orbiter Laser Altimeter) instrument—and then used these two images to "prove" that the Face on Mars is just another Martian hill.

According to the story, "The laser altimetry data are perhaps even more convincing than overhead photos that the Face is natural. 3D elevation maps reveal the formation from any angle, unaltered by lights and shadow. There are no eyes, no nose, and no mouth!"

The reality is that it is highly unlikely that anyone would recognize a picture of their own grandmother if it was stretched horizontally, flattened, compressed and shown upside down. So of course it doesn't look much like a Face.

There was one more major problem with NASA's argument: The MOLA instrument they were relying so much on has a resolution of *150 meters per pixel*. NASA was basing its entire "it's just a hill" argument on a MOLA "image" that is *six times worse* than the twenty-five-year-old Viking data. At that resolution, an object has to be about the size of three baseball stadiums to even show up. Frankly, trying to make their argument based on this "non-image" (which they *still* had to distort) was laughable and desperate.

In fact, NASA's logic and analysis was not only categorically wrong, but also *so* wrong that it could only be the result of a deliberate attempt to mislead the public about the true nature of the Face.

Because it can produce neat, colorized views like the one shown in the article, most people assumed that MOLA is a high resolution instrument. Some of our critics had even gone so far as to suggest that since MOLA has a "vertical" resolution of only a few centimeters, it is in fact a "better" and more accurate visual instrument than the MGS camera. In fact, it is nothing of the kind—and this sort of ignorance among the lay public and the press is precisely what NASA depended on to keep the Face from being seen it its true context. If that is not enough, as we have seen, they are not above flipping it upside down and backwards even in their "MOLA" presentations to enhance this deception.

What the MOLA really does is send out a series of pulses (ten per second) of laser light that "bounce" off the Martian surface and are reflected back to the instrument's receiver. This results in a single, circular "pixel" (or picture element) of data that is some 160 meters in diameter. Since the spacecraft is traveling at about 3.3 km per second, the next "dot" is some 330 meters down track (since MGS is in a roughly polar orbit), leaving a gap in the image track about 170 meters wide. While the spacing is greater at the equator and less at the poles, it still requires multiple scans to accurately define any object visually. MOLA has been operating continuously since MGS went into the primary science phase, except when it was turned off during solar conjunction. So what the MOLA has produced is one continuous string of data, consisting of a series of 160 diameter spots, with 170-meter gaps in between them—winding around the planet like a ball of twine for over two years. Sounds kind of cool, doesn't it?

It is—but what it is *not* is very specific or accurate on the scale of a mile or so. In other words, it is certainly *incapable* of "imaging" any individual object as small as the Face, nor is it anywhere near the spatial resolution of the MGS camera, even at the latter's worst. Some people have been confused by the stated "vertical" resolution of the MOLA. One particular critic at the time seemed to be utterly incapable of grasping just what the twenty to thirty centimeter vertical resolution of the instrument actually means. He even went so far as to suggest that because of that, the MOLA instrument is "better" for resolving features on the Martian surface than the MOC camera. If that were the case, they wouldn't have even bothered to put a camera on the spacecraft—thereby saving perhaps a hundred million dollars over the course of the entire mission.

Within that 160-meter diameter "dot" that we keep talking about, the MOLA can discern almost no detail. Its ten quick pulses hit the ground in the area in question and return the timing data to the instrument. MOLA then takes the *average* altitude of the spacecraft above the ground within that 160-meter pixel and assigns a value to that pixel based on the average. As a result, every bit of detail within that pixel is reduced to a single point, a single value: the average spacecraft altitude above the ground being "pinged." All of the individual stuff within that 160-meter circle is completely lost. That average value for the area in question *is* accurate to within about one meter with respect to the distance below the spacecraft, but that's it.

So just how big is "160 meters?" Just how much is missed by this "precision" instrument? A lot, it turns out. 160 meters is a huge pixel diameter. It is, within about five feet, the diameter of the Tacoma Dome arena near Seattle. The Tacoma Dome can hold upwards of 23,000 people, not to mention the playing field, the facilities, locker rooms, concession stands, press facilities, plumbing, miles of wiring and enough concrete to build a fifty-mile-long highway—and to MOLA, it would just be one big blob. A dot. It *would* be able to give a very close estimate of the average distance the roof of the Dome was from the MGS spacecraft, but that is it. It could discern no *details* about the object whatsoever.

The argument has been made that 160-meter resolution really isn't that bad, that it is "only" three times worse than Viking. But remember, 160 meters-per-pixel vs. Viking's fifty meters-per-pixel, is a *150%* difference. And when you consider the scale of the Face itself, it becomes obvious just how much crucial detail is missing from these "precision" MOLA scans.

To give you some idea of the scale involved here, we have placed the Tacoma Dome—approximately to scale—next to the Face on Mars [Fig. 8-5]. As you can see, the pixel size of the MOLA is so large that a feature like the eyeball in the eye socket (which is about the same size as the Dome) would be *completely missed*; assuming that by some miracle the MOLA scan actually ran across the feature in the first place [Fig. 8-5]. In fact, one MOLA "pixel" is about the size of the pupil feature itself.

By contrast, the MGS camera, at its maximum resolution of 1.5 meters per pixel, would "see" an area thousands of pixels "square" in this same 160-meter circular space. Objects as small as passenger cars could be made out without enhancement. And each of those pixels has a specific color value assigned to it from 256 available shades of gray. Enhancement processes can use these color values to bring out even more detail, effectively increasing the spatial resolution (under certain conditions and assumptions) even further.

So to argue that there are "no eyes, no nose, no mouth!" based on such a crude instrument (MOLA) is not only scientifically absurd—it is scientifically *dishonest*. The simple truth is that MOLA is incapable of resolving a feature as small as the (Tacoma Dome sized) eyeball. And the old-fashioned MOC camera, with its 1.5-meter dimensional resolution, and a "mere" 256 shades of gray-scale resolution, is *thousands of times* more accurate.

Which brings us to the next problem with NASA (and our critics) using MOLA data to "debunk" the Face. There is, again, a general misunderstanding about just how the MOLA works. Because planets are so large, and individual features like the Face are so small by comparison, the chances of MOLA actually tracking *directly* across the Face in the course of its two-year "nominal" Mission were very small indeed. Most critics assumed that MOLA, like some kind of "scanning camera," completely blanketed the Face in a tight grid-like pattern. In fact, once again, this is completely wrong.

In looking at the unprocessed version of the new Face image, we see the CCD "pixel dropout" lines. These less sensitive pixels, of the CCD "line camera" that makes up the heart of the MOC itself, represent the actual geodetic track around Mars that the MGS took over the Face, as it acquired this new image. In an ideal circumstance, MOLA would have tracked down right across the center of the Face along those darker "scan lines" (which are offset from true north by about 5°). The actual data pixels as they would have been acquired by the MOLA in an ideal "centerline" scan would be roughly every 160 meters with the 170-meter gaps between the "pulses."

This is far different than the idealized notion that there were hundreds of MOLA data points taken across the Face. At best, there could only be between fifteen and twenty points.

But wait a minute, why couldn't MOLA have made multiple passes across the Face, and gathered enough data to accurately measure the height of the entire object in its mission around Mars? After all, hasn't MGS been in orbit for *years*? Yes—but that had only amounted to about 10,000 orbits since MGS began the Mission Mapping Phase in March 1998. This might seem like a lot of orbits, but since Mars is such an enormous place (with a surface equal to the land area of *all* the continents on Earth combined), it means that MOLA has only covered the planet sufficiently to date to leave 1.5-mile gaps between the "twine" (at the equator). At the latitude of the Face (41° N), the distance between tracks is somewhat less—probably about 0.80 miles. Since the Face is only about 1.2 miles wide, it is highly unlikely that any subsequent parallel tracks actually scanned across the formation more than twice. Since there had only been one direct overhead MOC shot of the Face released by Dr. Malin to this point—the one taken in June 2000—there could almost certainly be no more than two samplings of MOLA data taken across the Face in the course of the entire mapping mission (because the first MGS image, taken in April 1998, and the latest one were taken "off nadir," so MOLA was not used). And not only that, the June 2000 example did not track accurately across *the middle* of the Face, but was offset to one side.

In fact, we can test all this rather easily. If we assume that the unprocessed version of the latest off-nadir (~25°) Face image has *not* been cropped, then the (bore sighted with the camera) MOLA scan (*if* the instrument was actually turned on) would have been pretty much right down the center of that frame. When we drew a simple line down the center of the unprocessed frame, it became obvious that the best MOLA track would have been off to the East side, and clearly would have missed the tip of the "nose"—which is the highest Facial point. This notion—that the MOLA scan NASA used in its (mis-) representations of the Face, missed the Nose *completely*—is further reinforced by the claim made in the NASA hit piece: that the Face is "only" 800 feet tall. Previous estimates, made from reliable methods like comparative stereo images and measuring trigonometric shadow lengths, have shown that the Face is actually some 1,500 feet high at the nose tip. This discrepancy can be easily accounted for when you see that the MOLA scan that NASA actually used had to have tracked to the side of Face's highest point (the Nose)—*completely missing* the tallest feature on the Face.

Jim Frawley, the contract scientist who is credited (along with NASA's Jim Garvin) as having created the "MOLA" image used in the NASA hit piece, admitted as much in an e-mail. When asked directly if there were only *two* MOLA passes over the Face, he responded "Your [sic] right. I found just two."[120]

So, that's *two* passes: each a series of fifteen to twenty dots, *160 meters* in diameter, with absolutely *no* discernable detail about the "Tacoma Dome-sized" areas that MOLA scanned. How could NASA, from this meager data, decide that the Face was "800 feet" in height, and generate the supposed "3D mesh" to create their now infamous "MOLA image" for the hit piece? How could they further decide, from just *two* scans that missed all these *crucial facial features*, that there was, as Garvin is quoted as emphatically stating, "No eyes, no nose and no mouth!"

They couldn't.

The fact is, there is *no way* for Garvin and Frawley to have created the "images" presented in the NASA "hit piece" from the available MOLA data. Further, it is equally impossible for them to have made *any kind* of accurate determination regarding the fine scale ("Tacoma Dome-sized") features—like the "eyeball" strikingly visible in the June 2000 image and in the April 8, 2001 second detailed image. These facts are in stark contrast to how the data was portrayed in the NASA hit piece (which was reprinted and treated uncritically on Space.com and other media outlets).

The image they were passing off as "MOLA-generated data" is nothing more than a deliberately "de-resed" version of the MOC image itself. Once again, confronted with this information, Frawley admitted to the truth. "You're right on this too. Image is 99% MOC. It's made with an 'inverse imaging' program I wrote some time ago. MOLA is used for constraints." "Constraints" simply means that he used the available MOLA data to make sure he had the height-to-width proportions correct when he made his shape-from-shading image—and in reality, it's more like the image is 99.99% MOC.

But why quibble? The key point here is that NASA had made *outright false claims* about the image they presented to the press and public as specific MOLA data, and compounded that lie by pretending that the instrument could resolve more than it actually could. When that wasn't enough, they flipped the image upside down and stretched it to ensure it was totally distorted. To be fair to Frawley, he simply produced the image he was requested to produce. He had no control over how Garvin and the NASA hierarchy used and distorted that information to serve their own partisan political purposes.

And make no mistake; this article was all about politics. As we have shown, there was no science in Garvin's MOLA claim at all. In fact, like his other statement comparing the Face to Middle Butte Mesa (and then not even producing an image of the mesa to support his claim), Garvin has been shown to be lacking either the intelligence or the integrity necessary to carry out his duties as director of NASA's Mars projects. He is either ignorant of the capabilities of his own instruments, or was engaging in a deliberate deception. Either way, the reality is that the MOLA claims were not only false, but they were calculatingly designed to "scotch this thing for good," as one unnamed JPL scientist put it after the 1998 Catbox fiasco.

Why does an open, honest agency that is *so sure* that the Face is "not exotic in any way," need to create the Catbox three years previously at all? And why did they need to embargo this new 2001 image for almost *two months*—while they built up a carefully orchestrated smear campaign against it? And why would they try to pass off data that is six times less precise than the original twenty-five year-old Viking images, to make their case?

The answer is; they wouldn't. But by this time, we had long since given up on the idea that NASA was open or honest.

Yes Virginia, it Really Is a "Catbox"

"And it [the temple] was made with cherubims and palm trees, so that a palm tree was between a cherub and a cherub; and every cherub had two faces. So that the face of a man was toward the palm tree on one side, and the face of a young lion toward the palm tree on the other side: it was made through all the house round about." – Ezekiel 41:18-19

Regardless of NASA's latest duplicities with the new image, we were still confronted with a fundamental problem; what did we learn from the new image and what did our new conclusions tell us about Cydonia that we didn't already know? That the Face was meant to represent two distinct species, one human and one feline, we were now relatively certain of [Fig. 8-6]. The implications of this startling new confirmation—not only for the reality of this object as a structured Martian Monument, but for its ultimate "message" to humanity at large—were overwhelmingly profound.

At the same time, we are confronted with the same quandary we ridiculed SPSR over on the "letters" on the D&M. Did the presence of a Man-Lion—on Mars, of all places—not by necessity imply that the builders, presumably ancient Martians, knew all about these two Earth-bound creatures? Didn't that imply some even more potentially preposterous answers to the questions already raised? Did lions once roam the Martian deserts? Were the original Martians humanoids, like us? And if these two things were true, what was the monumental fusion of the two on the Face trying to tell us?

Before we could address any of these metaphysical questions, we had to first confront the technical and scientific issues of the new image. This was crucially important, insofar as the Face represents the starting point for countless new or casual followers of this long scientific controversy. But simultaneously, we did not want to lose sight of the fact that the Face, at a more fundamental level, had almost become a secondary part of this debate. Hoagland's Geometric Relationship Model for Cydonia—with its potential for quantification and testing of the foundations of the "intelligence hypothesis" itself (in the form of specific predictions made by the hyperdimensional physics theory derived from that alignment model) had clearly stepped to the forefront of the debate over the artificiality of Cydonia. Because of this quantifiable basis for the model, the Face itself had been relegated to a secondary, "confirmatory" status—rather than the lynch-pin around which all decisions *vis-à-vis* the artificiality of Cydonia must (or should) be anchored.

The reason for this was that the Face, no matter how good an image we obtained, was always subject to interpretation. No matter that the new image showed fine structures that appeared to be supports for an artificial edifice, no matter that there were eyebrows and pupils and curved lips right where there should be, it was always vulnerable to one simple objection—"it doesn't look like a face to me."

Fortunately, we have an impartial arbiter that transcends the biases or *a priori* objections of any particular discipline of science. It is called the scientific method. The cornerstones of this method are specific measurement, and specific prediction.

There is a common axiom in science that reads, "you do not have a science without prediction." This is a modern (but no less correct) play on the axiom advocated by early 20th-century astronomer Sir Arthur Eddington, who inserted the measurement side of the equation into the method with his simple statement "Gentlemen, you do not have a science unless you can express it in numbers."

In this case, our prediction had been about the visual make-up of the eastern half of the Face. It stated boldly that the Face is really *two* faces: One human, one feline. But still, even if the presence of a feline side of the Face on Mars were universally accepted, that in and of itself would be meaningless without Hoagland's decade-old prediction. And in the end, there was no real way to quantify a visual interpretation. So we were left to debate the issue at a lower level: was it a Face, or wasn't it?

In a way, perhaps the name itself unfairly raised expectations that we would see a friendly, all-American, symmetrical human visage when we finally got a real good look. But we never expected that. And we said so repeatedly over many years.

With depressing unanimity, however, the news articles critical of the Cydonia investigation (in the *New York Times,* the *Washington Post, USA Today,* CNN, etc.) relied on a flawed recitation of previous claims made about the Face by other Cydonia researchers over the years. NASA's own position, highlighted by the hit pieces, is that all of the Cydonia researchers have consistently claimed that the central feature at Cydonia would be "a symmetrical humanoid Face."

In response to this long-awaited image, the independent research community responded by rolling out every excuse they could think of for why the Face wasn't totally symmetrical. These excuses ranged from declaring that the eastern half, or Cliff side, was "more eroded" than the City side to describing it as "more irregular," or even partially melted, *anything* apparently to keep from admitting that it's feline.

Some even tried to make the case that the entire Cliff side of the Face shows evidence of collapse (however, why would such an internal process be restricted to only one side...?). Reconstructions of the Cliff side eye socket and mouth area appear to support the notion that they may have once been more similar to the City side, but have now slumped inward. Subsequently, the beveled base around the upper and lower Cliff side has slumped outward slightly, from a proposed accumulation of material that pushes outward underneath the substructure. If this is the case, then it is possible that the Face did have a much more uniform left/right appearance at one time. Still, in the absence of a specific engineering analysis or, especially, a prediction that this process would produce the resulting asymmetric appearance, this after-the-fact reconstruction has little weight behind it.

Another serious problem is that whatever material the surface "casing" of the Face is made from should show serious signs of fracturing, if it has generally fallen in on the eastern half. Such a dramatic cave-in would have produced a chaotic, shattered appearance quite unlike the smooth and non-fractal appearance that we actually see on that side of the Face.

In truth, it is simply wrong that the eastern (Cliff) half is more eroded than the western (City) half. It is equally wrong that the Cliff side is also more irregular. These are clearly coping mechanisms put forth by those that expected to see a symmetrical "human" face. The reality is that the eastern half is simply less familiar than the more commonly seen western half—and, since it is decidedly feline, it is less consistent with many of the hopes and expectations of seeing a familiar, friendly human countenance staring back at us from the Cydonia plane. In reality, the new image showed that the eastern half is significantly less eroded and appears to have more of the original "casing" on it then the more weathered western half. What the problem really came down to is that the Cliff side confirms our model and not "theirs"—and that was a new scientific and political reality that many long-time researchers of this decades-long puzzle were having difficulty coping with.

The real test should have been whether the feline predictions stood up against the details revealed by the new Face image, and whether or not the Face could now validly be viewed as an eroded remnant of a once much grander Monument. The symmetrical beveled base, the rough facial symmetry and specific corresponding features (the left and right eyeballs and eye sockets, and the nostrils) all argue that even if we were wrong in our feline interpretation, the damn thing still looks an awful lot like a *Face*. And again, it is surrounded by a crucial *context*—all that other "weird stuff" also on the ground at Cydonia.

Our own preliminary analysis of the single high-resolution image NASA released May 24 had also revealed provocative evidence of structural detail. In other words, as opposed to being "carved"—like a Martian "Mt. Rushmore"— significant portions of the Face on Mars seem to be composed of highly eroded *manufactured* elements. There literally appear to be a series of still-detectable geometric *rooms* and complex supporting structures, nakedly exposed on the heavily eroded western platform of the Face.

Writing in *Monuments* in 1992, Hoagland—based on Carlotto's previous revealing fractal imaging analysis—noted that the appearance of "a Face itself" might be due to the "sophisticated placement of shadow-casting pyramidal substructures on [the] underlying mesa." In other words, that when sufficient optical resolution was achieved, the Face would be found to be a highly complex, *constructed* object whose former sophistication would now be evident by its repeating arrays of *geometric* ruins. The close-up from the May 24 image is striking confirmation of that major prediction.

Publicly, at least, the authors got very little support for our model from the independent research community. Ultimately, however, it is the predictive aspect of the "feline model" that gave it a leg up on the general collapse concept. But, in the absence of a good way to quantify our model, we were left to argue our position on the Face on much softer ground—the traditions of archeology and anthropology—rather than on the rock-solid footing of Eddington's numbers. And we had to address the biggest questions first.

What was the Face, exactly? A partially collapsed representation of a Pharaoh? Or, a Pharaoh/Lion hybrid split down the center? We obviously thought the latter—but if that's truly the case, the next (really loaded!) question must be answered: What is a terrestrial feline "half-Face" doing on a half hominid monument—on *Mars*?

That depends on how weird you want to get. Ultimately, you cannot argue that the Face is a possible monument on Mars, without spending some time studying the possible cultural significance of it *as a monument*. All monuments that we're familiar with are meant to impart a certain message—to pay homage to an epoch, or a person or an event—as a lesson or example to those who would come after. So it is with the Face on Mars.

If we can show that this alien artifact has a fundamental terrestrial connection, both in form and fact, to the practices and rites of ancient cultures here on Earth, then we can go a long way to explaining how a "Lion/Pharaoh Monument" ended up on a nearby planet. Our new model, shared (and inspired by) other researchers like Michael Cremo, Robert Bauval and Graham Hancock—is that all of the ancient advanced cultures on Earth ultimately sprang (in the form of refugees) from the same *pre-diluvial*, truly advanced root civilization. This golden age of science and technology was called the Fourth World by the Maya, the Zep-Tepi (the First Time) by the Egyptians and Atlantis by the Greeks.

So as we look to these ancient civilizations, we must question whether we see any similar examples in monumental architecture or cultural precedent to what we're now—unmistakably—seeing on Mars. It turns out that the Maya, one of the most advanced (and in some ways the most mysterious) of these early post-catastrophic civilizations, did indeed have *exact* examples of these split-faced gods.

We've found (with the invaluable research of George Haas and his colleague William Saunders[121]) that there are indeed innumerable terrestrial examples of precisely such "split faces" among the Maya—in ceremonial masks, monumental architecture, even in the classic "Mayan glyphs." In many cases, these split faces are precise *man/animal hybrids* (like the man/jaguar image)— just as Hoagland long ago proposed for the Face on Mars.

And, as the extraordinary quotations from the Old Testament at the beginning of this segment demonstrates, there is also an ancient Hebrew text describing Ezekiel's vision surrounding the statuary that would someday adorn the rebuilt Temple in Jerusalem. "So that the face of a man was toward the palm tree on one side, and the face of a young lion toward the palm tree on the other side." This quote dictates exactly *the same* Man/Lion split face imagery that we now see at Cydonia.

So, there is a major human tradition—across not one, but several human cultures—that reinforces the notion that the apparent asymmetry of the Cydonia Face is in fact *intentional*. But we think even more important is the specific nature of that union—the Man/Lion hybrid—for it uniquely speaks to a very sacred, very ancient human religious tradition.

The most obvious Earth-bound affirmation of the Man/Lion hybrid tradition is the Great Sphinx at Giza. With the head of a Pharaoh and the body of a Lion, the Great Sphinx is the ultimate terrestrial architectural expression of this deep "connection" to the ancient mysteries of antiquity—and, apparently, to Mars. Recent geological research has shown that the Sphinx most probably dates to a *much* earlier epoch than had been previously assumed, to a time when its gaze to the east would have let it bear direct witness to the rising of the sun in the constellation of Leo (*the Lion*)—to which the Sphinx is inextricably linked and identified. Most startling, the timing of this particular alignment, 10,500 BC, predates by literally thousands of years the existence of any accepted "advanced" ancient human civilization.

The constellation of Leo and the Sphinx itself were considered by the Egyptians to be one and the same. They were also both identified with a particular god of ancient Egypt, Horus.

As we've already learned, Horus was the son of the Egyptian gods Isis and Osiris, two Egyptian deities whom we have shown inexplicably appear over and over again in the mythical symbolism of the folks who took us to Mars—NASA. Horus represents the notion of "rebirth and resurrection" to the Egyptians, since he grew to manhood and defeated his uncle Set who was the murderer of his father. Afterwards, Horus re-established the good kingdom of his father Osiris to ancient Egypt, and according to Egyptian belief he was in essence "the first Pharaoh" —since all later Pharaohs descended directly from him and ruled as Horus themselves. What's even more provocative is that the Egyptians also identified Horus directly with the planet *Mars*—as they both shared a name; "Hor-Dshr," literally "Horus the Red." Graham Hancock also discovered that in its early history, the Great Sphinx at Giza was painted red— in honor of this specific Man/Lion-Horus/Mars connection. And the headdress, the one we are so used to seeing on images of Egyptian Pharaohs, is designed to represent the mane of a lion.

This Pharaoh/Lion connection even stretches into our own modern Christian traditions. Besides the startling Hebrew testimony of Ezekiel, there are additional "Man/Lion" connections at the very foundations of Christianity.

Elsewhere in the Old Testament, one of the great prophets was Daniel. During the first year of Belshazzar's rule in Babylon, at about 556 BC, Daniel had his own series of "great visions"—featuring four "great beasts." One of those eerily echoes the *same combined imagery* we've now confirmed on Mars.

> "*The first [was] like a lion, and had eagle's wings: I beheld till the wings thereof were plucked, and it was lifted up from the earth, and made stand upon the feet as a man, and a man's heart was given to it."—Daniel 7:4*

Jesus, the central figure of Christianity, had a lineage directly connected to the "House of David"—the first king of the tribe of Judah (Israel). The line that was prophesized to one day produce the "Messiah" was described in the Old Testament thus:

> "*Judah, you are he whom your brothers shall praise; your hand shall be on the neck of your enemies; your father's children shall bow down before you. Judah is a lion's whelp... the scepter shall not depart from Judah."—Genesis 49:8-10*

For this reason, Jesus was specifically known by the messianic title "Lion of Judea" in the last book of the Biblical canon—titled the "Apocalypse of Jesus," but better known as "Revelation"—Jesus' crucial role is prophesized at the End of Days:

> "*So I wept much, because on one was found worthy to open and read the scroll, or even look at it. But one of the elders said to me, "Do not weep. Behold, the Lion of the tribe of Judah, the root of David, has prevailed to open the scroll and to loose its seven seals."—Revelation 5:4-5*

In the Apocrypha (books no longer accepted into the Biblical canon), this dual imagery—Man and Lion—is also echoed... the Gospel According to Thomas contains this remarkable passage: "Jesus said, Blessed be the Lion, which eaten by man, becomes man. Cursed is the man, whom eaten by the Lion, becomes a Lion."

The Sacred City of Jerusalem itself—site of the famed Temple, and controversial modern crossroads of three of the world great religions, Judaism,

Christianity and Islam—flies a flag emblazoned with the lion image—a tribute to the symbol of power and authority behind all three. Many other official flags, such as the flag of Scotland (below, right), contain identical lion images of power and authority. One must now wonder...

Jesus was also known as the "King of Kings"—as good a description of supreme authority as you will ever find. Do *all* these Earthly "symbols of authority" extend back across an immensity of space and time to an eroded, monumental "Human/Lion" image lying on the rusted Martian sands...?

There are many extraordinary parallels between the Horus of the Egyptian tradition, and the historical Jesus. Indeed, even the traditional depiction of Mary and Jesus as "Madonna and Child" derives from earlier images of Isis and Horus.

How all of this terrestrial esoterica relates to a possible "monument" discovered by a ritually-bound space agency on Mars is ultimately to be found in the true *meaning* of the Face on Mars. The now unmistakable Pharaoh/Lion connection at Cydonia—and identical dual imagery long present here on Earth—was obviously intended to express some deep, fundamental message for the human species. Even the NASA hit piece astonishingly acknowledged that the Viking view of this Cydonia enigma bore a strong resemblance to "an Egyptian Pharaoh." Our own conclusion was that this monument was intended to be exactly what it appeared: A "Martian Sphinx"—the first Horus.

This unique redundant symbolism is now overwhelmingly apparent, the connections crystal clear—if you want numbers, the tangent of the Face's Cydonia latitude (41°) on Mars is precisely equal to the *cosine* of the Sphinx's latitude at Giza (30°).

The message of the Face on Mars is that of Horus here on Earth. It is either as a true "one-to-one" epic recreation of a specific personage on Mars, or a Monument to an idea: that the Golden Age may be long gone, but it still lives ("the King is dead, long live the King"). The literal recreations of the redundant "Man/Lion" message here on Earth—copied in increasing likelihood from their immensely ancient template at Cydonia—speak to a time of great human accomplishment and enlightenment.

A time "someone" has ever since apparently been patiently seeking to recreate here on Earth. Witness the extraordinary monumental civilizations of Egypt, Sumer, the Mayans and the rest. These attempted "recreations," however, obviously came long after whatever series of unimaginable catastrophes erased that Epoch Time, not only from two worlds, but almost from human memory itself. "Something" happened. That is increasingly obvious. Something destroyed (apparently not once, but several times) what was once a vast and far reaching solar system-wide human civilization, a civilization that left its calling cards on at least two worlds, anchored in the *identical* Pharaoh/Lion symbolism we've now identified on Mars.

The message of Cydonia (as Hoagland termed it years ago) is now apparent: we are supposed to ask, "how is this monument related to us?" and ultimately go back to Cydonia to find the answer.

And what will we find? Our own all-but-forgotten past amid the reddish sands? Or, something even more essential: a window on our coming destiny? Or simply this one essential truth: *we* are the Martians.

Chapter Eight Images

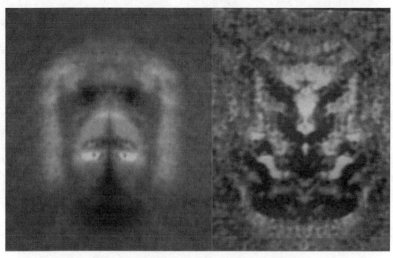

Fig. 8-1 - "Hominid" (L) and "Feline" (R) halves of the "Face on Mars'" – folded over symmetries – from 1976 Viking Cydonia data. Originally discovered and presented by Hoagland at the United Nations, 1992 (JPL/Hoagland)

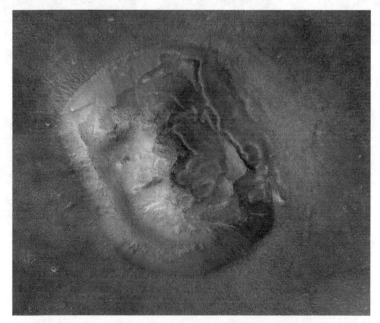

Fig. 8-2 - MGS image E03-00824. Eastern half is seen to be narrower and other features out of symmetrical alignment because of improper ortho-rectification by NASA. (MSSS\Laney)

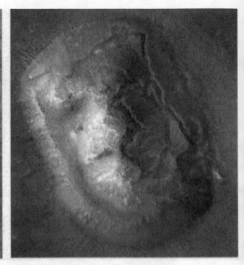

Fig. 8-3 – Middle Butte Mesa in Idaho (L) and the Face on Mars (R). Jim Garvin, NASA Chief Scientist for Mars exploration, believes there is a resemblance between the two.

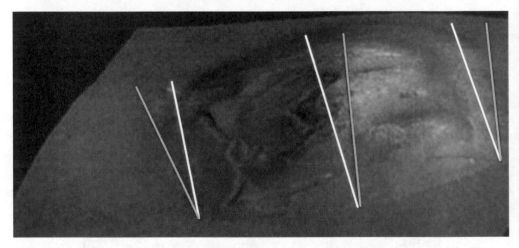

Fig. 8-4 –NASA version of MGS image E03-00824 from NASA web article titled, "Unmasking the Face on Mars." Face image has been turned upside down and horizontally stretched from normal orientation. Dark and light lines represent degree of divergence from proper rectification at three key points along the image (NASA)

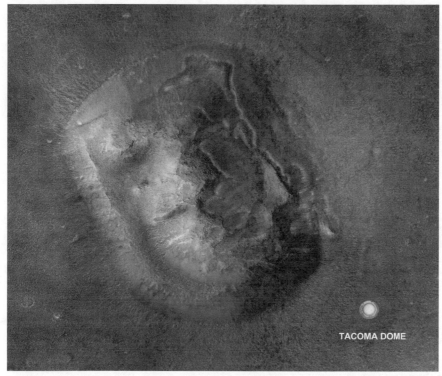

Fig. 8-5 – Scale comparison of Face on Mars to Tacoma Dome in Tacoma, Washington. Tacoma dome represents a single pixel from MGS' MOLA instrument, which NASA claimed proved the Face had no eyes, "no nose, and no mouth!"

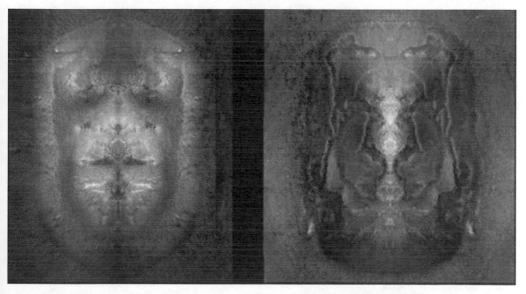

Fig. 8-6 – "Hominid" and "Feline" symmetries of the Face done with E03-00824. (Hoagland)

Chapter Nine
2001: A Mars Odyssey

Within a few days of the release of the new overhead view of the Face and NASA's "unmasking" of it, the agency began to make somewhat curious announcements about revisiting Cydonia. NASA's Lead Scientist for Mars Exploration, Dr. Jim Garvin, publicly promised that not only would more pictures of the Face be taken by MGS, but that Cydonia and the Face would also be a target of the 2001 Mars Odyssey spacecraft, due to go into Martian orbit in October, 2001. This seemed a little strange to us, considering NASA had now officially decreed the case of the Face "closed;" had gone to great lengths to debunk the possibility that the Face was artificial, and that it cost upwards of $400,000 to target any specific feature on the planet. More images from MGS would certainly be useful, but getting time on the Odyssey mission to look at Cydonia was an unexpected boon to our efforts. Odyssey had the potential to unlock even more secrets of Cydonia than the Global Surveyor had.

Odyssey carried not only a gamma ray spectrometer, which would be able to detect the underground hydrogen from any Martian reserves of ice or water (crucial for a manned Mars mission), it also carried a combined visual/infrared high-resolution camera called THEMIS, for THermal EMission Imaging System. (Interestingly, in the annals of Greek mythology, Themis was a Titan, one of an ancient race of gods whose origins are unknown but who ruled the Earth before the gods of Olympus. She was the ancient Greek goddess of justice, who was also the opposite of Nemesis. Where Themis was order, Nemesis was chaos.)

According to Garvin, this unique instrument (which had a spatial resolution capability about twice that of the original Viking cameras) would be able to distinguish the "Pyramids of Giza" from the background noise. It would do this by separating the unique spectral signature of the materials making up the Pyramids from their natural surroundings of the Sahara Desert. The infrared capability of THEMIS might also allow us to see "below the surface" of Cydonia for the first time, using the ground-penetrating capabilities of the near IR wavelengths. Certainly, we felt that if THEMIS could make out the Giza Pyramids, it could deduce the mysteries of Cydonia below the sands.

In the meantime, a new Martian controversy had broken out. Back in June of 2000, NASA announced through Michael Malin and Ken Edgett of MSSS that it had discovered evidence of water on Mars.[122] Their discovery centered around the notion that certain features found at latitudes above 30° north and south (in other words, from the those locations poleward, away from the equator) indicated the ejection or runoff of liquid water very near the surface of Mars. The ejection points were also curiously facing away from the sun. They proclaimed themselves to be completely baffled by this finding, since in the conventional model water almost certainly cannot flow at such high latitudes and out of the sunlight. According to the accepted view of Mars, the planet is so cold that any water should be frozen solid to a depth of at least six miles. Further, if there was liquid water near the surface, it should only appear near the equator, where it is much warmer, and in areas illuminated by the sun's warming rays. What perplexed them was that these findings were exactly the opposite of the ideal conditions for water on Mars. They were also surprised that the features were so geologically "young." Even though they admitted that their own model was inadequate to explain the phenomenon, they put forth an idea that the water was bursting forth from semi-permeable liquid pockets in crater walls. Yet they provided no examples of any mid-latitude, "poleward"-facing seepages.

Then, just a few days later, the authors shared a discussion about an image found by Hoagland in the vast image libraries that Malin had put on the internet. Hoagland was at first perplexed by the odd looking streak he had found on MOC image SP2-33806 [Fig. 9-1]. But in the course of the conversation, Bara was adamant that it was water. Bara based his reasoning on the visual resemblance to a hose trickling water out on a dry, dusty slope. The image shows a liquid seepage that exactly fits the Malin/Edgett model, with liquid clearly seeping from a dark crack in the crater rim and running down the slope of the crater wall that faces the direct sunlight. Add to that the fact this crater sits at about 10° N, which is precisely the kind of mid-latitude location for their model to work. Note also that the darkest portion of the flow is in the middle, where the largest amount of water would be in such a scenario.

But there was one real problem with this model. The water was so dark, so obviously fresh, that it begged the question of just how long ago this burst had happened. Water should evaporate rapidly in the exposed Martian environment. So rapidly in fact, that unless the water flow causes a destructive action on the crater wall, i.e. a groove or channel, there should be no real evidence left that the water was ever there. In this case, there is no such "destructive force" visible in the flow. It's just a dark, apparently wet patch. This means that it must have been a very gentle flow, and it must have happened only hours or even *moments* before MGS snapped this image. Not only had we confirmed that there was flowing liquid water on Mars, *we'd caught Mars in the act!*

There were, as always, immediate objections to our assertions. The first was that there was insufficient atmospheric pressure to allow for liquid water at the surface of Mars, never mind the temperature. This supposedly guaranteed that any water ice on Mars that melted in the heat of the day would instantly vaporize. However, we soon discovered a paper from Dr. Gil Levin, the principal investigator of the Labeled Release Experiment on the Viking 1 and 2 Landers. He cited several sources that confirmed that there were vast areas of the planet where atmospheric pressure and temperature exceeded the triple point of liquid water.

The final proof of this came ironically not from Odyssey, but from the venerable old Mars Global Surveyor. Surveyor has carried an instrument that up until this point had been pretty much an afterthought, called the Thermal Emission Spectrometer (TES). One of the most stunning (and stunningly ignored) results from this instrument was its finding that during the summer on Mars (remember, Mars' year is about twice that of Earth's), the regions of Mars even above 40° latitude warm to a ground temperature of over 60° F. Obviously, this is well above the threshold at which water can exist in a liquid state, and resoundingly destroyed the last objection to the seeps as *liquid* water.

We soon discovered that some amateur investigators were interested in the water seeps as well. Working with them, we found that there were quite a number of these "seep images." NASA, in the guise of Malin and Edgett, stepped up to propose that they were "dark dust streaks," simply the result of a rock becoming dislodged (by wind or tremors) and tumbling downhill, exposing darker material underneath the dry surface. Of course, "darker material" under the ground on Earth is usually darker because it's wet, but they didn't mention that in their paper.[123]

Hoagland then began an exchange of ideas with Effrain Palermo, one of the brightest of the amateur researchers. Under his mentoring Palermo had collected an enormous amount of data on the seeps. At Hoagland's suggestion Palermo and his research partner, Jill England, then proceeded to systematically map the locations of these "seep" images relative to Mars surface coordinates, to see if there was a global pattern to their distribution. As a control, they also mapped randomly-selected "non-stain" images until a representative and statistically valid sampling had been completed.

Immediately, two striking global patterns emerged: both pointing to present day liquid water as a source of the "stains" or seepages. In the first pattern, the map showed that seepage images appeared preferentially near equatorial latitudes, mostly between 30º north and south; none were found above 40º north and south. This implied that the phenomenon is restricted to warmer areas of Mars, which would be expected if these were truly water flows. An equatorial pattern is also completely inconsistent with the "dust avalanche" model put forth by Malin and NASA as an explanation for these features. After all, winds or tremors that shook the ground loose would not be restricted to the warmer equatorial regions.

The second, more important pattern discovered was that the water flows seemed to cluster preferentially around two pronounced geological features on the Martian surface: the Tharsis and Arabia mantle uplifts, or "bulges." The curious thing about these bulges, though, was their location: 180º apart.

It was Hoagland who first realized the significance of this distribution. "I've got it," he said in a phone conversation between the authors. "They're anti-podal bulges, Mike. Tidal bulges." His conclusion was simple, elegant and indisputable. The Tharsis and Arabia bulges on Mars were 180º apart, on opposite sides of the planet. Such bulges are commonly seen all over the solar system, on Jupiter's moons Io and Europa, on Saturn's moons and even in our own Earth-Moon system—and they are always, 100% of the time, caused by tidal forces between two orbiting bodies. The scars of her former life as a tidal-locked companion of a mysterious, long-forgotten parent planet told the story. Mars was not always a planet. It was a moon. A moon that had once been in a tidally locked relationship with her parent, just as the Moon was with the Earth.

Immediately, all sorts of implications fell out of this inevitable conclusion.

In our model, this relationship went on for millions of years, perhaps hundreds of millions, and was broken only when "Planet V" (named for the missing planet in Van Flandern's Exploded Planet Hypothesis) was destroyed in a cataclysmic collision with another body, or a gargantuan internal explosion. The resultant debris bombarded not only Mars, but also a large portion of the solar system. Mars, as a close-by satellite, was the hardest hit, as the devastating impacts ripped away most of her atmosphere and blasted the planet with rubble. It is this bombardment that accounts for the well known "crustal dichotomy" of Mars, where the southern hemisphere has a crustal thickness nearly twice that of the northern lowlands in some places.

And, the stains were indeed pockets of water. They were fossil remnants of a former Martian bi-modal tidal ocean. Vallis Marineris, so inexplicable in conventional terms, became a water-eroded, tidal bored scar—and the smooth-planed northern hemisphere was further (and subsequent to the planetary bombardment) massively re-sculpted in this process, by the sudden and catastrophic release of waters of the oceans. This newly released "double ocean" flowed north from the Tharsis Rise and Arabia Terra, completely flooding the northern lowlands in the process.

Although we tried to get our work published at established Mars conferences, we were told in no uncertain terms that our theory would not be allowed to be presented because of our work on Cydonia.

But it didn't really matter, because we now had a completely new piece of the puzzle. We now knew why it was so imperative for NASA to keep Cydonia suppressed. It wasn't just that confirmation of artifacts on another world would cause cultural upsets or even panic. What was really scary was the answer to the question of what happened to the civilization that built them. For such a technologically advanced civilization to be so utterly devastated would be terrifying. If Mars was once not so different from Earth, then the same thing could happen to Earth.

It didn't take long for one prediction after another of our model to be confirmed. Odyssey's data showed an abundance of hydrogen (probably water) in the northern and equatorial regions, right where our model said they should be. Palermo and England's distribution pattern was confirmed by work at Brown University, and new outflow channels were found implying that Valles Marineris had once been filled with water—and no one had even addressed the anti-podal tidal bulges of Mars.

Now, however, with Odyssey in orbit, we had more pressing issues to deal with. It was only a matter of time before infrared images of Cydonia would start beaming back.

It's Only a "Whole New Mars"–to Them

On January 21, 2002, a story by Leonard David appeared on Space.com.[124] In it, David quoted Steve Saunders, project scientist for the 2001 Mars Odyssey mission, as saying that the spacecraft was ready to begin the science phase of its operational life, and that the Face on Mars would be one of the early "high priority targets." Obviously, as we said above, if NASA really believed their own propaganda about what the ground truth of Cydonia was, they wouldn't be bothering to spend precious Odyssey resources on such an endeavor.

This heads-up prompted us to consider just what they might be doing behind the scenes at Mars Odyssey headquarters. Unlike the Mars Global Surveyor, Michael Malin or JPL would not control the THEMIS visible light camera directly. Dr. Philip Christensen, a relative newcomer to the Mars programs, would run it out of Arizona State University. This gave us hope that we might actually get some real data for a change. It was possible that Christensen was one of the insiders who believed that the public should get the straight scoop on Cydonia, rather than the part of the group that believed in continued repression.

For purposes of our own internal discussions, we had taken to calling these two groups the "Owls" and the "Roosters," two terms used inside the intelligence community to refer to groups that advocate suppression of a given issue vs. revelation of a given issue. We got the strong impression that the "Roosters" were winning, as President Bush had replaced Dan Goldin with his own man, Sean O'Keefe, shortly after the "Unmasking the Face on Mars" MOLA fraud had been exposed. Further, Deep Space told us that the order to take the May 2001 Face image had come directly from the office of Vice President Cheney.

We took all this with a grain of salt, of course. We didn't even know for a fact which side Deep Space was on, much less what the White House agenda might have been on this issue. Still, things seemed a bit more hopeful, since some of the previous cast of characters had been swept aside for this mission.

We strongly suspected it was this change in leadership that had led to the change in attitude that would make the Face an early target of the Odyssey suite of instruments, along with the promise of an "immediate" release of the data.

The more we learned about the THEMIS instrument, the more promising it looked. THEMIS is actually three instruments in one: a visible light camera, a thermal imager, and a multi-spectral imager—or infrared camera. The infrared camera can scan the planet's surface at 100 MPP resolution, but with a sensitivity of one degree difference in temperature. The same infrared instrument, by scanning in *nine* different regions of the IR spectrum, would allow determination of the surface composition of the objects that it scans at that same 100 meter/pixel resolution. The total image would actually be several hundred "pixels" square, allowing us to make precise comparisons of temperature *and* material composition variations over the entire surface area of the Face, at the resolution of a football field... on a structure over a *square mile* in total area. So despite the relative dearth of spatial resolution from the IR camera, in many ways it would tell us more about Cydonia and the Face than all of the visible light images taken so far.

As Odyssey approached Mars, the FACETS initiative paid off once again. Peter Gersten received a call from the office of Dr. Jim Garvin, requesting a phone conference with him and Richard Hoagland. The call was quickly arranged. During the course of the call, Garvin promised that FACETS would get the data it wanted from Odyssey, and that a future mission, Mars Reconnaissance Orbiter (MRO), would be *the* mission in terms of answering all of our questions about Cydonia. He also encouraged us to submit papers to the various Mars conferences and publications, promising that they would get a fair hearing.

Getting a call from NASA's Mars point man was weird enough. Having him completely reverse his position and encourage us to participate in the public scientific process was downright bizarre. Still, we had little choice but to take the man at his word, and hope for the best. The climate seemed to be changing for the better.

Then, early in 2002, Hoagland got a call from Deep Space. He told us that the early returns from the nighttime infrared were stunning, and had sent shock waves all throughout NASA. He assured us that Cydonia had already been targeted successfully, and that the data was described as "amazing." He encouraged us to demand nighttime IR data, especially of Cydonia.

On February 26, a new story appeared on Space.com by Leonard David,[125] in which NASA project scientists described the early data from the science phase of the mission as "amazing"—exactly the word Deep Space had used. Steven Saunders added that the data represented "a whole new Mars." While the story was short on details, it was clear by reading between the lines that the data that was causing such a stir internally was from the IR camera, and that raised a whole series of interesting questions. Some of those, we learned, might be answered in a press briefing scheduled for the following Friday, March 1, 2002.

As we contemplated what might be revealed at the press briefing, it occurred to us that we had no reason to think it might involve Cydonia or the Face, or for that matter even address the question of artificiality in any meaningful way. But if the data itself was honest—and given the "new tone" we'd been experiencing, it might be—then we still had reason to hope that something interesting would be revealed on the following Friday. But the important thing we decided to keep in mind was that the way NASA spun what they presented was increasingly irrelevant. What would matter is what the data *showed*, not what NASA decided to emphasize politically. Even after the "unmasking the face" hit pieces and the MOLA fraud, an MSNBC poll[126] had found that among those that had changed their mind, the vast majority had decided it was *more* likely the Face was artificial after viewing the May 2001 image.

Those of us longing for some sort of official "disclosure" on artificiality or even a radical geologic theory like the tidal model were missing the point. "Disclosure" is *not* going to happen with the president sitting at his desk in the oval office, with the paned windows and pictures of his family behind him, dourly reassuring us that it is worth going to work tomorrow despite the stunning revelations of the "last twenty-four hours." Disclosure had been happening *all around* us, and right in front of us, pretty much unabated since the magical ball in Times Square turned us all toward a new millennium. We were being given the data, in bits and pieces, and pretty much allowed to make up our own minds what we thought of it. NASA's own relevancy in the greater scheme of things was increasingly precarious—perhaps even by design—as it became more and more obvious to those of us that were paying attention just how truly wonderful and strange a place Mars, and our solar system, really was.

It was almost as if someone high up in the agency (or outside it) was pulling the strings, trying to get this particular rooster to get its message across without so much as a single crow.

We'd started to get a hint that "something" was coming about two weeks prior to the late February Space.com piece. A series of images suddenly appeared on the web, purportedly from Mars and leaked from a company called IEC. They included color, infrared and radar imagery of an "Anomaly 502" that was supposed to be of some Martian ruins just below the surface. While we immediately had serious doubts about the legitimacy of the image, this was exactly the sort of "trial balloon" that a "Brookings" pattern of disclosure would mandate—put out an image, see what the reaction is, and decide whether to go ahead with your disclosure based on the response. So while we were somewhat unimpressed with the execution, the timing and content of the release jogged our memories and got us to thinking.

Why a couple of weeks before the Space.com article and the press conference? And why emphasize infrared and other ground-penetrating technologies? Why not just forge a doctored-up surface anomaly? Then we remembered: Mars Odyssey would *not* be the first probe from Earth to use ground-penetrating technologies in Mars orbit.

The Russians had sent two probes, named Phobos 1 and 2, to Mars in the late 1980s to study the surface and atmospheric properties of the planet and the composition of one of its two moons (Phobos). Phobos 1 failed along the way, but Phobos 2 made it all the way to Mars and operated nominally for a period of several weeks. Its disappearance has become the stuff of UFO lore, but in the process the spacecraft made numerous valuable observations of both the Moon Phobos and Mars. One of the most curious discoveries was that Phobos' density was found to be extremely anomalous. According to a paper published in the October 19, 1989 issue of *Nature*, Phobos had a bizarre density of 1.95 g/cu.cm ("19.5" anyone?), meaning it was almost 1/3 hollow. Since both Martian "moons" are actually captured asteroids, this finding is extraordinary. There is virtually no way that a solid object like Phobos can be "hollowed out" in this manner naturally, leaving a really big question—just who hollowed it out, and why?

But things got even more interesting when Phobos 2 was rotated to look at Mars itself. The probe carried an infrared spectrometer, a device not too different from the infrared thermal imager on Mars Odyssey. While it lacked the resolution of Odyssey's far better THEMIS camera, the infrared device on Phobos 2 also gave the Russian scientists the capability to discern buried objects just below the surface of the planet (covered with sand or dust) via their relative rates of cooling.

In 1989, just after the loss of Phobos 2, a program appeared on England's independent Channel 4 revealing the discoveries of the Phobos 2 probe. Among them was a tantalizing infrared image, taken in the Hydraotes Chaos region (0.9º N, 34.3º W), showing what seemed to be a fairly mundane landscape in the visible light spectrum—but when the IR filter was applied to the same area, an astonishing rectilinear pattern appeared just beneath the sand [Fig. 9-2]. This regular, highly geometric pattern (across an area the size of Los Angeles) is strongly indicative of a cityscape just under the surface. Although some of the rectilinear features seem to be aligned with the scan lines of the image, others are unmistakably *not* aligned, and are also curved and somewhat geometrically irregular, as they would be if they were wrapping around uneven topography. Clearly, they are incredibly similar to some sort of buried (regular/geometric) construction or tunnel system.

There was no question that a lot of people noticed just how weird this all was. The program featured comments from Dr. John Becklake of the London Science Museum (and a very sober guy), and he left no doubt about what he thought of the images. Interviewed in front of an exhibit that was obviously prepared with the help of the Russians (remember, it was still the Soviet Union then), Becklake was unequivocal: "The city-like pattern is sixty kilometers wide and could easily be mistaken for an aerial view of Los Angeles."

The program went on to show excerpts from a Soviet Space Research Institute press briefing in which the anomalies were discussed. And yet, all of this high level interest in this story by scientific heavyweights was virtually ignored in the United States.

So maybe, just maybe, Phobos 2 had given us a preview of what would be shown in the March 1 press conference. If the infrared images from Odyssey were in any way similar to what Phobos 2 found, NASA was going to have a hard time spinning the data.

We had a sneaking suspicion that we were guessing right on this one. Keeping in mind that NASA is an agency steeped in ritual and bound by a Brookings-like code of behavior, we couldn't help but notice one odd little coincidence: The Phobos 2 images of Mars were taken on March 1, 1989, exactly thirteen years to the day before the upcoming NASA press conference.

Based on this, we expected that NASA would produce an IR image, most probably taken of or near Hydroates Chaos at their press conference. We had only a few days to wait and see if we were right.

Just as we suspected, we were dead on about NASA's proclivity for ritual and symbology. Exactly thirteen years to the day after Phobos 2 had imaged the Hydroates Chaos region of Mars, NASA released a new series of IR images from Mars Odyssey 2001. The undisputed star of show was an image of the Hydaspsis Chaos region of Mars. If that sounds familiar, it should, as it is just a few miles from Hydroates Chaos (0.9º N x 34.3º W vs. 2º N x 29º W), the location of the stunning Phobos 2 image. So, as we expected, we got an IR image of virtually the same area exactly thirteen years to the day from the moment Phobos 2 took its picture [Fig. 9-3].

And what a picture. Almost from the beginning, many of the scientists at the briefing seemed a bit nervous and uncertain. Dr. Phillip Christensen, especially, seemed edgy as he presented images from the infrared camera.

As Christensen haltingly displayed the image to the assembled media, he made no mention of the stunning *regularity* of the terrain—especially considering it is dubbed "Chaos." What was most striking initially is the incredible consistency of the channels between the "sand covered mesas" (how *does* sand stay on top of a flat "mesa" that is buffeted by 300 MPH winds from time to time?) in the image. The channels all seemed to be about the same width, and remained incredibly consistent for miles. The mesas themselves were shockingly geometric, not really what one might expect from a fluvial erosion process—but the Devil of course, is in the details.

Close-up enhancements showed some very unusual features of these "mesas" [Fig. 9-4]. There were regular, geometric "notches," or even openings, in some of them and others showed signs of being buried foundations for larger objects. A little further up in the image was a strange looking "crater" plastered on top of a dark rectangle with incredibly square edges. Considering that this was not aligned with the image scan, but *is* aligned with actual north/south, we were inclined to doubt this was an image artifact or natural formation.

We were also pleased to get more confirmation of our tidal model at the press briefing, in the form of Gamma Ray Spectrometer data. The GRS has a very coarse resolution and wasn't fully deployed yet, but the GRS team, led by Dr. William Boynton, gathered some data anyway while the instrument was still in its "parked" position.

Of the data, the most crucial was the information gathered on "high energy neutrons." These types of neutrons are typically absorbed by hydrogen—a key component of water and water ice—so by measuring areas with little or no "HEND" return, it is possible to determine just if, and how, water was distributed on Mars.

As it turned out, there was a hell of a lot of water, or water ice, on Mars. Most of it was clustered (as expected) in the south polar cap, but huge amounts of it are also present on two specific (and bi-modally *opposed*) areas of the planet—the Tharsis and Arabia bulges.

Obviously, this was a flat confirmation of our previous work on the tidal model. Because of our *specific* prediction more than six months previously in our paper, that Odyssey would find exactly this anomalous (beyond our model) water distribution, we could state categorically that these findings constitute absolute, inviolable confirmation that our model is correct. It also proved that the work done by Palermo and England is correct, since the water distribution corresponds precisely where they have found what were now proven water stain images.

What was stunning was that neither Boynton nor any members of the assembled press, either because of ignorance or timidity (is there any American institution that has fallen into a greater state of disrepair than the Fourth Estate?), made the slightest note of the fact that the water—inexplicably for the established conventional models of Mars evolution—was distributed on two such prominent features of the Martian landscape. This does not change the fact that there is nothing in any conventional model that can account for this distribution. The only explanation that fits the data is our tidal model.

Even after they had hyped the data to be presented as a "whole new Mars," and "tremendously exciting," not one of the scientists bothered to explain just *why* any of this constituted a "whole new Mars"—and nobody asked.

This astounding lack of curiosity by the press as to just what made this data so exciting played right into the NASA strategy we'd come to expect. NASA's policy was clearly now one of disclosure—but *unacknowledged* disclosure. With so few science beat reporters actually having a science background anymore, they have become completely dependant on the agency for their material. The scientists at NASA have become priests, virtually unassailable.

We posted our thoughts on all of this that night on the Enterprise Mission website. The reaction—from all sides—was swift. In fact, few stories over the years generated as much negative comment as our web posting on the new Hydaspsis Chaos IR image.

With unanimity (and predictability), our critics steadfastly ignored the flat-out confirmation of our Mars tidal model that was included in the piece, and instead preferred to focus on the infrared image. During the course of reading through these nasty comments, it became clear to us that most of these "armchair geologists" not only did not understand what they were seeing, but also completely misunderstood (or deliberately misrepresented) just what it was about this data we found so fascinating. In many cases, they accused us of immediately claiming that the geometric block-like features in the IR image were artificial, which we flatly had not done—to that point.

For the most part, the attacks focused on our (perceived) inability to recognize "simple geology." The attackers also criticized us for not providing a context visible light image of the Hydaspsis Chaos region in our piece. One web article even blatantly accused us of being "unscientific" for not including such an image. Of course, this critic made no mention of the fact that NASA, at its initial Odyssey press briefing, also failed to provide such an important context image. This, in spite of the fact that the THEMIS folks had their remarkable nighttime IR image for over a week before the press conference, not to mention the budget and staff to easily do a search for such a visible companion image from the Viking archive. We did not have the luxury of those resources—certainly not in the few hours immediately after seeing what the THEMIS team presented, and the posting of our initial article, which we clearly characterized as a "preliminary assessment" in any case.

We found this reaction, from the alleged "anomalist community" to be quite disturbing. They were the ones supposedly most interested in finding proof of extraterrestrial artifacts, but when we confronted them with solid evidence, they reacted angrily. We began to wonder if maybe "Brookings" had been right, that the science and engineering buffs and professionals would have the hardest time with "proof" of extraterrestrial intelligence.

In any event, when we read through the criticism, it became obvious that the critics literally had no idea why we found this nighttime IR image so fascinating, and so potentially important. They assumed that we were expecting to see blatant evidence of artificiality in the broader context visible light image. We weren't.

Key to understanding the bizarre nature of this area is the *buried formations*— the underlying structure beneath the Martian surface features, revealed for the first time at this roughly "Viking" resolution by Odyssey's IR imager.

Just for the sake of argument, let's review one natural geologic model for this kind of formation for a minute. One of the arguments consistently advanced to explain such regular patterned formations is the tired, old, "frost wedging" model.

In typical examples of frost wedging here on earth, the forms are caused by repetitive freeze and thaw over many years. The ground eventually forms cracks, or weak points, sometimes in vaguely polygonal patterns like you see above. The melting snows find their way into these fissures, slowly wearing away the ground in between the harder blocks and creating water-filled channels for runoff. This is a slow process, and the wedges are very shallow— but the important thing to remember is that the softer soil is pushed away by the flowing waters, which eventually shapes the harder stuff into the polygonal forms we see above. In other words, the runoff flow creates the shapes. Now let's consider Mars.

In our tidal model, what happened to Mars was quick and cataclysmic. In a very short period of time, starting with the fateful day of the destruction of Mars' parent planet, "Planet V," the Red Planet took a beating almost unparalleled in the history of the solar system. That first day she lost better than half her atmosphere, experienced floods of Biblical proportions, was bombarded with literally miles of Planet V's debris, and lost her ability to sustain higher forms of life.

Within a few months, if not weeks, virtually all of the remaining water on Mars was either frozen in place as surface or subsurface water ice, or sublimated to the poles. So whatever sculpting took place was rapid and intense. Since frost wedging would have taken decades to accomplish its erosive process, it seems unlikely that it can be attributed to what we see here.

Look again at a close-up of this nighttime IR data from Mars Odyssey [Fig. 9-4]. Remembering that dark is "cool" and bright is "warm," this image makes no sense in terms of conventional geologic models. Casting aside for the moment NASA's silly explanation that the "mesas" are "dust covered," which is why they are "cooler" in this image (dust collecting deeply on the *tops* of flat, windswept mesas, on a planet where the winds can exceed three hundred miles per hour), take a good look at the edges of these tantalizingly regular "cells." Note that they are all brighter by a factor of at least ten than the interior of the "mesas" they encircle. If, in fact, the walls of these mesas were made of the same stuff as the interior (rock?), then they would be expected to have pretty close to the same heat-emissive properties as the interiors, even if there were a little dust on top. Instead, we see a dramatic difference in the amount of heat retained by these precisely defined, amazingly geometric "walls," compared with their very cool interiors. For conventional geologic models, this is a big problem.

How can a natural ring around a mesa be made of something completely *different* than the interior of the mesa itself—since mesas are carved (in natural geology) out of pre-existing bedrock? And how can the (shadowed) channels between the mesas now be brighter (warmer) than the windswept (exposed to wind *and* sunlight), rocky tops? The channels should collect and trap significant amounts of sand and dust over any interval of time—which should then act as insulation, producing dark (cold) channels. In fact, the entire situation is reversed—to NASA's own admitted bafflement.

Enter the Mars tidal model. In our take on this remarkable region, Hydaspis Chaos was a "dumping ground" for some of the huge volumes of water that were released from the sudden severing of Mars' prior gravitational relationship (orbital lock) with Planet V. As this catastrophic flood unfolded, the sudden orbital release dumped trillions of tons of water, massive boulders, dirt, silt and sediment all over this (and every other) low-lying area on Mars. The rushing tidal waves would eventually slow to a trickle, but not until they left literally trillions of tons of miles-deep debris, mud and sediment in their destructive wake. The finest sediment—initially deposited in catchments between more resistant areas of rock—would later be worn away by the remaining flows before the planet then literally froze. After a period of drying, the incessant Martian winds would then continue to erode—but at a much slower pace, and for literally *millions* of ensuing years—what those last flood waters only started.

Anything standing above ground before this almost inconceivable cataclysm would have been simply obliterated by the sudden floods. As these raging waters ebbed, the remaining flows have found their way to narrower and narrower rivers and streams, constrained somewhat by the previous geology, eventually carving out the shapes we see here by removing much of the initial sediments deposited between the "mesas."

OK, so at this point there's no difference between our model and conventional processes, right? Well, our model assumes that the area Odyssey and Viking imaged is covered now with a (relatively insulating) "hard mud" to a significant depth (at least several kilometers). The initial flood of water could not have gone on too long after the initial cataclysm (because of the sudden loss of atmosphere, and the freezing temperatures), so there would have been relatively little post-catastrophe erosion from remaining, flowing water before those waters froze. Mud and sediments, unlike exposed bedrock, do not retain heat, so it would be expected (in this model) that these deep sediments would now show up in any nighttime IR image as a "dark" or "cool" area, just as we see here over most of the nighttime IR image—but if simple geologic theory were not enough, there is further proof in this same image that we are looking at a huge "mud flow," rather than exposed planetary rock.

Just above the "mesas" in this nighttime image (to the west, since north is to the right) are two modest impact craters, each a couple of hundred feet across. In both cases, in the infrared they have brighter (warmer) rims than their surroundings, and dark (cool) interiors. This is as it should be—if the crater rims are (warmer) exposed outcrops of underlying bedrock, uplifted by the initial impact process, and the bowl-shaped interiors have trapped (cooler) blowing dust and sediments, which now appear dark because of their thermal insulating properties.

Yet something's not quite right with this standard geological picture. If you compare the brightness of the crater rims (the "exposed rock" in the impact model), they are only slightly brighter than their surroundings. They are certainly not as brilliant as the (presumed) exposed mesa rock walls of the Chaos region just to the east. It's as if the craters were created, not in bedrock, but in an ancient layer of insulating *mud*. If the impacting objects had struck a hard-pan surface, presumably the underlying rock would have been shattered and raised, exposing it as brightly as the edges of the nearby eroded "mesas"— but it has not. The IR signature of each crater rim seems curiously dull—as if the thermal emission was coming from something with a much softer, far more insulating nature than bare rock (which is exactly what these nighttime IR images are supposed to reveal). It in fact appears like clods of friable (and thus insulating) *sediment* instead of deeply excavated (by the cratering process) bedrock.

Thus, these completely unexpected IR crater signatures neatly reinforce our model—that this region is in fact a deep flood plain of overlying mud and sediments, with the ancient bedrock now buried miles below the surface.

So these simple craters reveal remarkable, independent confirmation for our basic massive flood scenario. These cannot be simple exposed, rocky mesas, now covered with accumulated windblown dust, but rather, that this whole area (and the other "Chaos" regions?) is one massive mud flow from the ancient cataclysm that rent Mars from its mother planet about sixty-five million years ago from its orbit of Planet V.

If this area is indeed covered to some depth in ancient sediments, then the simplest reason the interior of the mesas are now dark is not because they are composed of the same material as the much hotter walls and are covered with dust on top, but because they are literally catch basins filled in with those same ancient muds still contained within the current IR bright walls.

The latter, then, are simply composed of some much more heat-retentive material arrayed in a stunningly regular, geometric pattern of amazingly uniform thickness—and stretching for literally hundreds of collective miles around each mesa.

Thus, these Hydaspsis Chaos mesas do not appear themselves to have been shaped and eroded by the vast tidal floods gauging out pre-existing bedrock into intricate geometric forms, but by the tidally-released sediments flowing around and over pre-existing sets of resistant walls composed of a heat-retentive something, walls clearly now already in place *before* the catastrophe itself.

So then, the next question must become: What kind of natural geologic process forms such walls—with regular, uniform thickness and intricate (and opposing) geometric patterns—that can be filled in by massive mud flows of the type we now see here? The short answer is: none.

There is, however, a perfectly viable alternative scenario. Let's consider what would happen on a planet like ours if a sudden cataclysm on the scale of what happened to Mars were to occur.

If the Earth's rotational axis suddenly shifted, or if the collapse of the Ross Ice Shelf resulted in massive amounts of water suddenly inundating habitable coastal areas, everything above ground would be swept away almost instantly in the resulting massive floods. Large buildings would be crushed and tossed aside, leaving only their foundations below the ground. If this happened in an area the size of say, the Los Angeles basin, then all one would expect to see would be the remnants of these former rising structures, arranged in a hauntingly geometric pattern indicated by their remnant foundations.

If you then flooded that whole area with miles-deep sediment, the hollowed-out cores of these former skyscrapers and other structures would also fill with sediments—and the whole L.A. basin would become one big, featureless sea of mud.

On Earth, however, as rain and wind began to erode away these newly-formed deposits, eventually the eroding sediments would sink low enough to reveal the preserved foundations of the former artificial structures. At this point, the waters (and wind) would take the path of least resistance, as they always do, and flow *around* the massive base foundations of these former structures. Yet, protected from this erosive action by the basement walls, the interior sediments would remain where they were initially trapped in order to dry, and form a series of essentially level "mesas" to any aerial observers, separated by the former streets between the buildings.

Thus, our hypothetical post-cataclysm L.A. basin would strikingly resemble (in the infrared) exactly what we were seeing in this nighttime Odyssey Mars image (though on a much smaller scale)—even down to the geometric crenellations on the mesa walls.

It was increasingly apparent that these bright, intensely geometric outlines are, in fact, the exhumed foundations of former massive artificial structures—all that remains after the ancient raging floods completely leveled the original structures' towering upper levels.

So what we'd found in the Odyssey infrared of Hydaspsis Chaos are the ancient foundations of something far more crucial to our eventual understanding of this entire shattered planet, far more relevant to the search for who we ultimately are, than mere frost wedging.

It is precisely these remarkable structures, if not answers, that NASA was obviously looking for when they took and then released this specific Odyssey IR data as their "first nighttime image," precisely thirteen years to the day after the Russians took their own provocative Phobos 2 IR images just a stone's throw away from this same area. Is it truly an accident that this first IR image—from a mission named literally after Arthur C. Clarke's famed epic of extraterrestrial intelligence's formative involvement with mankind—should be a set of buried artificial structures which (if we had now read the Odyssey IR correctly) can teach us so much about Cydonia itself, if not the history of Mars destruction.

The last nail holding in place this increasingly tattered assertion, that these intricate, highly ordered Hydaspsis Chaos forms are somehow the product of a "simple Martian geology" was then removed not by us, but by NASA itself.

At two subsequent Mars Odyssey public lectures, one held at JPL's Von Karmen Auditorium, the second at Pasadena City College (in Room 333, of course), Dr. Roger Gibbs, the new Odyssey Project Manager, provocatively showed this same baffling IR Odyssey image, and flatly admitted that he and the entire Odyssey team had "no idea how to interpret it."

"Why does a channel have no dust, and the top of a mesa does?" he asked rhetorically. He then asserted that the working model was that the channels had somehow "collapsed around the mesas."

Again, this is silly. Clearly the exposed channels were literally scooped out by an erosive process that attacked the softer, drying mud, but left the hard, geometric foundations—each containing its own reservoir of trapped sediments—of the ancient arcologies essentially intact.

Somehow, Gibbs and his colleagues can't publicly consider a non-natural explanation for these impossible features, even while blandly admitting, "It [the IR image] really asks more questions than it answers." Clearly, if the explanation for Hydaspsis Chaos was "simple geology," as our critics had asserted, it seems to be "geology" well beyond the leadership of the Odyssey Mars mission, if not of NASA's best planetary minds.

Yet still we had one last hurdle to clear. Some of the critics then raised the similarity to even older and similar geologic anomalies on Mars, like Mariner 9's so-called "Inca City." They pointedly used the MGS images of "Inca City" in an attempt to dismiss Hydaspsis Chaos as an obviously similar "natural set" of features—but there is very little natural about the "Inca City" at all.

As you can see, "Inca City" [Fig. 9-5] compares very favorably (in terms of straight geometry) to well-known archeological ruins on Earth, like the Anasazi structures in Chaco Canyon. Even though Inca City is on a much larger scale and completely unexcavated, this does not mean that artificiality should be ruled out *a priori*—certainly not based simply on existing visible light data. To expect ancient, highly eroded archaeological "ruins" on another planet (and ones of potentially immense scale) to look as clean as a recently excavated dig on Earth is not only naïve in the extreme, it's not even logical.

As noted previously, an IR study of this formation is crucial to determining the underlying structure that (even heavily "mantled") presents such a compelling geometric comparison to our own terrestrial monuments. It's bad enough to dismiss the sand and frost-covered "Inca City" as "natural" without a closer study of the entire area, but to then cite it as some sort of "proof" that the Hydaspsis Chaos images have a similar "natural" explanation, is almost criminally idiotic.

After all, what lies just beneath the surface sometimes tells a completely different story.

Do Geologists Dream of Windblown Sheep?

After another month of waiting, on Friday, April 12, 2002, we got the first Mars Odyssey data to cover the Cydonia region of Mars. Unfortunately, we did not get what we had hoped for (a multi-spectral nighttime infrared image of the Face and surrounding structures). Nor did we get a full-color image of the Cydonia complex in the promised five-band color. What we did get was a nice grayscale strip from the spacecraft's visible light camera [Fig. 9-6]. At nineteen meters-per-pixel, the image was substantially better resolution than the fifty meter-per-pixel images we got in the Viking era. Still, it is an order of magnitude lower than the four to five meter-per-pixel images we got in the best of circumstances from Mars Global Surveyor.

This did not prevent it from being useful. It was still a better overhead view of the center of the Cydonia complex than we had before, allowing us to view the specific objects of interest in context and at a resolution we have not generally seen (the D&M, for instance, had been almost completely missed in the publicly released images from Malin Space Science Systems). Previously identified objects of interest that were visible in the new Odyssey image strip included the Face, the D&M Pyramid, the Fort... and some surprises.

The first thing we noticed was that the so-called "massive tetrahedral ruin" adjacent to the symmetrical mesa just south of the Face was much more clearly defined than in the previous images. This object was first spotted on the infamous Catbox MGS image strip. It had been stylistically interpreted from previous images as a "dolphin," or various other absurd pictogram shapes, with one amateur pixologist even claiming he saw a "trailer park" at the base of the object. Part of the illusion was the extreme forced perspective of that original MGS image—taken 45° off-nadir—which effectively distorted the shape of the ruined pyramid. We could now see clearly from directly overhead that there were two distinct faceted walls that once made up this tetrahedral structure. The object lies just south of the suspiciously symmetrical mesa and—conveniently—lies at exactly 19.5° off the base symmetry axis of the D&M. This same axis passes right between the eyes of the Face.

The real prize, however, was the good, higher-resolution image of the D&M, which we could now compare with the Viking data.

One of the most controversial objects in the entire Cydonia artificiality debate (because of its status as the lynchpin of the Hoagland/Torun Geometric Relationship Model), the D&M had always been key to deciphering the correctness of the original Cydonia observations from two decades ago. The most crucial question surrounding the D&M has been that of the Pyramid's five-sided symmetry, which was suggested strongly by the original Viking data and was an issue of some considerable debate in years past. Vince Dipietro, for one, had maintained for years that the D&M was only a four-sided object, objecting to the implication of a "fifth buttress" in the shadowed side of the Pyramid on the Viking data.

This projected five-sided symmetry not only held up extremely well in the new image [Fig. 9-7] (note especially the four clearly defined faceted sides to the pentagonal pyramid meeting at a central apex—exactly as observed by Hoagland and Torun[127] on the original Viking data in 1989), but we could now substantially more detail on the lower section of the damaged right hand side. We also (despite deep shadow) could verify the existence of a "fifth buttress" to the northeast—the final piece needed to complete the pentagonal form and reconstruct the object's original, undamaged shape. The buttress seemed to be pretty much the same length as the other visible buttresses (the southeastern buttress is mostly buried under debris flow from the mild collapse the object has endured) and verified the predicted geometric form proposed by Torun in 1988 perfectly. Obviously, such a "hit" was way beyond even "the power of randomness," and is a compelling confirmation of the validity of Torun's original work.

We could also see additional evidence of the internal bulge in the Pyramid as noted back in the original investigation. It was once argued—by Hoagland in *Monuments*—that the "crater" to the right of the Pyramid seemed to be some sort of "entrance wound" for a possible projectile that may have accounted for the bulge in the upper right hand quintile. This new Odyssey image gave far greater detail of this area, and it seems to substantiate earlier 3D shape-from-shading work done by Mark Carlotto that indicated it might actually be a horizontal "entrance point," either from a projectile or as an actual architectural entrance. Additionally, the new image—at nineteen meters per pixel resolution—revealed new structural details of both southern "buttresses" that further reinforce the idea that these are essential architectural elements of a massive artificial edifice.

The conventional geologic model of these features is that they are due to "slumping." Basically, the idea is that loose material from the top of the D&M tumbles down the slopes of the Pyramid and preferentially piles up at the corners of the object. Forgetting for a moment that none of these geologic experts seem to have noticed that the mass-wasting has put piles of debris precisely at the five *corners* of a 1.5 mile high, bilaterally symmetric, *pentagonal* Pyramid, our model is that these are actually reinforcing "buttresses" that have a specific architectural and mathematical function. Close examination of the new image—remember, twice as good as the Viking data—shows that these buttresses are indeed just that. They have a very "boxy," geometric look—and the southwest buttress even seems to have two rectangular openings (doors?) in the base. The simple truth is that no "mass-wasting" process produces rectangular box shapes—certainly not ones with several-hundred-foot-wide *doors* in them.

The other major discovery from the new image was that the entire structure could now be seen to be placed atop a huge rise (or platform) much like the Giza Plateau on Earth—and this newly discovered platform, which was not resolvable in the Viking data, seemed to have a geometry all its own.

When you rotate the image of the D&M from the way we are all used to looking at it, we can see that there are two distinct but partially buried edges to the plateau that the D&M rests on. These two edges, not visible in the original Viking data, meet in an apex point that is exactly aligned with the SE buttress on the opposite side of the structure [Fig. 9-7].

What this new perspective on the D&M now allows us to see for the first time is that the pyramid rests on a 2D, seven-sided platform (or base) upon which the massive 3D five-side "Rosetta Stone" structure was constructed. It also revealed an additional, second alternative line of symmetry for the object, which conversely produced a second, bi-laterally symmetrical four-sided geometric figure.

When these two shapes (the seven-sided platform and the five-sided pyramid) were superimposed upon each other, they once again reinforced the quintessentially tetrahedral message of Cydonia [Fig. 9-8]. Indeed, one of the new internal angles generated by the new figure is none other than the ubiquitous 19.5°. Not only did this new data flatly validate the original Torun reconstruction and analysis of this enigmatic object, but it also demonstrated the correctness of the geometric relationship model derived from it (as if the numerous successful predictions of the hyperdimensional physics model hadn't already done so).

The new image also allowed us to do an actual side-by-side-by-side comparison of the Face from Viking, MGS and Odyssey. Immediately, several things became obvious. First, the Odyssey image confirmed that the MGS April 2001 image had been poorly ortho-rectified, as the Face platform was much narrower (from Odyssey's more direct overhead perspective) than it was on the Surveyor image. Also, the "nostrils" from the Catbox image had returned after being nearly invisible on the April image, and the "lion"-side eye-socket appeared to be better aligned with the opposite socket.

Of course, none of this prevented the usual nay-saying from NASA. While almost all images released of anything other than Cydonia are not accompanied by captions or headline stories, images of the Face and Cydonia invariably are. This image was no exception. The caption, posted on the Arizona State University THEMIS web page, was a laughable mish-mash of airy homily, inappropriate comparisons and scientific contortions that seemed way too anxious to describe the indescribable in terms that make sense to the limited minds of the geologists, who were trying to close this issue once and for all. It began by declaring that "nature is an imaginative artist" and then proceeded to inform readers that we are always seeing things we have "dreamed up" in our imaginations. They then compare the Face to Arizona's Camelback Mountain (shades of NASA's previously failed "Middle Butte Mesa" comparison), and "Sleeping Beauty" near Ludlow, California, among several other places. They naturally fail to mention that all of these locations only take on the visages they are famous for *as profiles from the ground*. This is, as has been pointed out to the NASA regulars on many occasions before, completely unlike the Face, which is designed to be seen from overhead. And, as is par for the course with these kind of assessments from the "scientific community," they tried to deal with the Face in isolation, completely ignoring the preponderance of very anomalous structures scattered all over the rest of the image—like the bi-laterally symmetrical, pentagonal D&M Pyramid.

They finally attempted—not very well—to claim that the Face's appearance is the result of wind-based erosion, or a combination of Aeolian processes and a bizarre "pasting" process that they didn't even attempt to explain thoroughly. This is just a variation of a discarded "differential erosion" argument that NASA had floated several years before. This recently rediscovered notion is frankly nothing more than the authors' substantially uninformed (or deliberately obtuse) opinion, which seems to permeate all levels of NASA and its affiliate academic institutions, and not only ignores the substantial history of predictions and observations of various independent researchers, but simply does not hold up under the most elementary critical scrutiny. They fail to note the incredible variety of "mesas" in the Cydonia region. How exactly did the Martian wind decide to change directions, at very precise 85.3° and 69.4° angles, just around the apex of the D&M every few thousand years to sculpt the precisely mathematical Cydonia Rosetta Stone and Pentagonal D&M shape? Or perhaps the wind only blew a few miles to the north, where it "sculpted" the Face out of our dreams? Perhaps some completely different process was responsible for forming the D&M only a stone's throw away? Now we're willing to engage in an exchange of opinions, but for these "experts" to couch theirs in the guise of stipulated fact is not "science," any more than any other sweeping and unsupported assertion would be.

The reality is that if the Face and the rest of Cydonia were artificial, then geologists, of all people, would probably be the last group on the planet to recognize or acknowledge it.

The reasons for this are myriad and complex, but revolve chiefly around the major problem confronting modern scientific archaeology. Initially, from mere appearance, archeologists and geologists can't even agree on whether a fairly obvious Earth artifact is artificial or derived from a completely natural process. It is therefore hardly surprising that a potential artifact on another world has been a source of intense, continuing debate for decades.

A case in point is the set of recently-discovered pyramids below the Pacific Ocean just off Japan, on a submerged undersea island called Yonaguni. Archeologists were quick to point out the various characteristics that identified these structures as artificial, monumental architecture. Geologists countered with a series of equally reasoned arguments based on the biases of their science—that natural processes could explain all these data points just as readily. To most observers, the winning argument was made patently obvious simply by looking at the pictures. The underwater pyramids were blatantly artificial. Yet the debate has raged on for years.

It was only when a team of divers sent by the Discovery Channel discovered a stunning, underwater Face, complete with a Mesoamerican headdress (or lion's mane?), that some of the geologists relented. These new finds, joined with some key observations by Japanese marine geologists,[128] finally broke the dam and forced the admission of the obvious—somebody built the stuff. But even this single, redundant, unifying theme—faces amid the ruins, both here *and* on Mars—can leave some geologists searching for some remotely plausible natural explanation.

This sort of debate is exactly why we have always argued that planetary geologists are not well equipped to solely evaluate Cydonia—or any other putative artificial structures on the planet Mars. If the Face and other objects there are artificial, then they should be looked at first and foremost thru the lens of archaeology, and geology only secondly (for their natural context)—if at all. Geologists tend to look at everything as the product of one of their familiar natural processes, since their training has told them (a planet full of artifacts on Earth notwithstanding) that this is the process by which everything in the solar system has been formed. This is how you get the simplistic "sand dunes" argument when geologists are confronted by anomalies like the Glass Tunnels of Mars. The simple fact is that when dealing with a problem outside their own discipline, and therefore limited expertise, most scientists—from whatever field—will resort to any familiar explanation from that field, and then cling to it at all costs, no matter how silly or ultimately contradictory it turns out to be. This is what their training has taught them to do (go for the familiar first). It is also what their training limits them to.

This is why the Brookings Report was so concerned with the impact that the discovery of artifacts might have on scientists of all stripes.

Still, despite the limitations imposed on the image by the ASU geology department, there were some genuine geologic observations to be gleaned from the new image. For instance, there was some evidence of wind erosion at Cydonia, including a predominant "wind direction."

There is a clearly predominant "wind direction" at Cydonia. We could see from the earlier close-ups of the Face that the city side (and the city side only), showed some evidence of minor pitting as the result of wind erosion processes. But the Cliff side was clearly protected from this process—it showed no sign of pitting—and quite obviously the minor wind erosion we observe was not a significant factor in the overall shape of the Face, this last observation meaning it was either very light (unlikely on a planet with occasional gusts for several weeks in the 300 mph range) or very *recent*. The new image of the D&M only confirms this single predominant wind direction. The three fully exposed buttresses are the ones that are most exposed to these presumed "winds of Cydonia." This is because this wind has simply blown away the dust and sand that might have once covered the structure. The two partially buried buttresses are on the opposite side of the structure, away from the howling winds of the Cydonia plain and presumably protected by the towering structure itself.

From these simple observations we can easily deduce that which seems to elude the best minds that NASA can throw at the "Cydonia problem." Wind was not a significant factor in the geomorphology of either the Face or the D&M. To ascribe the shape of any object of interest at Cydonia to this process flatly ignores the observations that contradict it.

It would be easy, even comforting, to dismiss this as merely rationalization—to assume that the staff geologists at ASU, so used to explaining things in comfortable terms, had simply chosen to ignore the frightening paradigm shift that the Face, the D&M and other objects represent. To somehow convince themselves that, yes, the wind *could* make those turns...

Chapter Nine Images

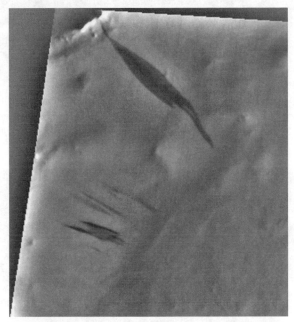

Fig. 9-1 – MGS image SP2-33806. Prime evidence in a 2000 Enterprise Mission press release announcing the Enterprise discovery of *liquid water* – flowing across the surface of Mars (JPL/Enterprise Mission)

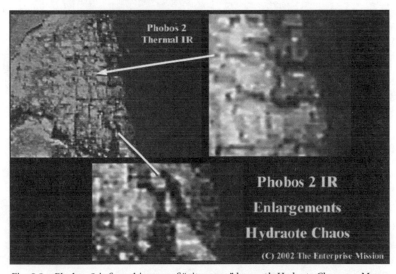

Fig. 9-2 – Phobos 2 infrared image of "cityscape" beneath Hydoate Chaos on Mars.

Fig. 9-3 - Mars Odyssey thermal Infrared night-time image of remarkable "geometric structures" – in the Hydroates Chaos Region of Mars. NASA's explanation: "windswept mesas" (ASU/Enterprise Mission Enhancement)

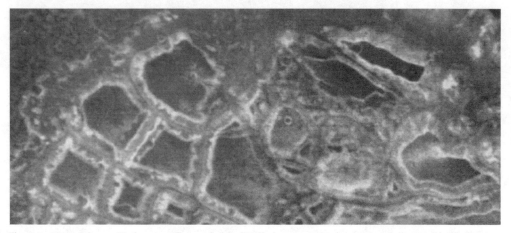

Fig. 9-4 – Mars Odyssey IR close-up of remarkably detailed symmetries on geometric "mesas" in Hydroates Chaos (ASU/Enterprise Mission Enhancement).

Fig. 9-5 – Comparison of "Inca City" in South Polar region with Anasazi ruins in New Mexico.

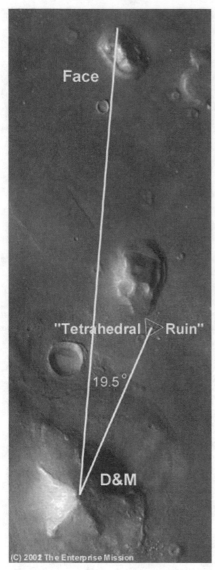

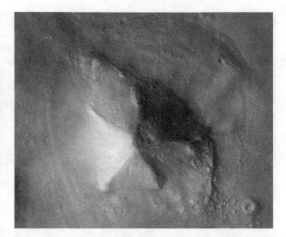

Fig. 9-7 - The heavily-eroded but still remarkable symmetries of the "D&M Pyramid" – from visible light Mars Odyssey 2001 THEMIS image with ~20-meter resolution (THEMIS-ASU)

Fig. 9-6 - Mars Odyssey 2001 visible light image of Cydonia, showing the Face, the D&M pyramid, the "symmetrical mesa" and the tetrahedral ruin. April 2002 (THEMIS-ASU).

Fig. 9-8 – Mars Odyssey THEMIS close-up of D&M seven-sided base platform, with five-sided original Torun symmetry superimposed. Note internal D&M 19.5-degree angle (ASU/Torun/Enterprise Mission)

Chapter Ten

Mars Heats Up

"There are unimaginable wonders [out there], for those who can remove some of truth's protective layers." —Astronaut Neil A. Armstrong, on the occasion of the Twenty-Fifth anniversary of the Apollo 11 landing

"The first thing we do, let's kill all the lawyers." —William Shakespeare, Henry VI, Part 2

As the Spring of 2002 drifted into Summer, we were more optimistic than we had been in twenty years that we were on the brink of a breakthrough with Cydonia. Odyssey's chief project scientist, Steven Saunders, continued to make positive statements to members of the Enterprise Mission's bulletin board service regarding the possibility of getting multi-spectral color images of the area around the Face. Deep Space informed us that the THEMIS instrument had taken both daytime and nighttime IR images, and that we needed to push for release of them. Art Bell had returned from his abrupt "retirement," and in concert with him Hoagland used the airwaves to apply public pressure to release the data. However, Deep Space had also issued a disquieting warning: After viewing the initial data from Cydonia, "the lawyers" had taken over inside JPL.

As we tried to grapple with the meaning of this admonition, we got the news: Saunders announced that the Cydonia IR data would be released a few days later, in late July 2002. We posted a "Captain's Update" on the evening of July 23, congratulating the "crew" on the BBS for having extracted this latest bit of data from NASA with their public e-mail and fax campaign. We knew that if this dataset was honest, we might possibly have a "smoking gun" that would end the debate about Cydonia's artificiality once and for all. Although we had already made what we considered an incredibly strong case over the years that certain objects at Cydonia—including the famous Face on Mars—were artificial, the latest THEMIS data was on a totally different level. Victory, it seemed, might finally be at hand. If only we could get the data to stand still.

When the image was finally released, on July 24, 2002,[129] we had to admit to disappointment. Instead of the full five-band color, fully processed multi-spectral image we had expected, we instead got a series of nine grayscale image strips, lined up side-by-side on a black background canvas. Still, we downloaded the data and began to work with it.

As a deliberate choice, the authors decided early on to stay completely out of the "processing phase" of this investigation. We instead elected to utilize the skills of two outside volunteers in image processing—who freely offered invaluable technical assistance: Holger Isenberg and Keith Laney.

Neither were full-time members of our long-standing team of other imaging and geological experts, whom we have previously turned to over the years, but rather were highly qualified and completely independent recent "interested persons" in our work. Holger is a graduate engineer for Applied Computer Science at Dortmund University, as well as a Unix System and Network Administrator at a German software company. Keith Laney is a digital imaging and software applications specialist and MOC image processor for the NASA-Ames' MOC MER2003 Landing Sites Project. We felt this would give the maximum degree of credibility to our eventual results.

Since the other key factor in gaining credibility in scientific circles is reproducibility, we also decided that Keith and Holger should aim at accomplishing the same tasks, but work as separately from each other as was practical.

Keith, at first, was not very enthusiastic about any aspect of the "project." Hoping to see a full-color, multi-spectral image (per Dr. Saunders' preview web announcement, made just prior to the July 24 release), Laney was so displeased with what was eventually published on the THEMIS website (a set of grayscale, side-by-side, multiple-band image *strips*), that he didn't even bother to download the composite THEMIS image the first day (July 24, 2002), flatly declaring that "it sucked" when asked about it in the Enterprise Mission BBS and on other message boards across the Net.

Because of this highly visible position, Keith (and a few others who had also publicly criticized the quality of the new image) then began to get a series of interesting "responses" in the Enterprise BBS and in private chats. These responses came from two new visitors to our site: "Bamf," and one "Dan Smythe." The former had begun posting about a month before the image release; the latter appeared simultaneously with its release. Both of these newfound "friends" seemed to know quite a bit about infrared image processing, and pretty much *goaded* Keith into finally going back and processing the full-size TIFF file—which (according to his own computer log) he had downloaded on July 25, at 10:27 p.m. EDT.

We were, from the get-go, more than a tad suspicious of both Bamf and Smythe, and that suspicion only deepened when we discovered that "Bamf" was in fact none other than Noel Gorelick—who claimed (in his on-line Enterprise BBS profile) to be the manager of Arizona State University's Mars Computation Center. Gorelick went on to further claim that he was in fact *the person* who had hand-assembled (over a period of four days) the infrared Cydonia image posted on the THEMIS website.

While from the beginning publicly acerbic and disparaging of the very reason for the existence of the Enterprise Mission website, privately "Bamf" and Smythe provided Keith and Holger, and anyone else with the smarts to work with the new image, a virtual on-line "how-to" tutorial in infrared image processing.

Using the image he had downloaded on the twenty-fifth, and working with a state-of-the-art software suite that included ENVI 3.5 (which provided a crucial "decorrelation-stretch" tool), Keith began to achieve astonishing results once he got the hang of it. His initial color ratio composites—from the nine strips of drab B&W imaging he had downloaded from the official THEMIS website—were nothing less than stunning.

In addition to a whole new IR perspective on the long-familiar features of Cydonia—the Face, the D&M, etc.—Laney found a troublesome, quasi-regular pattern of easily discernable pervasive rectilinear markings running almost the entire length of each band of the IR Cydonia strips. When he queried Gorelick about the annoying "blockies" (as he called them), "Bamf" grudgingly admitted, "so, you've found our dirty little secret." The ASU Manager acknowledged that the THEMIS team had also seen this peculiar "noise," and had decided *not* to create the normal, multi-spectral *color* images produced of other areas on Mars from the same THEMIS camera because of this disfiguring pattern. Remarkably, he also admitted that, despite their best efforts, they could not seem to remove the "noise" or find it on other published IR THEMIS images.

It seemed unique to the Cydonia THEMIS IR images...

Keith then began feverishly working on his own methods to remove this offending pattern, even sharing a couple of destriping tricks with Gorelick (who was quietly soliciting outside assistance on the problem) in the process—but, ultimately, he had no more luck than the official THEMIS team.

Around the same time, Hoagland (who was following this unique exchange between Bamf and Laney via e-mail) decided to suggest a radically different process for eliminating this peculiar noise, a well-known technique for "noise averaging" used for decades in astronomical photography—called "luminance layering." Hoagland outlined adding the earlier Odyssey visible light image (released April 12, 2002) to the stunning three-color IR composites Keith was creating from the latest THEMIS release. Laney, remarkably, had at about the same time independently decided to do the same thing. Once Keith did this—essentially adding the visible light image to the color stretched IR data—things rapidly got *very* interesting.

What both men found to their surprise was that the pattern not only wouldn't go away, it actually got *more pronounced*. It suddenly struck Hoagland that what he and Laney (not to mention the THEMIS team itself) thought was "noise" in the Cydonia images might really be *the signal*. Working on this radical assumption, Hoagland switched gears and quietly urged Laney to start trying to amplify the pattern in his on-going ENVI processing of the raw data, rather than continue efforts to remove it.

At the same time, taking the composited VIS/IR "decorrelated" versions Laney had already completed, Hoagland began a few experiments of his own.

That's when the "cityscape"—the mile upon mile of individual, clearly artificial structures hiding beneath the dusty plains of Cydonia—became completely obvious.

By simply increasing the image saturation of the composite VIS/IR image, what had been an annoying and mystifying block pattern was suddenly revealed by Hoagland to be a stunning, highly detailed *city,* silently sleeping for untold ages beneath the timeless sands of Mars (see Color Fig. 17).

It quickly became apparent the cityscape was real. For one thing, the pattern did not follow the scan lines of the imager, but rather true Cydonia north and south—exactly as terrestrial cities most often do. In addition, there were a wide variety of individual "structures" which did not conform to this north/south grid at all. We were completely confident that these structures are not noise, since they show, in both layout and design, clear evidence of architectural intent and organization—and even the smallest structures were at least an order of magnitude (ten times) above the very low noise threshold of these superb VIS/IR composites, both in terms of resolution as well as intensity.

Without any further enhancement, massive individual structures literally "fell out" of the data by simply turning up the color saturation. Previously identified features, like the "tunnel" running out of the Fort, could now be seen to continue out across the Cydonia plain, literally *under the soil*.

It was first noted by us because of its similarity to the Glass Tunnels Hoagland found several hundred miles west of Cydonia two years previously, and that got Arthur C. Clarke so excited in 2001, we said at the time that this "linear Fort feature" was consistent with our model that the Glass Tunnels were some kind of subterranean transport system. The exposed portion of the "Fort tunnel" is clearly visible even in the Viking data from over twenty years ago, and runs out from underneath the Fort at a roughly 60° angle to the Face-side wall. However, in the visible images, the tunnel runs for only a few hundred feet before it disappears beneath the Cydonian sands.

In the more recent B&W visible light MOC images we have received from MGS (Mars Global Surveyor), the close-up view of the tunnel revealed that it has a distinct *rib cross-structure*—strikingly similar to the other Martian "tunnels" we had found (and which our NASA critics, with boring predictability, dismissed as simple "dune trains").

On the Cydonia VIS/IR composites, we found more (see Color Fig. 18).

One new transport tunnel emerges from the base of a building we are now calling the "train station" in downtown Cydonia and runs straight into the "mesa" just south of the Face. In fact, it intersects the "mesa" at the exact lateral center of that massive above-ground structure. Close examination of the other end of the tube shows that the tunnel terminates at the base of the "train station" directly in front of a series of striking *open archways* in that building.

What was absolutely clear from the new THEMIS enhancements is that the "tunnels" are indeed just that. The THEMIS image allows us to see that the tunnel running out from the Fort is not just a simple depression in the landscape, but a coherent, tubular feature that runs for many miles beneath the Martian desert and across the Cydonia plain into the heart of the buried "Cydonia city." There are also clear structural cross-members—massive in scale—that run 90° to the tunnel at various points along its traverse, as well as its companion tunnel leading to the "mesa" south of the Face. Architects and engineers will again quickly recognize these features as obvious evidence of intent and design. The simple fact that the tunnels can be seen to continue *below* the ground, is clear evidence that they are not natural formations of any kind.

And there were the other, even more unimaginable wonders on this THEMIS image, the most prominent being a stunning, overwhelmingly architectural structure we called "The Temple" [Fig. 10-1].

You can clearly see [Fig. 10-1], without even straining your eyes, the individual cells in the roof and what appear to be *decorative buttresses* all around this multi-story structure. This object, just east (to the "Face side") of the Fort, is an architectural marvel—clearly a structure of significance and, with the exception of its apparently missing roof, in nearly pristine condition. It is also enormous, taking up an area the size of an entire city block all by itself. We quickly detected numerous other completely obvious buildings of various types scattered all across the image.

Obviously, we were excited by our new find. This data fit exactly the kinds of descriptions we had been getting from Deep Space, and even Saunders. It was very tempting to release the data immediately—but before we could go public, we had a few issues to resolve. First, if there truly was an underground city at Cydonia, how did it get there? Was it older or newer then the eroded ruins above the ground? And even more pressing, how could we see it at all? IR images had some degree of ground penetration capability, but these images seemed to be visualizing objects well beyond the (published) capabilities of the camera.

Longwave infrared imagers have a limited ability to penetrate underground by detecting surface "hot spots" conducting heat upward from below, but they typically cannot show the kind of detail we were seeing. The one possibility was that we were remotely sensing our "city" through a layer of fluffy, micron-sized dust essentially transparent to the THEMIS wavebands; probably covering another thicker layer of protective and also infrared transparent material, namely ice.

Exactly how that ice got there, and how we were seeing the city below it, quickly became secondary considerations, however. Just as we were ready to "go to press," we suddenly became aware that the original THEMIS daytime IR image of Cydonia posted on the website may have been altered.

The first hints came when Holger Isenberg (remember, the other imaging associate working from Germany), was unable to duplicate our key results. Holger was not working with precisely the same software suite as the rest of us (ENVI 3.5 was unavailable to him, gratis, unlike Keith's copy, in Europe), which was both a good and bad development, as it inadvertently gave us a chance to compare the results when two different, but functionally similar sets of imaging tools were applied to the same IR data.

The first assumption was that there might be something that either he or we were doing "wrong" that gave us differing results. We eventually discovered that Holger was using the PNG version of the THEMIS image, as opposed to the full size TIFF that the rest of us were working with—but that did not account for the entire discrepancy that we were seeing. PNGs are a fundamentally solid file format, completely lacking the "lossy" compression problems of JPGs and GIFs.

Beyond that, the objects that we were finding were so big, so far beyond the noise level of the image and so easy to bring out (once the compositing of the IR bands had been done properly), that even with a compressed web image file format, Holger should have seen these objects literally jump out at him. Instead, there seemed to be a distinct lack of information in the images he was processing.

And all of this came on the heels of the dramatic and curious developments of the preceding week or so. Our new-found friend, Bamf, *aka* Noel Gorelick of ASU, made some emphatic statements that seemed to fly in the Face of his previous attempts to help our efforts. On or around August 20, 2002, Gorelick made what seemed at the time to be a bizarre and arrogant claim. He insisted he had the power—and indeed the *right*—to flagrantly alter the data posted on the official THEMIS website... *at his whim.* To quote him verbatim:

> *"The daily images website is a service I provide and support entirely because I'm a nice guy. There is no NASA mandate or contractual requirement for me to provide these images to anyone before they're delivered to the PDS [Planetary Data System]. I do so because the public is interested in what's going on with the mission, and it's good for public relations.* Accordingly, if I feel like degrading the data before I post it, I'm certainly free to do so. If I want to scribble on the images with a crayon before posting them to my website, I'm probably free to do that too. *The 'government data' that the public paid for is being well cared for while it's being prepared for delivery to the PDS." [Emphasis added.]*

Tagged on to an earlier, equally strange statement in the same vein made by Gorelick—claiming that "no one can tell if an image has been fundamentally altered"—we began to become more than a tad concerned. This, in turn, led to us withdrawing from our original plan to post the infrared Cydonia THEMIS data and our preliminary analysis on August 22, 2002—while we rechecked our data's fundamental integrity and lineage.

That prompted an immediate response from ASU's Dr. Philip Christensen, the principal investigator for the Mars Odyssey 2001 THEMIS instrument. He sent an e-mail to Bara late in the evening on August 22, as if he had been waiting either at home or in the office for our appearance on *Coast to Coast AM:*

> *"Dear Michael,*
>
> *"I am confused by your statements regarding the THEMIS IR data and your decision not to release your findings. The data were calibrated [sic] our standard processes and in the same way that is done for the THEMIS science team. I am not sure why you are suggesting that Noel, or anyone else on the THEMIS team, has done anything to alter the data—he was simply questioning how you have treated the data and how you have validated your methods and processes.*
>
> *"I look forward to your release and to a detailed description of exactly what you have done to the data once you have downloaded [sic] from our website. Hopefully you will provide an accurate description of these techniques and the methods that you have employed to validate everything you have done to the data.*
>
> *"Sincerely,*
> *"Phil Christensen "*
> *THEMIS Principal Investigator"*

At this point, we made a conscious decision to pretend in public that we were more than satisfied with this response. The truth was, it only deepened our concern. Not only does Christensen's e-mail make a blatant misstatement— Gorelick certainly *did* imply he might have altered the data (and more than once), and was not simply "questioning how you have treated the data." If threatening to "scribble on the images with *a crayon*" doesn't strike Christensen as a reason for us to imply that Noel might have done something untoward to the data, well, Phil, what *would* qualify?

The most disquieting part of the e-mail was Christensen's inordinate interest in our "methods and processes." Why was he so interested in what *we* had done to the data, when he himself and his own team had provided *nothing* to the public (which, Gorelick's assertions to the contrary, owns the data) about what *they* might have done to it? The THEMIS team hadn't provided any description as to how the data had been handled, treated or possibly "enhanced." They hadn't provided the slightest hint of what filters may have been applied, a paper trail of who had handled it, or even the most rudimentary ancillary data (spacecraft geometry, lighting, etc.) for *any* of the released Odyssey images. If they weren't even willing to provide the date, time and orbit number of the THEMIS Cydonia image, why would they be so interested in us providing an "accurate description" of what *we* had done?

We didn't quite know yet, but we knew we needed to find out—so we went back through the data stream and tried to find any discrepancies in either our processes or tools. The one thing that kept coming out was that Holger Isenberg just did not seem to be getting the same results as Keith Laney. In fact, he wasn't even coming close. Granted, he didn't have full use of ENVI, but that alone didn't seem to fully address the striking differences. At this point our "separate but equal processing strategy" seemed to have effectively run its course.

Keith was non-plussed by this problem as well. After initially thinking that Holger just wasn't "getting it," he finally decided to retrace his own steps, go back to the THEMIS website and download the image one more time and start from scratch, working with Holger step by step. When he did so, he—to put it bluntly—got a major shock.

When he brought up the two images side-by-side—the one he'd downloaded on the evening of July 25, and the one he just downloaded again from the exact same official THEMIS website on the evening of August 25—*they didn't match.*

Keith's original July 25 version is noticeably different—even to naked eye inspection—from the one currently posted (as of this writing) on the official THEMIS website. In Keith's version, you can plainly see the block pattern, or part of it, even in the so-called "raw" data—without *any* computer processing. In the official THEMIS posting, these "blocks" are completely absent.

The differences become even more striking when standard IR enhancement processes are applied. In Keith's version, he could easily replicate his previous results (with ENVI) and get down into the data. The result is a beautifully clean IR composite image with very little noise. When he uses precisely the same processes on the "official" website version, the result is absolute, unadulterated "noise-filled crap" (see Color Fig. 19).

Some readers in the BBS were confused because in the raw, the official version seems noticeably sharper. In fact, this is because it had a sharpening filter applied to a visual image overlaying the IR. The "real" one had a mild (0.42) Gaussian blur applied to it. Keith mistakenly applied this to the original image in an attempt to enhance what initially appeared to him to be a fuzzy, noisy image. Fortunately, he saved an unblurred copy of the original, which he used for subsequent processing. In any event, this should not be a major cause for concern.

As the ENVI software manual points out, in a true multi-spectral IR image, you cannot destroy or add to the data with a simple filter, like a Gaussian blur. You can only enhance it, because each individual IR pixel retains the same original base intensity data. On the other hand, if you reduce the resolution with resampling, overlay a visible light image and then apply a sharpen filter (while adding an unknown amount of pure white noise)—as apparently had been done with the official website version—then you will get what we see on the degraded official version. The aforementioned "noise-filled crap" (see Color Fig. 20).

So what was happening here? Our conclusion is that the image Keith had obtained was a genuine—or at least close to genuine—Mars Odyssey 2001 THEMIS IR image. What was officially posted on the THEMIS website on July 24, 2002 (and remains there to this day) when the vast majority of interested parties downloaded it, is a highly-modified, highly-degraded image of unknown origin. In other words, a highly diluted fake.

Indeed, if you read the abstract of Dr. Christensen's own recent paper on the THEMIS results,[130] he makes it quite clear that the instrument is more than capable of achieving excellent "ground penetration." To quote from the abstract:

"Regional 100m mapping has revealed the presence of channel systems in ancient crater terrains not detected by Viking and not mapped by the high-resolution camera on Mars Global Surveyor."

In other words, "We're seeing stuff with the THEMIS instrument below the ground that can't be seen on the visible light Viking or MGS images."

Faced with this discrepancy, we elected to hold off publishing on Thursday night, August 29, 2002, and instead simply put the source data out side by side so that it could be evaluated. We waited to see if there was a response from Christensen, but when none was forthcoming, we elected to take the initiative and initiate an e-mail exchange, which was also ultimately fruitless. In the absence of absolute proof one way or the other, we were left to go over the points which make it overwhelmingly clear to us that the image Keith was working with is the real deal. Our analysis resulted in four separate *proofs* that led us to this conclusion. In addition, there are several other, finer points that reinforce our stance.

First, we know that Keith's image and the official image probably don't come from the same source. The header tag in Keith's version shows that it was converted from a PNM format, a standard conversion format used inside NASA to process remote probe images. In Holger Isenberg's opinion (remember, he's a UNIX administrator for a German software company), Keith's image was also processed on an older UNIX-based computer—typical of the type of equipment used by universities and the government. The "official" version, on the other hand, appears to have been processed on a Windows-based machine and shows no sign of having been converted from a PNM. So, there is nothing inconsistent in the header tags with Keith's image being the genuine article.

Since we are all used to working with data in the visual light spectrum, we have been taught to assume that sharper is better, which it is—if you *are* working in the visible light bands. However, visual resolution does not equate to infrared sensitivity or information richness. The real image, while it may look a bit blurry, in fact contains far more data than the official version, no matter what the official version looks like on first (naked eye) inspection. That's why the processing and enhancement tools used for multi-spectral imaging, to accomplish exactly that, are *crucial* to extracting that signal from this set of IR images.

When you do a visual inspection of the two sets of raw data side by side (or, in our case, one over the other), there are additional visual clues that will tell you that the "real" version is true THEMIS IR data, and that the "official" version is at best a poorly rendered copy.

The first proof of this can be found with a simple visual inspection of some specific features in the "raw" versions of both the "real" and "official" versions of the THEMIS Cydonia IR data.

In Keith's real version, there are subtle but distinct differences in specific features from frame to frame. This is what true IR data should show, because each signal return in each individual IR waveband is going to find (in the "dusty surface model") a slightly different bottom on the planet below. The longer wavelength signals will penetrate more deeply than the shorter wavelength signals. As an inevitable result, there will be subtle changes in the appearance of certain features—but only if they are "real" features on (or just below) the planet's visible dusty surface.

Conversely, in the official version of the Cydonia IR dataset we found no difference whatsoever from frame to frame as you go up the infrared bands. All you can see is steady, overall brightening. This proves that the top layer of the official version is not only *not* real IR data, but that someone must have simply brightened the entire image, or methodically darkened the individual image strips, in a blatant attempt at misdirection—a blatant fraud made up to look like real IR data to the uneducated eye.

The second major proof validating the real version is obtained by a comparison of specific features in earlier visible light images of Cydonia. There are numerous areas on the Odyssey, MGS and Viking visible light images that can be inspected and compared to the real IR data. When we do so, we would expect to see some of the block features that appear in the infrared to also appear in the visible light images, assuming that some of these buried structures are actually near the top of the dusty, icy layer.

There are numerous examples on MGS visual images of block-type features matching up with the IR blocks. We posted several on our website at the time. If these were processing artifacts (or scanner marks, as one person actually suggested), then there would not be *any* correlation between visible features and the IR blocks. The existence of even one correlation constitutes a proof that the blocks represent real features—on or just beneath the Cydonia dusty plain.[131]

The next proof is in the analysis of the noise floor.

Using an ENVI-generated data table comparison, Keith's "real" version was much clearer, showed far more detail, and simply contains far less noise than the official version. The "real" image had a much wider range of spectral data in it, owing to the enhancement capabilities of the ENVI 3.5 decorrelation-stretch technique and the *much cleaner source data*. The official version may have actually been generated from the same original, but the spectral range of the image—in essence "the signal"—was been *deliberately compressed,* resulting in the extremely noisy and scientifically worthless (by comparison) infrared image. In short, if you were working with the official version, you couldn't see the proverbial forest for the trees, or in this case, the artifacts for the Artifacts...

This second major point—which numerically validated for us the entire reality of Keith's version—is that it is *entirely consistent* with the kind of quantitative results one would expect from the exquisite THEMIS instrument, while the "official" version is entirely inconsistent with what that amazing instrument is capable of producing.

According to a JPL document[132] (which was subsequently removed from the JPL website after we downloaded it), the THEMIS instrument is accurate to plus or minus 0.001K (one thousandth of a degree Kelvin) temperature measurement.

That means that it is capable of differentiating temperature differences—instrument thermal "noise"—in increments that are incredibly precise.

What that translates to visually is brought out when you decorrelation-stretch the images—as Keith has done with ENVI 3.5—essentially separating the thermal data from the compositional data, to allow you to see more detail in the "composition bands." And that is why, when we added the visible light overlay to the "good" THEMIS data the screen literally explodes with rich detail and stunning clarity. In short, our version of the THEMIS data behaved exactly as *real* THEMIS data should—and the "official version" posted on the ASU website does *not*.

The final and most conclusive *proof*—that Keith's version of the Cydonia IR image is the *real* one—comes, from of all places, the very instrument on *another* spacecraft (MGS) that was fraudulently used by NASA in a failed attempt to debunk the Face back in 2001.

In an ironic twist, we can use published MOLA (Mars Orbiter Laser Altimeter) data from Mars Global Surveyor to ultimately clinch our "buried city" model.

The laser altimetry data from a single pass over the Face provides an excellent test of our "dust covered ice" model. The MOLA instrument operates at the 10.6 micron wavelength, essentially smack-dab in the middle of the THEMIS sensing band. The altimeter is not capable of penetrating solid ground at that wavelength, but it should not have a problem piercing a fluffy layer of micron-sized dust and an underlying long-wave IR translucent material, like ice. When we looked at the altitude scan, we could see that the majority of the ground truth at Cydonia seems to lie *below* the mean datum of the visible plain (in some areas many *thousands* of feet below). In addition, there are no corresponding visible troughs or geologic features to account for this anomalous penetration—but there is our stunning "cityscape."

So what the MOLA data is exquisitely confirming is that Cydonia is not a flat, relatively featureless plain with some mildly eroded mesas scattered about. It is, in fact, a deep, dust-covered plain which lies atop a thick layer of ice that has preserved and concealed—literally for countless millennia—a once highly advanced, plainly technological and now mysteriously absent civilization. All it takes to peel back "truth's protective layer" is to go there, brush away the dust and literally start drilling.

Another reason we are so certain that this new Cydonia data was the real deal, and the official version is a hoax, is that the data from Keith's version of the IR data is consistent with what we already know about what secrets Mars holds under the surface.

In case any of you have forgotten, this THEMIS Cydonia city pattern is a virtual twin for the discovery made by a similar thermal IR scanner aboard Russia's Phobos 2 spacecraft in 1989. A side-by-side comparison clearly demonstrates that "Cydonia City" is not an isolated settlement, but rather part of what appears to be a planet-wide, far-reaching civilization [Fig. 10-2].

There can be no middle ground here. Architects and engineers will instantly recognize the form, fit and function of their craft in the objects scattered all throughout the THEMIS image. And even more, some of the features can be seen, just barely covered over, in the Odyssey visible light image and even on the original Viking data. How "noise" could occupy the same area and location on images taken twenty-five years apart, by completely different instruments and on completely different spacecraft, is going to be a tough hurdle for the usual suspects to clear. In addition, the MOLA data confirms that a great deal of Cydonia exists *below* the visible surface precisely in the basin occupied by our frozen city. The laser can't lie.

So Keith's version of the IR image—the only one which is consistent with *all* the other observations—must be the closest thing to the "real" data returned by the THEMIS instrument we have now in the public view.

There is one final point. We had hoped, based on Hoagland's discussion with NASA's Dr. James Garvin—the head scientist of NASA's ongoing Mars programs—to avoid the usual "cloak and dagger" nonsense that had so permeated and tainted our various interactions with NASA over the years. However, we were now forced to reluctantly conclude, based on this "bait and switch" maneuver apparently cooked up by *someone* inside ASU's official THEMIS team, that nothing had really changed in this regard. As we considered what might have happened had we gone ahead and published our results on August 22, 2002 as originally planned, it became very obvious that someone in NASA/ASU was trying very hard to set us up.

Had we gone forward with our original announcement, and revealed the images without the background story concerning the two conflicting datasets, we would have naturally called upon NASA professionals and experienced IR imaging experts in the private sector and academia to replicate our work. They would have then gone to the official THEMIS website, downloaded the "official" version of the THEMIS IR data—and been totally unable to accomplish that, even given the proper multi-spectral software and experience.

We would have been thoroughly and publicly discredited, certainly in the eyes of the curious middle of this debate, and any nascent interest in the possibility of a former, now-extinct Martian civilization beginning to stir among some of the honest professionals inside NASA or in the press would have quietly died, as *someone* clearly intended.

We now understood what Deep Space was referring to with his cryptic warnings about the "lawyers" being in charge. In spite of the growing signs that the "Roosters" were winning the internal war, this whole incident reminded us that this does not mean that the losers, the "Owls," would quietly go away.

So, we knew a few things for certain. We knew that someone inside JPL had tried to set us up, tried to sucker us into releasing data that couldn't be confirmed independently by using the official data. We knew that Keith's data was real, because it had passed every test and was entirely consistent with what THEMIS data should look like. What we didn't know was just who had been behind this rather clumsy and desperate attempt to discredit us.

In order to separate the good guys from the bad guys, our first order of business was to determine just how and why Keith Laney was apparently specifically targeted to receive this data, and just who pulled the necessary strings to implement this part of the plan—and, most importantly, who gave the orders.

In order to reconstruct that trail, we had to go back to the two main players in this little psycho-drama, Bamf and Dan Smythe. Both showed up in the Enterprise BBS within about a month of the release of the Cydonia IR data (Smythe only surfaced after the image was released). This was shortly after Hoagland's seeming "breakthrough conversation" with Jim Garvin, and before any of us had a hint that this Cydonia IR data release from Odyssey was coming.

In retrospect, it is now obvious that this was a reconnaissance mission, an attempt to sort out just who among our crowd might have the knowledge and expertise to work with the real data once it was released. Keith Laney would be an ideal candidate for this kind of operation. He was reasonably independent of the authors, worked for the NASA/AMES Marsoweb Project, and had quite a bit of experience in visual image processing. What he lacked was any experience working with multi-spectral data. Into this void stepped Gorelick and Smythe, as we've documented already.

Once we had a heads-up from Stephen Saunders, the Chief Project Scientist for the Mars Odyssey Mission, that this data was going to be released the week of July 26, we looked back to our "ritual calendar" and tried to make a prediction as to what day it would actually be released. Since the original nighttime IR data had been released of Hydaspis Chaos exactly thirteen years to the day after the taking of the Phobos 2 image of the same region, we felt very strongly that this data release would follow a similar pattern. Very quickly, a date jumped out at us as perfect for the completion of this NASA ritual.

July 25, 1976, was a very important date. It was on that day, at 3:26 p.m. GMT, that a NASA probe named Viking 1 passed over a northern plain on Mars and snapped a photograph that would forever change the course of the space program, if not of history itself.

It was on that date, precisely in accordance with the now familiar "Osiris ritual" that we have documented so many times in NASA's forty-four-year history, that Viking snapped the first image of the Face on Mars, frame 35A72. Immediately, we realized that in order to complete this arcane "ritual," to stay consistent with the bizarre pattern of behavior that we had documented over and over again, they would *have* to release the new THEMIS Cydonia IR data on Thursday, July 25, 2002, the thirty-first anniversary of the taking of 35A72—the famous first Face image. No other date fit the pattern.

Imagine our surprise when, for the first time in memory, they broke the pattern and released the image in the early afternoon of July 24, 2002, not on the twenty-fifth. As we contemplated this break with tradition, we wondered if perhaps this was a hopeful sign, and indication that the ritualists inside NASA had lost control of the agenda. We now know better.

What was released on the twenty-fourth, and what the vast majority of the press and public downloaded, was not the real data, but the not-so-cleverly constructed fake. Keith Laney, almost certainly their number one candidate to receive the "real" IR data, took a quick look at the public release, blurted out "it sucks" and didn't even bother to download it. Hoagland quoted Keith that night on *Coast to Coast AM* and, following that, Keith got into a series of pointed exchanges with both Bamf and Smythe. Eventually, after much prodding, Keith went *back* to the THEMIS website and downloaded the image we now *know* is the "real" data, on July 25, 2002 at 10:27 p.m. EDT. In fact, as we studied the logs of the discussions, Smythe and Bamf *badgered* Keith into going to the site and downloading the image.

So, in essence, someone *did* complete the ritual: by getting the "real data" to Keith—and thus to us and to the American people—on the correct ritual date, July 25.

There is one other ritual point to be made. In the course of processing the "real" data, Keith Laney discovered that when you crop and rotate the individual image strips (remember, created by Bamf) to the vertical, and examine the pixel count, it turns out that the image strips are *1947 pixels* by *333 pixels,* down and across.

Or, 19.5 x 33, if you are tracking this sort of ritual. We took this as yet another hint from Bamf that this was the true Cydonia data that they had been working with on the inside, the data that had so amazed the JPL scientists in the Space.com articles.

Once Keith had downloaded the "real" image, Bamf proceeded to provide enough assistance to Keith (and Holger, among others) to enable Laney to eventually figure out how to properly process the data he'd been given, and *exactly* what software he would need to carry out the job. Bamf interlaced this crucial private e-mail tutoring in multi-spectral imaging with his previously mentioned bizarre public comments on our investigation, along with other categorically false assertions on the THEMIS Cydonia data themselves.

Among these were unambiguous statements that "the IR data on the website were useless" because they had "not been calibrated," and that "the Face is not any different than any *other* object at Cydonia... is not made of anything unusual, like metal or plastic..." Not only are both these statements *wrong,* but they are also flat out contradictory.

A JPL paper on decorrelation-stretch techniques[133] flatly states that calibration is not necessary to make use of decorrelation-stretched IR data, because of the nature of the algorithm and the quality of the process itself. All calibration ultimately does is identify *which* colors relate to *which* materials— assuming you know what materials you are dealing with in a totally alien context in the first place. There is, obviously, some question on that matter raised by this particular Cydonia data.

And as for Gorelick declaring "he knew what the Face is composed of," after he already claimed that the data "wasn't calibrated," this simply made no sense. Forgetting that it conflicts with not only the image caption that he (presumably) posted, it also conflicts with Christensen's e-mail in which he (Christensen) restates that the data *is* "calibrated." So for Gorelick to presume he knows what the Face is made of, when by his own claim he is working with uncalibrated data, was and is nonsense.

And even if the data is calibrated, as the THEMIS website and his boss now claimed, he still couldn't possibly know what the Face or anything else is made of, because as of the moment, nobody knows what "Cydonia City" is composed of. Unless of course there is a special "spectral library" on alien construction methods and materials available inside JPL.

All of these conflicting signals left us with a few certainties, and many more unanswered questions. We knew Keith's dataset was valid, and we knew that Gorelick had tried to help us to some degree. We also knew that somebody was hoping to sucker punch us by changing the data on the official site, in order to create profound confusion when we tried to have our own analysis verified by reference to that "official" data set.

So this left us with two questions: just who were the "good guys" and "bad guys" inside the ASU THEMIS team, and how did they get the right data to Keith Laney—and *only* Keith—for their "ritually perfect" July 25 ritualistic window?

What then became clear was that the answer to one question should provide the answer to the other. Keith Laney has a cable modem connection to the Internet. What that means is that he has a "static" IP address. Every time he goes online (or even has his computer turned on and physically connected to the web), he broadcasts *the same identifier* to the whole world. It would certainly be a piece of cake for any reasonably savvy computer network expert to "ping" Keith's system, determine his static IP address and lay a trap for him.

Once Bamf and Smythe had convinced Keith to go back (on the twenty-fifth) and finally download the THEMIS Cydonia image, it would be an easy process to wait for his static IP address to "log in" to the official THEMIS website. At that point, Gorelick (Bamf) who ran the THEMIS website could then redirect him, *and only him,* to a *duplicate website* that looked exactly like the official THEMIS page, but which instead contained the "real" data Keith now had on his computer. And that, we can say with certainty, is exactly how "they" did it.

Our confidence in this assertion is based primarily on the fact that this exact process—the act of redirecting someone to a counterfeit server or webpage, with or without their knowledge—has a specific name amongst the world's computer geeks. In the jargon of the techno-savvy, of which Noel Gorelick is undeniably a world-class member, it's called, simply:

Bamfing.[134]

So what Gorelick was doing all along in his months-long stay on our BBS was apparently seeking out just the right person to leak this crucial IR data to, finding the perfect character to "Bamf" the *real, stunning Cydonia images*.

So just who were Gorelick and Christensen, anyway? Heroes or villains? Friends or foes? Both or neither? It could be that they worked together to help set us up, or it could be that Gorelick leaked the real data to Laney because he was tired of all the cover-ups and nonsense behind the scenes; or it could be that both men were pressured, put upon by a powerful outside "Owl" element, to participate in this fraud. Unfortunately, we tend to think that this latter scenario is the more probable at this point.

Don't forget that Deep Space, our inside source (who, in turn, is fronting for a host of other sources), told us weeks before that "the lawyers are running everything at this point" inside NASA, in response to our own continuing political and legal initiatives. This sort of smarmy trick, this kind of underhanded double dealing on this key new Cydonia data, is *exactly* the way such individuals tend to think and operate—especially when they're desperate.

There was a firestorm of attention when we finally revealed this whole story to the public at large. Stories ran on MSNBC's website (Alan Boyle is one of the few truly curious reporters out there) and on several local newspapers in and around ASU. Many of our own backers in the anomalist community saw the images and promptly panicked. After spending years debating the artificiality of Cydonia, when confronted with proof of it in the form of Keith's images, they reacted angrily. We were surprised to read in our own BBS that many of these supposedly curious souls were accusing us of having faked the "real" data. They were apparently more inclined to believe this than to believe that we had actually been leaked data that proved Cydonia artificial.

Unfortunately, one of the nuttiest members of our board was Holger Isenberg. Once he heard that there were two sets of data (and we released them side-by-side on our site), he ceased all attempts to process the data. No matter how hard they tried, neither Keith nor Hoagland could get him to simply take Keith's version of the raw data and process it as he had done with the "official" version. After we posted our analysis of the data and laid out the events surrounding Keith's acquisition of it, he then turned on us and accused Keith of creating "fake" images and Hoagland of "lying" about how Keith had obtained the raw data. He then went on to declare that Bamf was "his friend," and that his new-found friend (whom he'd never even met face-to-face) would never do anything like post degraded data in place of the real stuff on the ASU-THEMIS website. Finally, he alleged that the "real" data was created by scanning a print copy of some "fake" images created in Photoshop (an idea we later learned he was fed by Gorelick in a private chat on our BBS). When we pointed out to Holger that he was the one who verified that the header data on Keith's file confirmed it came from a NASA file format and was processed on a UNIX machine (there is no Photoshop for UNIX version), he broke off all contact with us.

We found this process exasperating, but sadly confirmatory of the "Brookings" model. Ironically, "Cydonia" derives from a Greek root meaning "enlightenment"—or, in this context, "to illuminate the True History of All Mankind"—but the data in this case only seemed to create fear and confusion. Even friends like Art Bell were not immune.

Art was having a hard time grasping the idea that the images Keith was getting *couldn't* be faked. He was convinced that Gorelick or Christensen had somehow embedded the images of various buildings and structure in the data. As we had pointed out in our web story, the meticulous process of creating thermal IR images prevented any such fraud.

Keith was essentially working with *nine different individual images*, each slightly different than the next due to the fact that they each came from a different IR filter wavelength, and from the motion of the spacecraft. Keith then cropped the nine images out, rotated them to "vertical," and hand-aligned and assembled them into single frames, one layer on top of the other, two at a time. In other words, he took image strips one and two of the nine, and laid them on top of each other, then he took three and four and laid them on top of each other, and so on. This makes eight possible layer combinations for every individual image strip, adding up to a total of seventy-two different possible wavelength filter combinations. Next, he "color-ratioed" the combined images, assigning a color value (and weight) to individual elements in the image. This means that, as an example, titanium will appear "red" in the given image, if he assigns that color to that material. However, without calibration, i.e. knowing what the materials are in a given image (pretty tough on Mars!) all you can do is assign a color value to a material. It tells you nothing about what that material is, just whether it is the same as other pixels around it. The same rules apply to heat differences in a given image. You won't see a temperature, but you will see a temperature differentiation.

Next, Keith used a decorrelation-stretch tool on the images. This is a critical tool that exaggerates (not creates) the color/material/heat signature differences in the color-ratioed images. Then, as a last step, he added the luminance layer of the visual image.

In short, there is no way that Christensen, et al., could have possibly painted "fake" buildings on each of the nine images and guess how he'd combine each layer, what color ratios he'd use, what settings he'd use in the decorrelation-stretch tool, how he would rotate, align and crop the images, and whether or not he'd then combine the visual image on top of that. It's just flatly impossible. So the objects we were seeing in Keith's processing were definitely "there" on the real IR image.

Next, Keith endured a series of attacks from various amateur anomaly hunters who claimed it was he that "created" the blocks and buildings on the images. Further, they argued that nobody "knew for sure" that "Bamf" was really Noel Gorelick (despite "Bamf's" evident encyclopedic knowledge of infrared image processing) and that he might have led Keith into creating artifacts with the imaging suggestions he'd made.

At this point, we had had enough. It was arranged for Keith to brief Art in more detail, and show him a comparison between the earlier Phobos 2 IR image and the new Cydonia data [Fig. 10-2]. That did the trick with Art on the images, but everybody still wanted proof of who "Bamf" was—so one of the authors (Bara) called him.

It took no time at all to find Noel Gorelick in the ASU phone directory. Bara called him the afternoon of September 6, 2002. In the course of the conversation, Gorelick freely admitted he was "Bamf," that he was responsible for all of the postings in the Enterprise Mission BBS under that name, and he reiterated his stance that the Cydonia IR image page had been untouched since he posted it on July 24, 2002. Bara then went on *Coast to Coast AM* that night to pass this information on.[135]

To some extent, these accusations and this sort of behavior were to be expected from the "usual suspects" on the various message boards. What we weren't prepared for, however, was an assault on our integrity from a former friend, especially one who should have known better.

Stretching the Truth

Shortly before we published our revelation of the existence of two THEMIS Cydonia IR images, on August 29, 2002, SPSR's Dr. Mark Carlotto put out his own analysis of the Cydonia multispectral image. From the beginning, we were concerned about the contents of the Carlotto analysis. For one thing, there were simple errors (for instance, the IR image is referred to as "E0201847.gif," which is the wrong file name for the multispectral Cydonia image, "20020724A"). In addition, there were numerous misspellings and other obvious mistakes—which gave the whole project an air of haste and sloppiness. It did not seem to be up to Carlotto's usual thorough standards—at least as Hoagland had remembered them from working with Carlotto years before.

As we got into the content of the article, it seemed to stray into even odder territory. Carlotto started out by comparing the Odyssey Martian THEMIS data to terrestrial Landsat images, a very inaccurate comparison to say the least. Landsat is a 1970s-era technology that produces primarily surface reflectance data in the visible region of the spectrum, as compared to THEMIS, which concentrates on extracting data primarily from intrinsic thermal infrared emissions from surfaces and objects. Landsat, in contrast to THEMIS, has virtually no "ground penetration" capabilities at all. A far better comparison would have been made to a relatively new Earth orbiting instrument (1999) called ASTER (Advanced Spaceborne Thermal Emission and Reflection), which has very similar near-infrared capabilities to THEMIS.

Carlotto went on to make some even stranger "errors" in his article. He proceeded to conclude that there were various clays making up the composition of the region surrounding the Face. However, Carlotto surely must have known that he could make no such conclusions, since the calibration data for either the official version or the "real" version of the Cydonia IR data had yet to be released. He also failed to perform a decorrelation-stretch, a step that is crucial for separating the thermal data from the composition information in the image. However, even if he had, it would have made no difference, since without the crucial calibration data his conclusions about the specific composition of Cydonia were meaningless. Beyond that, he completely ignored the overwhelming amount of noise in the official image and seemed more than satisfied with the poor quality of it.

Even as we tried to make sense of Carlotto's seeming "brain fade," our attention was drawn to another recent item posted on his new website. In it, he directly addressed the issue we had raised concerning the discrepancy between the two IR datasets, the "official" and the "real" one obtained by Keith Laney. Carlotto declared that clearly the "real" version we had posted was a degraded version of the "official" one.

We knew this, of course, to be utterly laughable. In his analysis of the two images, Carlotto had taken only a single unnamed band from the "real" data, and compared it to a single unnamed band from the "official" version. He had not done a full composite of the "real" image bands, not done any color ratios, nor performed a decorrelation-stretch to enhance the data. All he did was make a simple visual inspection of two grayscale bands without performing any of the accepted processes for enhancing false color thermal IR data.

This, as we have said, is flatly *not* how you handle thermal IR data. This is higher-order IR information, compared to a simple "pretty picture." Even though it is lower *resolution* data, it is information totally unavailable from even much higher resolution visual images. Which is what makes lower resolution thermal IR THEMIS data such a potential breakthrough on the thirty-year-plus Cydonia problem, even at around 100 meters per pixel.

It is this fundamental fact of infrared optical physics that made us believe from the beginning that Laney's July 25 "blurry" Cydonia IR image was the "real deal," and the July 24, apparently much higher resolution "official" image was, in fact, the "doctored" version.

The "real" image, while it may look a bit blurry, actually contains far more data than the "official" version, no matter what the "official" version looks like on first (naked eye) visual inspection. That's why the processing and enhancement tools used for multi-spectral (more than one band) imaging, to accomplish exactly that, are crucial to extracting that "hidden" signal from this set of IR images.

Carlotto, it seemed, was making the same elementary error that some of our readers were making—assuming that sharper is automatically better, as it is for visible light images. And curiously, he did not do the other crucial steps, which would have instantly shown him that the "real" (July 25) images contained far more, and far better quality, data than the official version did. This simple analysis in and of itself would have immediately disproved his core hypothesis—for how can "degraded" data produce quantitatively better results (in a superb program like ENVI 3.5) under proper analysis, than its supposed "source data," and with infinitely less noise?

The simple answer is that it can't—but that's not what really bothered us. What troubled us in the extreme was that Dr. Carlotto, a world-class imaging expert, and DOD contractor on a host of classified imaging analyses, *should have known all this.*

So, given this decidedly odd state of affairs—a well-known and respected imaging specialist, who at least *used* to be curious about Cydonia and suspicious of NASA, making not only a seemingly colossal error in judgment, but compounding that error by failing to simply put it to the test—we decided to consider our options. The authors and Keith Laney conferenced about Carlotto's article on the night of September 3. The general consensus was that Carlotto had effectively "polished a turd" and declared that he discovered a pearl—all without even considering the field of gems which had been placed right in front of him.

Hoagland, however, refused to buy into the notion that Carlotto was as incompetent as his analysis made it seem. He staunchly defended Carlotto's skills and professionalism, insisting that there must be some other reason for his reticence to properly process the THEMIS images. Considering Carlotto's membership in SPSR, we decided it was possible that Carlotto had simply fallen in with the "honest but stupid" crowd that forgives every NASA transgression, yet holds the authors to much higher standards. The decision was taken to reach out to Carlotto—if only to save him the embarrassment of a public response pointing out his lack of thoroughness. Bara subsequently e-mailed Carlotto, informing him he'd made several errors in his article, and advising him to pull it, at least until we published our own analysis.

Carlotto responded via e-mail and pointed out that he had plenty of experience working with thermal images, that the "real" image was "obviously" degraded, and that his paper had been peer reviewed (by Dr. Horace Crater, a statistical analyst who has no working experience we are aware of with thermal IR). Mike's response back was basically "suit yourself, but if *we* had peer reviewed your paper, it would not be published right now." Carlotto e-mailed Bara back, got Hoagland's phone number, and the two men had a chat on September 4.

According to Hoagland, what Carlotto seemed most concerned about was that his previously published paper would be made obsolete by our article. After a wide-ranging discussion, which included Carlotto pointing out that he'd written his own decorrelation-stretch algorithm, Carlotto agreed to take the "real" image, perform all of the proper steps (composite, color ratios and decorrelation-stretch) on it, and either call or e-mail Hoagland with his results. That never happened.

We have no way of knowing if Carlotto ever did the analysis he agreed to do, but after more than a week of waiting, Carlotto's only response was to publish yet another "update," in which he dug himself into an even deeper scientific and ethical hole.

Instead of following the proper protocols for processing thermal infrared data as he'd agreed to do, Carlotto decided instead to take the "official" version of the Cydonia THEMIS image and subject it to a series of contrast and blur filters in an apparent attempt to "prove" that the Laney image was generated by degrading the official one. He did this by taking only a single band image, not a composite, and he of course did not do any of the other tests he agreed to perform in his conversation with Hoagland. This led him to conclude "its similarity to the top left ["real"] image strongly suggests that the Enterprise image is an altered version of the ASU image."

He then goes on to claim that because the Laney image changes from band to band in the unratioed grayscale it is a "distorted" version of the "official" ASU release. What is truly disturbing is that Carlotto's "test" here is proving exactly the *opposite* of what he is claiming. Real multispectral data (and certainly thermal IR data) *does* change from band to band. What he is illustrating is exactly what true multispectral data should look like. He's not seeing distortion, but rather the expected shift in "return" to the camera from slight variations in the thermal signatures of actual features on the Cydonia plain. The other phenomena he seems to be describing, the "shift" in certain edges of some of the large features, is simply due to the fact that the various bands are not all taken at exactly the same time. There are significant shifts in the spacecraft's position as different filtered CCDs in the THEMIS camera record (in a rapid-fire sequence) of the actual imaging data for all the bands. This makes it a near impossibility to simply overlay the various bands when doing a composite. Of course, had he simply read the camera specs, he would have known this, and physically corrected (as Laney has successfully) for the minute geometric shifts.

That Carlotto made no effort to correct the alignment problem is not only a testament to his lack of thoroughness in this case, but an outright indictment of his methods and possibly even his motivations. Keith Laney, at our request, produced an image similar to the section that Carlotto had done, only with the various bands properly aligned geometrically. It took him all of five minutes.

All Carlotto would have to do, if he truly *wanted* to decide the question of which dataset is "degraded" and which is "pristine," would be to run the two images *side by side* through a quality enhancement tool, like ENVI 3.5 as Keith had done. Had he done so—as he promised Hoagland he would—he would have clearly seen that he got it *completely backwards.*

What makes this truly egregious is that Carlotto certainly knows everything we have described above, that a simple visual inspection of a single grayscale IR band is *not* a valid comparison of these incredibly information-rich datasets. He has evidently decided that it is better to try and cover his own mistakes by making up a "pretty picture" which only proves his talents as an artist, not as a scientist. What seems to have happened is that Carlotto is unwilling to publicly face the fact that his initial declaration "that the Laney image is the degraded one" is flat out wrong. Given the opportunity to admit his mistake, he has instead decided to cover his tracks with this absurd comparison. We were truly sorry that Carlotto had chosen to take a politically defensive stand, instead of the scientifically courageous one.

The absurdity of this position is underscored by an e-mail Laney received from Research Systems, Inc. While refusing to get into the middle of the "which image is real" controversy, the communication made several points totally inconsistent with Carlotto's "analysis." Said Keith's RSI ENVI representative:

"I must admit, this has made quite a stir in the astronomy community! At any rate the images look awesome! As I tell anyone who asks about RSI's stance... 'RSI does not have an opinion either way, we just want to provide the best software to scientists so that they can do their own best work.'"

How could someone crudely "degrading" the official Cydonia data create "quite a stir in the astronomy community?" Given the scientists that RSI routinely deals with, and their level of multi-spectral expertise, any simple "degradation" of the official ASU THEMIS website image into the one Laney has been working with (with RSI looking on) would most certainly have been caught if it was as simple as Carlotto was now claiming. Also, if Keith's image was a "hoax," why would the RSI representative go *on record* saying that the images produced from it "look awesome?" Wouldn't the better part of valor be to simply refrain from *all* comment until the lineage of the "real" image was determined?

In effect, Carlotto had simply parroted the position of Dr. Phillip

Christensen of ASU by declaring that the Laney data is "degraded," when all parties involved (as the RSI e-mail underscores) certainly know that the official version is far inferior to the Laney data. By refusing to put that data to the true scientific test—whatever the reason—Carlotto and SPSR are reduced to being nothing but mouthpieces for the NASA party line. As we said before we released this data, people were going to have to take sides... and SPSR and Carlotto evidently had.

However, the attempt to bring Carlotto back into the fold was not all wasted. A reader, Wil Faust, made a truly inspired suggestion to us. Why not compare, he said, Carlotto's own seminal work—his *fractal* analysis of the Cydonia region from the Viking data—to our own IR results?

So we did.

It turns out that when you use Carlotto's own methodology, and take a single band of the IR image strip from Keith's "real" version of the data to compare it against Carlotto's own fractal analysis[136] of Viking frames 35A72 and 70A13, you get quite striking results. Not only are the THEMIS blocks—which Carlotto now flatly claims are "enhancement" or "filtering" artifacts—clearly visible on *his own work* from twelve years ago, but they match up very precisely with the blocks on the Laney image, *literally one for one* [Fig. 10-3].

It is now incumbent upon Dr. Carlotto, who is so sure that the Laney image isn't valid (or is at least "degraded"; he'd parsed his words pretty carefully) to demonstrate just how "filtering artifacts" can not only line up with features in Odyssey's THEMIS *and* visible light images—but also with direct non-fractal "hits" in his own dataset.

Significantly, these non-fractal patterns—to show up in the visible images at all—also have to be caused by geometric structures buried just beneath the ground. Clearly, Dr. Carlotto's methods—which have led him to dismiss all of these converging anomalies as just "degraded data"—had led him off the edge of the paper.

It is one thing to make honest errors in a piece of scientific work. It is quite another to compound those errors and miss the entire forest by hiding behind an incompetent "peer review" and obvious political propaganda, without even checking your own previously published work.

We have no desire to pillory Carlotto any further here. His own demonstrable lack of true scientific curiosity has more than accomplished that. We simply challenge him again in this volume—as we did in 2002—to produce and *publish* the composites, color ratios and decorrelation-stretch results from the appropriate multi-spectral analysis he promised Richard Hoagland he would do on the "leaked" THEMIS image. Until he does this, sadly, we can no longer endorse his methods, his competence or even his intellectual honesty on any issue pertaining to this continuing extraterrestrial artifacts investigation.

Night and the City

As the summer of 2002 drifted into fall, we had just about run the course on options over the IR data. We still hoped for a legitimate nighttime IR image, since that would provide far greater contrast than the daytime image and should theoretically reveal the "buried city" in even more dramatic relief than the "real" daytime IR had. Just when we reached the point of giving up, however, it was Bamf to the rescue. Shortly after our publication of an article highly critical of the deceptive statements he had been making in our online BBS,[137] he finally delivered that which we had been asking for—sort of.

On October 31, 2002—Halloween no less, the same pagan holiday that marked the "birth" of JPL—ASU released what they alleged was a nighttime infrared image of Cydonia. Not only did this image appear on that most pagan of pagan ritual dates, but the caption claimed it was taken just a few days before, October 24, 2002, the one year anniversary of Odyssey's insertion into Mars' orbit.

These ritualistic manipulations aside, it was immediately clear that there were issues with the data. Rather than the full nine bands as we had been given in the daytime image of July 24–25, this time we got a cropped image showing a portion of a single nighttime image strip alongside a similar crop from (supposedly) the July daytime image. The official ASU caption read:

"This pair of THEMIS infrared images shows the so-called 'Face on Mars' landform viewed during both the day and night. The nighttime THEMIS IR image was acquired on October 24, 2002; the daytime image was originally released on July 24, 2002. Both images are of THEMIS's ninth IR band (12.57 microns), and they have been geometrically projected for image registration."

Yet a comparison with band nine from the actual July 24 release revealed the first of some troubling discrepancies. Band nine of the July 24 daytime IR is listed on the original ASU graphic as being "12.58 microns"; but on the October 31 comparison, it's cited as "12.57 microns." It turns out that depending on which official THEMIS document you reference, you get a different center wavelength value for the infrared filter strips on the THEMIS camera CCDs. This can be quite confusing, since it makes a difference in how IR bands are subsequently processed.

Even more troubling than this perplexing "filter change" was a major discrepancy in the new image itself. In the July 24 original, the image "footprint" cuts off two mesas at the top of the scanned strip; in the October 31 "version" of the same image, there is clearly significantly more surface detail captured northeast of those mesas. Two verifiably different image "footprints," for what had been categorically maintained by NASA, ASU and JPL for months as only one July 24 image.

Given the shrill accusations of "hoax" and "fraud" leveled at both Keith Laney and the authors for even suggesting that there could be two different versions of this image, the new posting was vindication. The publication of two demonstrably different "official" versions is elemental proof that we were right. It also opens even more possibilities; if there are no less than two "official" versions of the July 24 data, why not a third version that matches Laney's "real" data?

That being said, what—if anything—was to be gained by taking seriously the contents of this first nighttime Cydonia IR image—THEMIS image number 20021031A? First of all, just looking at the image, something was clearly wrong: the data itself is just too "noisy" for a Martian summer image.

Even a casual comparison with a nighttime IR we'd seen previously reveals that the noise level of this new nighttime Cydonia IR is comparable to that image (101180002) which was taken in the dead of Martian winter in the northern hemisphere, March 21, 2002. The current Cydonia image, by contrast, was supposedly acquired October 24—eight months *later*—just after the beginning of Martian summer in the northern hemisphere.

So why was it so noisy? One basic reason could be that Bamf had simply lied again. The truth was that, yes, this was a nighttime Cydonia IR image, but it was taken (probably as one of a series of unreleased Cydonia images) *much* earlier in the current Odyssey Mission than publicly admitted—when it was simply a lot colder at Cydonia. This effectively reduces contrast and increases noise. After the publication of our latest article on October 20, 2002 Bamf (and those managing his actions on the THEMIS team) hurriedly put out something on nighttime Cydonia IR just to shut us up. So he reached into the THEMIS "hidden Cydonia drawer" and pulled out an older, "colder" (and partially sanitized) "new" image, which he then simply labeled as one taken on October 24, 2002. At least, that was our working theory. We now set out to prove it.

Because we know the latitude of Cydonia (about 41° N), and the tilt of Mars on its own axis (about 25°), we can easily calculate—by observing the details in the image, and comparing them to the geometric position of the sun at any time of the Martian year—when this image had to have been acquired. Essentially, it's "Astronomy 101."

This is the simple equation that allows us to derive this crucial geometric information:

$$\sin D = \sin a \,/\, (\cos b)^2$$

Where D = max deviation (north or south) of the rising or setting sun from an east/west line; a = planetary obliquity (its tilt); and b = the latitude of the observer. "D" is also defined as the "summer or winter solstice": i.e. the longest days of summer, or the shortest days of winter, in that hemisphere.

Halfway between these two farthest excursions of the sun, north and south along the horizon, is the geometric position of the "spring and fall equinoxes." The word derives from the Latin, meaning "equal night": i.e. the length of the days and nights at that position in Mars' annual solar orbit (year) are approximately equal (Mars' orbit being decidedly elliptical).

Applying this calculation to the new Cydonia image, we could derive—with absolute scientific certainty—*when* this image *had* to have been taken, and we didn't need to "trust" We began with the orbits of Mars and Earth. Using a newly created "analog orbital computer," courtesy of Dr. Bob Zubrin, we were able to compare the two planetary orbits and graphically convert from any date and season here on Earth to its equivalent on Mars.[138] By using this tool we could confirm that Odyssey arrived in Mars orbit on October 24, 2001, just *after* the Martian winter solstice in the northern hemisphere.

By then advancing the dates to correspond to Bamf's claim that the new nighttime IR Cydonia image was taken on that one-year anniversary—precisely twelve Earth months later, October 24, 2002—we could see that the new image would have been taken over Mars just after the Martian *summer* solstice in the northern hemisphere.

Because he arranged both images—day and night—at the same but opposite symbolic angle to the Martian equator in his graphic (about 7º), it's a simple matter to match the illumination of the features seen in the images with the true Martian coordinates—and compare that with the calculated sun angles for any particular Martian season derived from the equation cited earlier.

When we overlaid (within an error of +/-3º) the calculated angle of for the maximum summer and minimum winter illumination angles (solstices) to an east/west line (which determines the angles at which slopes receive the last rays of the setting sun, and thus which will be warmest throughout the night)—a very interesting picture regarding when this nighttime IR image was actually taken begins to emerge.

In the "natural model," the only source of energy to warm the Martian surface is solar illumination. For objects on that surface to "glow" in the thermal infrared at night and be detectable by the THEMIS camera, therefore, requires that they be dense enough to retain a significant amount of solar energy for hours after sunset and be angled essentially "face on" towards those last warming rays of the setting sun.

If we examine the nighttime Cydonia landscape Bamf had now given us, several intriguing aspects of this model nicely come together. Apart from the Face, there is another distinctive feature at Cydonia, "the island." It's a roughly rectangular mesa, several miles due east of the Face, with a flat surface area of several square miles—standing a few hundred feet above the surrounding plane. Its two western, relatively steep vertical cliffs face northwest and southwest roughly in the directions of the summer and winter solstice sunset points [Fig. 10-4].

Examination of the "daytime thermal image" [Fig. 10-4, left] clearly shows the afternoon sun coming from the southwest, casting distinct cold shadows off the mesa's northeastern cliff (dark band, upper right). The top, northwestern cliff [Fig 10-4, dark band, upper right]. though illuminated, is almost as dark as the shadowed eastern cliff, indicating that sunlight is reaching it at "grazing incidence." There's also a distinct (colder) "darkening" extending back from the western "tip" along this northwestern cliff, indicating an "outcropping" at the tip, creating a distinct (thus colder) shadow extending about a mile.

Even though Bamf claimed (on the BBS and also in several private e-mails to some members) that this image was taken on May 5, one can confidently state from this solar geometry (mathematically extrapolated to the sunset point), that the July 24 image had to have been taken approximately halfway between the northern Martian hemisphere winter solstice (just before Mars Odyssey arrived, October 24) and the Martian spring equinox. In other words, on Earth, sometime in *January 2002... not* May 5.

Coincidentally, this was the precise timeframe of the final "tweaking" of Odyssey's Mars orbit (after aerobraking), which was advertised as a means to achieve the final mapping orbit for the formal science mission, which was to begin on February 18, 2002.

If Cydonia was (and is) a "hidden priority" for this entire mission—as certain official Odyssey mission press statements, the released Cydonia imaging data, and "Bamf's" own still-inexplicable, summer-long disinformation campaign on nighttime infrared capabilities at Cydonia all strongly suggest—January 2002 would have been the perfect time to take the image. While adjusting the orbit of the spacecraft for the main science mission, some "trims" could be included that would allow Cydonia to be quietly imaged very early in the mission. In fact, it could have been done even before the formal mapping mission (and unwanted press attention) began.

So what about Bamf's claim—that the *nighttime* Cydonia IR image was acquired on October 24, 2002?

In the right hand portion of Fig. 10-4, if you look carefully at this enlargement of the nighttime image you will notice that the brightest "thermally glowing" cliffs are once again facing *west*, toward the setting sun. This critical geometry is simply due to that being the last input of solar energy to these exposed rocks, before the onset of the frigid Martian night.

Remarkably, for when this image was supposedly taken, the most heated portion of "the island's" cliffs in this nighttime shot is the southwestern section and only a very small portion (that previously described "outcropping") of the northwestern-facing cliff.

If the sun were anywhere near the northern Martian summer solstice October 24, 2002, when this image was acquired, this entire northwestern cliff would have been directly heated by the setting sun and brilliantly glowing after sunset. And it is not.

From this simple (but irrefutable) solar geometry, the *truth* behind this image is now obvious. THEMIS image 20021031A—the nighttime IR image of Cydonia—could *not* have been acquired on October 24, 2002. The only time, according to this solstice geometry (lower green arrow) in the entire Odyssey mission this Cydonia image *could* have been taken, was in the same timeframe as the July 24 daytime IR image release, sometime in January 2002.

This now also explains the other, *totally inexplicable* discrepancy about this image—the nighttime Cydonia surface temperatures, reportedly recorded at the "height" of Martian summer.

In another March 21, near-Cydonia nighttime IR image (I01180002), the temperature range—from the coldest region in the image to the "warmest"—was cited in the Planetary Data System as "−56º Centigrade to −40º Centigrade"—a spread of about 16º C. This image was taken during the shortest period of Martian daylight, and on one of the *coldest*, longest nights around the northern winter solstice, and only about a hundred miles east of Cydonia itself.

By sharp contrast, the official caption for the "October 24" nighttime Cydonia image—taken at the *same* latitude and in the *same* geologic province and (ostensibly) just after the northern *summer* solstice—mysteriously reports *much colder* nighttime temperatures.[139] According to the official ASU October 31 release:

> *"The temperature in the daytime scene ranges from -50° C (darkest) to -15° C (brightest). At night many of the hills and knobs in this region are difficult to detect because the effects of heating and shadowing on the slopes are no longer present.* The temperatures at night vary from approximately -90° C (darkest) to -7° C (warmest)..." [Emphasis added.]

Let's get this straight: The *coldest* reading (-56º C) in the nighttime March 21 winter image was only 6º colder than the coldest daytime temperature (-50º C) in the "July 24" image—but the lowest surface temperature in the nighttime "October 24 image" (-90º C), taken near the *summer* solstice, was 46º *colder* than the coldest winter temperature on that March 21 image—and the *highest* temperature recorded in that new nighttime Cydonia image (-75º C)—again, reportedly taken at the height of the Martian *summer*—was 35º C colder than the peak nighttime *winter* temperature (-40º C), measured in that nearby March 21 winter nighttime image... huh?

Why are the readings taken on a "balmy Martian summer night," compared to the dead-of-winter nighttime readings taken just across the hill, so much (impossibly) colder?

Unless they're not *really* "summer" nighttime readings at all, because

Bamf (if not the entire THEMIS team behind him, from Christensen on down) just blatantly lied to us... again. That's what the objective *science* in these images now tells us—and from two totally independent disciplines, planetary orbital geometry *and* THEMIS' own radiometry of Mars.

Having satisfied ourselves that this new Cydonia data had been deliberately withheld, probably because it contained some critical new corroborative information on the "intelligence hypothesis" (otherwise, why bother to conceal it at all, and for so long?) we can now move on to consider what genuine anomalies might lie hidden here, even in this "sanitized" version of the latest "trick or treat." Remember—in order to be believed, some "truth" *must* be continually mixed in with all the lies.

The most striking "Cydonia anomaly" in the newest NASA/JPL/ASU release is the almost complete "disappearance" of the Face. In a side-by-side presentation, the Face—as seen in the July 24 daytime IR image (left), compared with its nighttime "October 31" counterpart (right), is almost completely missing [Fig. 10-5].

Remember, these are thermal IR scans. What we are seeing—in both the day and nighttime THEMIS images—is infrared radiation due to solar energy, being reflected back and/or emitted from the sun-warmed Martian surface.

In the afternoon close-up, the THEMIS camera is recording reflected long-wave solar energy (thus shadowed surfaces are extremely dark and cold), as well as re-emitted thermal radiation from the exposed sunlit portions of the "mesa," externally heated by absorbed radiation from the sun. In the nighttime image (right), the *only* radiation being picked up by the Odyssey camera is this *re-radiating* stored solar energy (this, of course, in NASA's "externally warmed, natural" model).

This being the essential physics of these images, why is the nighttime close-up of the Face, compared to its daytime counterpart, of such obvious lower quality and resolution? Part of the reason, we now know, is due to the fact that the image was taken *much* earlier than Bamf let on in his extensive BBS conversations. This image was taken literally at the *coldest* period of northern Martian winter—some time in January 2002. The frigid ground temperatures reported in its official caption, -90° C to -75° C, quietly confirm this, even if we didn't have the seasonal lighting geometry to cinch it.

In such a bitterly cold environment—and at night—the solar energy absorbed during the day, even by materials capable of efficiently retaining such solar radiation, is going to be relatively weak. So we would expect a nighttime IR "signature" to be significantly noisy if the image was actually taken in January 2002 (as we've now proven).

On the other hand, when we examined the pixel details of this nighttime Cydonia IR image, it was also obvious that "someone" carefully "added" a significant amount of noise in an obvious effort to obscure certain geometric patterns that were recorded. Fortunately, knowing this, it is possible to significantly reduce their final impact and amplify the real geometric patterns. In fact, in the official presentation of this data by the THEMIS team there are "coded" instructions in the caption for precisely what to do:

"Both images are of THEMIS's ninth IR band (12.57 microns), and they have been geometrically projected for image registration." [Emphasis added.]

In other words, *place one image precisely over the other* ("image registration"), as that (as every astronomer, physicist and imaging specialist knows) will significantly reduce the "noise."

Which, of course, we would have done anyway. But Bamf's careful preparation of both images, already precisely scaled and tilted by the appropriate amounts, made it *far* easier to carry out. Keith Laney prepared an "averaged" version of both images in a couple of minutes, thanks to Bamf's "helpful" presentation [Fig. 10-6].

The Face immediately pops out [Fig. 10-6, top, center], as well as a number of other, highly intriguing thermal anomalies across the image. The Face is revealed in the nighttime IR to be a perfectly rectangular "box," with lots of internal rectilinear geometry inside. Even in a simple contrast-adjusted and "Gaussian blurred" version of the "raw" image the striking geometric lineaments are plainly visible, as are the "squared off" proportions matching the daytime image.

One intriguing feature that appears in the nighttime version, but does *not* appear in the daytime image, is the "symmetrical extension" below the "chin." Clearly, something close to the surface, but underground, is warm enough in the nighttime IR to allow its heat to "leak" up through the overlying layers of sand and dust and reveal its symmetrical presence *underground.*

The brightest (warmest) sections of the Face are clearly those highest on the feature—the left "eyebrow ridge" and the "nose"—consistent with their retaining the most heat from being exposed the longest to the last rays of the setting sun. The rest of the mile-long "mesa"—remarkably, for a "rocky, eroded outcrop"—has almost blended into the very frigid Martian surface, clearly having cooled off very fast after sunset.

Interestingly, in the author's foreword to the fifth edition of *The Monuments of Mars*, Hoagland had predicted months previously precisely what we were now seeing in this latest nighttime image:

"The [Mars Odyssey] camera's long-wavelength ability to sense and image subtle temperature differences... will allow detection of cooler geometric artificial structures against the warmer natural background deserts (especially at night), the same way suspended bridges and skyscrapers on Earth cool first, *before surrounding landscapes.*" [Emphasis added.]

In addition to the highly anomalous internal geometric structure of the Face, which now strongly reinforces the non-natural, constructed model, the unnatural coldness of this structure must also be addressed. Why is the whole of the Face so cold compared to the surrounding background if it's just a "naturally eroding mesa?"

One obvious answer is that it's not a "natural mesa" at all—that the Face, as we've suspected all along, is composed of other types of manufactured materials and in a form that quickly allows the daytime heating from the sun to be dissipated after sunset. This requirement would be satisfied if the composition were some kind of "conducting material" like a metal, and if this material was significantly porous (perhaps "honeycombed," thus allowing very efficient night air cooling) to boot.

Clearly, something is definitely "anomalous" with this one feature, at least regarding the nighttime thermal properties of its external surface, which is precisely what we were expecting—if it's artificial. Close-ups of other "average mesas" on this landscape—some of which are, like the Face, located precisely according to the unique "hyperdimensional geometry" that's overwhelmingly "coded" here—also seem to be exhibiting "anomalous" thermal properties. Anomalous, that is, for mere outcroppings of erosively resistant rocks.

Is this why this image was apparently one of the first clandestinely acquired by the Odyssey mission, even before any other science was begun— and at literally the coldest period of the northern Martian winter, the best time to determine (with the low-noise thermal background) if "the lights were still on" somewhere at Cydonia?

Is this why the image Keith Laney was leaked on July 25, 2002 bears an eerie resemblance, as if it was a combination of a high quality daytime IR overlaid on another nighttime image? Is this why the D&M in Laney's multi-band version of the NASA July 25 image has this astonishing "transparent aspect," because it's a much more noise-free (averaged) version of what we've been discussing?

The only logical conclusion one can reach regarding this whole elaborate charade is that someone went to a lot of trouble to conceal something critical regarding this entire, early focus of the Odyssey mission on Cydonia, and on this nighttime image.

And, as always, there was more to the story. A few weeks after publication of a story on the nighttime IR on the Enterprise Mission website, Gorelick got involved in an online chat about it on the MarsNews.com forum. In the course of the chat, he was asked why he only posted a single band of the nighttime IR, instead of the full nine bands as they had with the daytime (both "real" and "official" versions). Bamf replied that "I could do a nine-band IOTD like the 7/24 image, but since we've already done of it [sic], I don't think we'll do another one."

The implications of Gorelick's off-the-cuff and arrogant remark were telling. Did he *really* mean to publicly imply that THEMIS has acquired all nine bands of nighttime infrared across Cydonia—in addition to the one he'd deigned to publish? His statement cannot really be interpreted in any other way. Gorelick couldn't tease us with his "I *could* do a nine-band IOTD..." unless the spacecraft had in fact acquired all of them. And that brought with it some major political implications.

A random check of about fifty (of the approximately two hundred nighttime) images then displayed on the ASU THEMIS archive website revealed only *one* other image which was acquired with all nine bands in the entire THEMIS library. If carried through, this is a rate of only two percent.

In other words, if Bamf was telling the truth in his online chat, a nine-band data set of nighttime Cydonia IR, according to ASU's own records, represents an almost singular occurrence in terms of other nighttime images of Mars.

Clearly, acquiring a full nine-band nighttime IR—of a region that has repeatedly been termed "scientifically uninteresting" by the entire THEMIS team—can only be additional corroboration of the clandestine nature of the Odyssey mission from the beginning. It is also further confirmation as to why there have been *so many* systematic lies even about when images are being taken of Cydonia. "Someone" is truly scientifically *obsessed* with the IR composition of Cydonia and was doing all they could to hide that increasingly obvious obsession.

So, did Bamf truly "slip" in revealing this new gem? Or was this another in an increasing number of fortuitously timed, carefully thought-out revelations about what was *truly* going on inside the Mission?

We decided, frankly, that we were no longer willing to speculate or wait for Christensen and his team to drop us another coded nugget. It was time to go on the offensive.

Caltech

In early 2003, most of the nation was trying to cope with the national tragedy surrounding the demise of the Space Shuttle Columbia. While there was a great deal about the disaster that made it look suspicious, there was equal weight to the arguments that it had simply been a terrible accident. Investigations revealed that NASA, in an effort to appease the Clinton administration's EPA, had switched to an "environmentally friendly" external tank insulating foam in 1996, and had been having trouble with it ever since. Once that decision was taken, it was only a matter of time before a Columbia-like disaster would befall the agency. We could not devote too much time to Columbia because, frankly, we were knee-deep in Mars.

In the spring of 2003, NASA released a new five-band color image of the Face on Mars. The image had the same date stamp as the nighttime Cydonia infrared image that was released the previous year. This immediately raised our suspicions. We had already proven that the nighttime Cydonia infrared image was not taken on October 24, 2002, as NASA claimed. That meant we could not be sure about this new five-band color image either—and, almost immediately, there were new questions raised about the visual light image.

This would have to wait for more detailed investigation, however. We had discovered that both Michael Malin and Phil Christensen would be attending a public event at JPL's sixth international Mars conference in Pasadena. We felt like this would give us a great opportunity to confront Christensen about the daytime infrared image that had caused so much controversy the previous year. So on July 23, 2003, Mike Bara dutifully attended the public event. Both Dr. Malin and Dr. Christensen gave slideshow presentations. Christensen could not resist showing the recent five-band color image of the Face on Mars. As he brought the image of the Face up on the screen, he smirked and asked the audience if they knew what it was. There was a nervous chuckle throughout the crowd. He then put forth a new idea about the natural geologic evolution of the Face. He suggested that the unusually bright reflectivity of the eastern side of the Face mesa was because of an accumulation of carbon dioxide snow on that side of the Face. He smirked once again when he asked the audience if they had any comments on his idea (there were none). However, a few moments later, when Dr. Arden Albee opened the forum for public questions, the smirk was wiped off his face.

As no one stepped to the microphone initially, Bara decided to ask a question. Dr. Albee suggested that the person asking the question give their name. When Bara approached the microphone and gave his name: "Michael Bara," Dr. Malin audibly groaned and proceeded to hide behind a stage curtain. Malin emerged from behind the curtain only after hearing that the question was for Dr. Christensen.

Christensen himself became highly agitated. He began pacing back and forth across the stage as Bara asked him if the original Cydonia daytime infrared image, which was posted on the ASU website on July 24, 2002 had ever been changed. At one point as he stumbled for an answer, Christensen dropped the battery pack for his wireless microphone and had to scramble to pick it up. Obviously, he was surprised and caught off-guard by this confrontation. When he did answer, it was in a nervous and halting tone, and he only occasionally made eye contact with his questioner.

Christensen defended the data on the THEMIS website. He claimed that once it had been posted on July 24, 2002 it had never been altered. He also went on to state that he had no idea how the "artificial stuff" had gotten on the images that Keith Laney had processed. At no time did he say that the Laney version of the image was an outright fake, nor did he accuse Hoagland, Laney or Bara of creating false data. His choice of the words "artificial stuff" was quite telling, because he could have just as easily used the word "fake." He pointedly did not.

Just to be sure, Bara asked for clarification on the most crucial point: had the data that was posted on the ASU THEMIS website been changed after July 24, 2002? Christensen again emphatically answered "no." What he did not know at that time was that at precisely that moment, we had him.

Unbeknownst to Christensen, there is a website called the "Internet Archive" (archive.org). Sponsored by the Library of Congress, its purpose is to document every page that has ever existed on the World Wide Web. Using a search engine called the "Wayback Machine," we had been able to plug in the URL of the daytime Cydonia infrared image. It had shown that, contrary to Christensen's previous e-mail statements and contrary to his latest public statements, the ASU website containing the image created and managed by Christensen's lieutenant Noel Gorelick had been originally posted on July 24, 2002, and had *subsequently been altered* in the early morning hours of July 26, 2002 [Fig. 10-7]. As you'll recall from the previous chapter, Laney downloaded the "real" version of the Cydonia daytime IR late in the evening (10:27 p.m.) on the twenty-fifth. In other words, just hours after Keith Laney had downloaded the "real data" from the THEMIS website, Gorelick, Christensen or someone else had changed the contents of the site.

So, all of Christensen's protestations to the contrary, there had been at least one change to the Cydonia infrared image website after Keith Laney had downloaded the "real" data. Within a few months of our obtaining the Wayback Machine data on the Cydonia IR page, the Internet Archive ceased tracking changes to any of the THEMIS ASU image release pages. This can only be done at the request of the website owner, who in this case would have been Noel Gorelick. So it was clear we would not be trapping Christensen or Gorelick in any more lies anytime soon.

It was hard to tell from Christensen's reaction whether he was so nervous because he was afraid he was going to get caught in a lie by us, or whether he was under some pressure from outside forces to keep the story straight even after this encounter. We're still not certain whether Christensen is a friend or an enemy of our independent investigation. On the one hand he has given us the data to prove our thesis beyond any reasonable doubt. On the other hand, he has participated (at least tacitly) in an attempt to discredit our investigation publicly by spreading false information and data to the public. Yet still, he had gone to some lengths to avoid accusing us of faking the "real" IR data.

So in the end, we simply didn't have enough information to reach a conclusion about Christensen the man, but we did have the new five-band color image of the Face, and that would prove to be perhaps the most significant piece of data we have received in the entire history of the investigation.

Chapter Ten Images

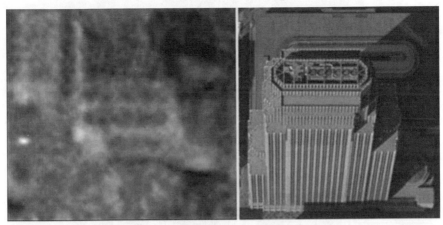

Fig. 10-1 - The "Temple" of Cydonia (L) – and a terrestrial comparison, Minneapolis, Minnesota (Bara\ASU).

Fig. 10-2 – Comparison of THEMIS Cydonia IR reflectance image (L) and Phobos 2 thermal IR image of Hydoate Chaos (R). Both images reveal completely anomalous geometric "cityscapes" beneath the Martian sands (ASU/USSR)

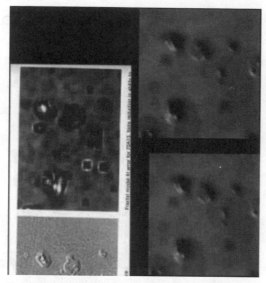

Fig. 10-3 - From "The Martian Enigmas: A Closer Look" by Dr. Mark Carlotto (left); and from Keith Laney's "real" version of THEMIS IR data. Laney image enhanced by contrast stretch only.

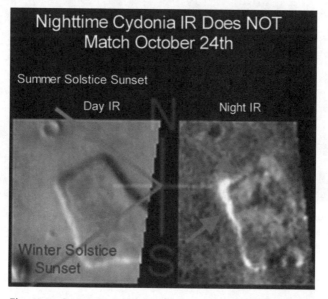

Fig. 10-4 - Comparison between Mars Odyssey THEMIS daytime IR reflectance data (L) and nighttime thermal IR emission data (R) – showing, via bright nighttime emission at mesa edge, that daytime sun-angle warming this Cydonia feature does NOT match the seasonal date claimed by the Mars Odyssey 2001 Project and JPL (ASU/Enterprise Mission)

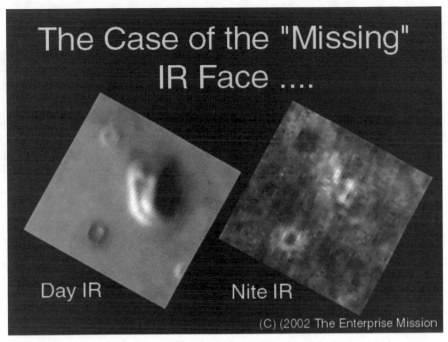

Fig. 10-5 –The Face on Mars mysteriously "disappears" in this THEMIS nighttime IR emission image (R) released by ASU, compared to daytime IR reflectance image. This is due to nighttime image being taken in deep Cydonian winter (contrary to official ASU statements), in addition to deliberately- added "pixel noise" (ASU/Enterprise Mission)

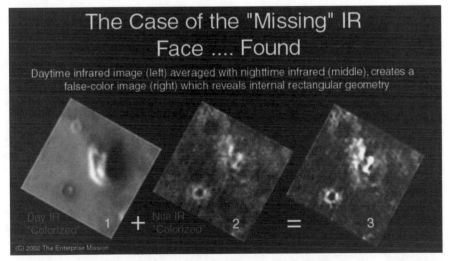

Fig. 10-6 – Averaged combination of daytime (L) and nighttime IR images (M) results in the Face being revealed (R) to reside over a *box-like, highly geometric substructure* – consistent with a *massive, manufactured foundation* (ASU/Enterprise Mission)

INTERNET ARCHIVE
WaybackMachine

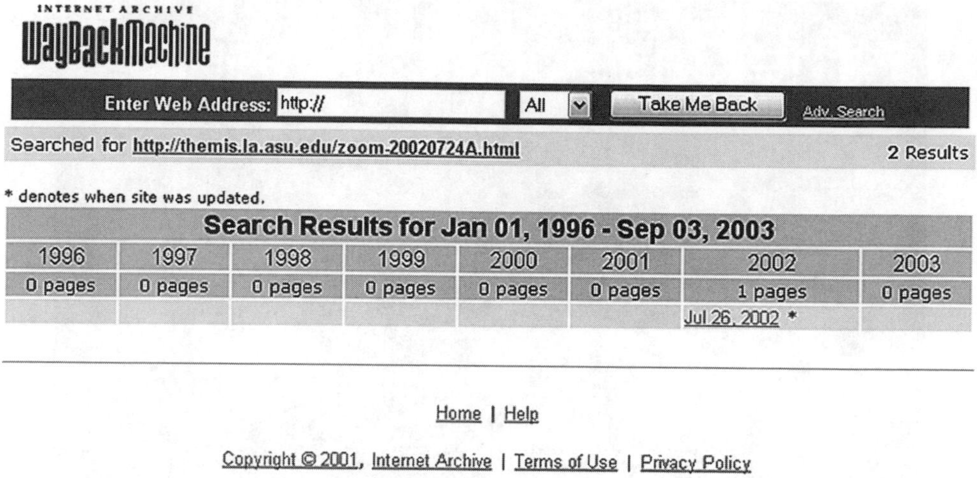

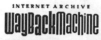

Enter Web Address: http:// All Take Me Back Adv. Search

Searched for http://themis.la.asu.edu/zoom-20020724A.html 2 Results

* denotes when site was updated.

Search Results for Jan 01, 1996 - Sep 03, 2003							
1996	1997	1998	1999	2000	2001	2002	2003
0 pages	0 pages	0 pages	0 pages	0 pages	0 pages	1 pages	0 pages
						Jul 26, 2002 *	

Home | Help

Copyright © 2001, Internet Archive | Terms of Use | Privacy Policy

Fig. 10-7 – Screen capture from the "Wayback Machine" – archive Website showing ASU-THEMIS page containing the "real" Cydonia IR image was *altered* just hours after Keith Laney downloaded the "real" IR data from the official ASU-THEMIS site. ASU and Dr. Philip Christensen (THEMIS-PI) have repeatedly (and officially) denied this crucial page was ever altered (Wayback Machine)

Chapter Eleven

The True Colors of NASA

The image shown in Fig. 11-1 is an enhanced, grayscale close-up created by Keith Laney and Richard C. Hoagland from a combination of three 2001 Mars Odyssey visual frames (of the five simultaneously taken by the Odyssey color visual camera) and the 2001 Mars Global Surveyor image of the Face on Mars (E03-00824). The Odyssey image release is officially designated JPL/ASU V0 3814003. The five frames—from the near "IR" end of the visible spectrum, to the "violet"—were acquired by the Odyssey spacecraft as it flew over the Cydonia region on October 24, 2002, precisely one year (Greenwich Meridian Time) after it arrived in Martian orbit. This date is the same one that was given for the Halloween release of the nighttime Cydonia IR image. As we have established in a previous section, it is highly unlikely that the nighttime image was actually taken on that date. Whether the new image was actually taken on that date is harder to determine.

What made this close-up so remarkable is that, for the first time in over a generation, a NASA spacecraft acquired multi-spectral images of the Face as seen in morning light—with the illumination coming from the east [Fig. 11-2]. What this unique sun angle revealed was nothing less than revolutionary. Even casual examination of the Face as seen in this "new light" revealed two new pieces of vital information:

    1. The eastern side, under even this pre-dawn illumination, is incredibly reflective, and;

    2. In lowered contrast images, the source of this anomalously "high albedo" is an inexplicable series of highly geometric "panels."

    Beyond that, the key parameter that made this new image so remarkable is "when" it was acquired. Carefully examining the "data block" for image V0 3814003 on the Arizona State University (ASU) THEMIS website,[140] we can immediately ascertain that it was taken by the Odyssey camera at 4:39 a.m., local Martian (Cydonia) time. Further reading of the table reveals that the "phase angle"—that is, the geometric relationship between the sun, the Martian surface directly underneath the spacecraft just east of the Face and Odyssey itself—was 90.3º. Since 90º (for a spacecraft directly overhead) would indicate the sun was literally on the eastern horizon, the slightly greater angle reveals that actually the sun was 0.3º below the horizon when the image was acquired (and even slightly lower at the location of the Face itself).

The last line in the table, "Description: Cydonia—Face at Night" confirms this geometry. Technically, then, this Odyssey dawn image was actually acquired just before sunrise, with the sun still hidden below the Cydonia horizon. This simple, inarguable geometry marks the high brightness of the Face's eastern side—before the sun has risen—as extraordinary. This, in turn, leads directly to the pivotal question: just what could make "an average Martian mesa" (to quote Carl Sagan) so incredibly *reflective*... even in the semi-dark, pre-dawn twilight of Cydonia?

A side-by-side comparison [Fig. 11-3] reveals the true incongruity of such a brilliant-surfaced object. The official NASA version of the Face from V0 3814003 (left) is totally "washed out" on the illuminated (eastern) side—even though the image was shot before the sun had risen. In the Laney-Hoagland rendition (right), after considerable effort to lower brightness levels, some surface details can just be seen beneath the glare.

Again, for this over saturation of the THEMIS imaging CCDs to have occurred, and under these really dim lighting conditions, something about the innate reflectivity of this Martian surface feature—at this geometry—must truly be "anomalous." Why, under pre-dawn lighting, is the Face—even allowing for the obviously increased gain settings in the Odyssey camera—so incredibly bright? And why is that inexplicably reflective eastern surface also arrayed in those stark, startlingly geometric patterns?

For the inevitable critics of the image processing and Face analyses, it is imperative for everyone to realize that these spectacular results were achieved by merely decreasing the brightness of Odyssey image V0 3814003, and then applying a 3X pixel over sampling (to smooth out "jaggies"). No other filters or enhancement techniques have been used. So the usual canard of "image processing artifacts" cannot be applied as an excuse for the geometry revealed by the brightness reduction.

By the time the brightness has been decreased to approximately ten percent of the official published image value, the astonishing 3D geometry of the Face's eastern side is overwhelmingly apparent [Fig. 11-4].

Given that we did not have access to the absolute photometric transfer functions of the Odyssey VIS camera, this technique can also be used to put some crude upper limits on the absolute reflectivity of the Face. If we equate the surface brightness of the soils around the Face with NASA's published estimates of the average albedo (reflectivity) of the Cydonia region (about twenty percent—according to a pre-Pathfinder NASA Conference),[141] we can then approximate the relative reflectivity of the Face in this new image, based on when the surface "disappears" in our brightness reduction experiment.

Keeping firmly in mind that these are only "ballpark estimates," the Face via this technique turns out to be reflecting—straight up—an amazing 99.9% of the surface illumination falling on it, compared to the ground's average of about 20%! Since no rocky surface can *possibly* have this degree of right-angle, natural reflectivity, the Face's eastern side *must* be acting like a set of coherent, artificial *mirrors*. By the process of elimination, we were inexorably forced to conclude that only some kind of manufactured, highly "directional" glass and metal surface on the Face—whereby the angles of all the surviving reflecting elements, despite the curving underlying structure, are aligned—could redirect the horizon sky glow coherently, vertically, at such a specific angle, and thus produce these startling optical phenomenon seen in Odyssey's VIS camera.

Further, the multi-colored, "prismatic" appearance of these striking 3D "panels" raises the serious possibility of semi-transparent, glass-like refractions in this material—from whatever is making up this structure. This, in turn, is totally consistent with Hoagland's predictions that the Face was a specifically constructed and completely artificial edifice. It is also completely consistent with Mark Carlotto's earlier fractal analysis, which identified the Face as the most unnatural object for over 15,000 square miles. Hoagland argued in *Monuments* that Carlotto's then newly-published results were likely due to a "sophisticated placement of shadow-casting [artificial] pyramidal substructures on [the] underlying mesa..." In other words, that major portions of the Face were composed of a collection of polygonal objects, much like modern 3D computer models are made up of thousands of polygons, now showing (through erosion) countless exposed elements of their internal geometric structure.

Of course, the first criticism to be leveled at this assertion is that there may be a perfectly prosaic explanation for the eastern half of the Face being so anomalously bright—like Dr. Christensen's aforementioned "pasted on snow" model. In actuality, this explanation is fairly easy to refute.

According to the Canadian Meteorological Center,[142] fresh snow has an albedo of about 83%. Old "aging" snow only scatters about 50% of sunlight falling on it. Since we've now quantitatively estimated the eastern side brightness of the Face at over 99%, this effectively rules out any natural high albedo, lambert-type (all-direction) scattering surfaces caused by snow or ice (whether water or frozen CO_2) as an explanation for the severely overexposed Face's eastern half. And, if "snow" was present on the Face when the image was acquired (hardly likely, given the official Martian northern hemisphere date of image acquisition—northern summer), why didn't this same snow fall on the shadowed side in the Odyssey view, or on the other, nearby mesas?

The truth is, "something" about the protected eastern half of the Face's surface (captured by Odyssey in an instant of precise pre-dawn sun/spacecraft illumination geometry) is capable of producing mirror-like reflections, bounced 250 miles straight up, even *before* the sun had actually risen over the Cydonia horizon. "Snow" simply cannot do that.

The other conventional explanation for the anomalous brightness of the Face in this pre-dawn image is that it simply isn't all that bright at all. That its brightness is simply a function of the camera's gain being turned up. A close look at the image itself [Fig. 11-2] proves that this is simply not correct. Comparing the Face with its closest "next door neighbor," a mesa located just to the southeast, shows that while the Face's eastern flank is totally overexposed, the mesa right next door is barely lit up. Yet the source of illumination for both objects is exactly the same—the pre-dawn brightening several degrees above the Cydonia horizon.

Further, this dramatic brightness difference is not because of any major differences in height. As can be seen in pre-dawn imagery, the primary landscape lighting comes from a large area of the sky—several degrees above the point where the sun will actually rise. Thus, the sky illumination of both features in V0 3814003—regardless of their intrinsic height—is essentially the same.

Significantly, the only other object on the color image that even begins to approach the Face in brightness (albedo) is the D&M. In the bottom left portion of V0 3814003, the Odyssey camera managed to catch the northeast quadrant of the D&M at about twenty meters per pixel resolution (roughly twice that of Viking)—and in color. Perhaps even more important, whereas all previous D&M images (Viking and Odyssey) have been taken with the sun coming from the left, in this image the Pyramid is clearly illuminated from the right—the direction of the not-yet-risen-sun. It is immediately apparent in this image that the D&M is as reflective at this viewing angle as the Face itself.

What was now clear was that in this revealing pre-dawn light, the light-scattering properties of both the Face and D&M were highly anomalous. This bizarre reflectivity supports the idea that the Face is at least made up of a series of highly polished (possibly by wind), geometrically aligned artificial panels. The reflectivity also powerfully reinforces the earlier findings of the nighttime infrared image, in which the Face seemed to all but disappear except for a box-like, geometric under-structure. For if the Face was actually constructed of glasslike or metallic panels, rather than being an "eroded rocky outcrop," it would not retain heat at all through the night. Its anomalous reflectivity in this image is confirmation of its equally anomalous lack of heat retention in the nighttime IR. In other words, the Face is anything but "a giant pile of rocks."

It's hard to imagine that Dr. Christensen failed to appreciate the lighting geometry of this image when he took it. Our suspicion was that he arranged this particular lighting because he expected to see something different and unique, but it hardly seems probable that he was looking to get a good pre-dawn image of carbon dioxide snow. Given that only this specific lighting could produce an image of the geometric panels on the Face's eastern side, how had Christensen been tipped off to take this photo? The panels didn't clearly show up in any of Dr. Malin's "better" visible light images, although they were faintly suggested by at least one.

In this Mars Surveyor image of the Face, compared to the same area from Odyssey, note the distinct, glowing, 3D quality of the eastern geometry seen in the Odyssey view. Then note hints of a similar rectilinear structure on the Face's surface in the white light image, at opposite lighting and higher resolution. This key similarity—but at two totally different scales—suggests that the brilliant reflecting elements seen in the Odyssey twenty-meter color version may in fact be larger scale, more massive interior structures, captured underneath the visible light features seen in the MGS five-meter view.

This would have been possible because of the unique illumination angle of this image. Pre-dawn sunlight, shining almost horizontally through a high-tech, though porous and still eroding eastern surface covering, photographed by Odyssey looking straight down at a 90° angle. A good analogy would be the view through a fine-mesh window screen at twilight, into a well-lit room—where the mesh is literally too small to be seen against the massive illuminated pieces in the room.

The overall effect in this Odyssey view would have been identical to an internal lighting system, producing an imaging effect almost like an x-ray, making the internal architectural structure of the Face on Mars visible for the first time.

So given that there was little evidence to suggest that an MGS image (at least a publicly released one) was the source of Dr. Christensen's curiosity, we were forced to look to the political arena for a reason he might have wanted to "get there first." After a little searching, the reason became obvious—ESA's Mars Express.

The Mars Express Orbiter was scheduled to arrive in Mars orbit in late December 2003. Because of the vagaries of the spacecraft's orbit, it would not be in a position to image Cydonia for a few years after its arrival. However, it did carry a color imaging camera that was vastly superior to the color camera on Mars Odyssey 2001. The on-board HRSC (High Resolution Stereo Camera) system could take images in color, stereo and at very high resolutions, down to about two meters per pixel. This meant that an image of the Face, if it could be captured under the same pre-dawn lighting conditions, would give us color and stereo images equivalent to the Odyssey color images, but at the spatial resolution of the best Mars Global Surveyor images.

Given that the mission was going to be run outside of the influence of NASA, we had some hope that we might get good data from it. At least we hoped that it would be more honest than what NASA had provided over the years—and in color.

The True Colors of NASA

The reason we were now so focused on color images was that from the very beginning of NASA's Mars explorations, the agency had a problem with getting color images correct. As we had gathered data over the years, it became obvious that this was not a technical problem, but a *political* one. For some reason, NASA had an aversion to showing the world the "true colors of Mars."

Perhaps the most infamous account is of the controversy that still swirls around the release by JPL of the first color Viking Lander image. Within a few hours of that historic publication of the first color photograph from the surface of Mars, another hurriedly revised version of this first color surface image was suddenly produced, supposedly correcting the initial "color engineering problems" in the first image.

Decades later, one of those personally present at JPL would relate a very different story of this incident. The witness is the son of the scientist in charge of one of Viking's three historic biology experiments, the Labeled Release Experiment. The LRE's principal investigator was Dr. Gilbert Levin. His son, Dr. Ron Levin, is now a physicist at MIT.

In the summer of 1976 (when Viking landed on that oh-so-familiar "ritual date" of July 20), Ron was a newly-graduated high school student, assisting his father at JPL during that incomparable Viking Summer. Hoagland was also present, covering the extraordinary Viking story for millions of readers of a major magazine and a couple of broadcast television networks.

Dr. Gil Levin's first-hand recollections of the whole affair are recounted in a recent book by science writer Barry DiGregorio.[143] In the book, Levin relates the remarkable overreaction by JPL that occurred in response to Ron Levin's naïve efforts to "correct" what seemed to him to be a deliberate distortion of the incoming Viking Lander data. According to DiGregorio's narrative:

"At about 2:00 p.m. PDT, the first color image from the surface of another planet, Mars, began to emerge on the JPL color video monitors located in many of the surrounding buildings, specifically set up for JPL employees and media personnel to view the Viking images. Gil and Ron Levin sat in the main control room where dozens of video monitors and anxious technicians waited to see this historic first color picture. As the image developed on the monitors, the crowd of scientists, technicians and media reacted enthusiastically to a scene that would be absolutely unforgettable—Mars in color. The image showed an Arizona-like landscape: blue sky, brownish-red desert soil and gray rocks with *green splotches.*

"Gil Levin commented to Patricia Straat (his co-investigator) and his son Ron, 'Look at that image! It looks like *Arizona*'" (see Color Fig. 21).

"Two hours after the first color image appeared on the monitors, a technician abruptly changed the image from the light-blue sky and Arizona-like landscape to a uniform orange-red sky and landscape. Ron Levin looked in disbelief as the technician went from monitor to monitor making the change. Minutes later, Ron followed him, resetting the colors to their original appearance. Levin and Straat were interrupted when they heard someone being chastised. It was Ron Levin being chewed out by the Viking project director himself, James S. Martin, Jr. Gil Levin immediately inquired as to what was going on. Martin had caught Ron changing all the color monitors back to their original settings. He warned Ron that if he tried something like that again, *he'd be thrown out of JPL for good.* The director then asked a TRW engineer assisting the Biology team, Ron Gilje, to follow Ron Levin around to every color monitor and change it back to the red landscape.

"What Gil Levin, Ron and Patricia Straat did not know (even to this writing) is that the order to change the colors came directly from the NASA administrator himself, Dr. James Fletcher. Months later, Gil Levin sought out the JPL Viking imaging team technician who actually made the changes and asked why it was done. The technician responded that he had instructions from the Viking imaging team that the Mars sky and landscape should be red and went around to all the monitors, "tweaking" them to make it so. Gil Levin said, "The new settings showed the American flag (painted on the Landers) as having *purple* stripes. The technician said that the Mars atmosphere made the flag *appear that way* [emphasis added]."

Hoagland had been at JPL that same afternoon, and vividly remembers a similar shock when the "Arizona, Mars" image initially flashed on the JPL monitors was suddenly transformed into a Martian "Red Light District." He now kicks himself for not asking many more questions. But, as he would later put it, "it was 1976—and we all *trusted* NASA back then."

One of the basic questions that should have been asked involves the physics behind JPL's abrupt color alterations. As Gil Levin phrased it in DiGregorio's book:

> *If atmospheric dust were scattering red light and not blue, the sky would appear red, but since the red would be at least partially removed by the time the light hit the surface, its [the direct sunlight's] reflection from the surface would make the surface appear more* blue *than red. There would be less red light [in the direct sunlight illumination] left to reflect. And what about the sharp shadows of the rocks in the black and white images yesterday? If significant scattering of the light on Mars occurred [from lots of red dust in the atmosphere], the sharp shadows in those images would not be present, or at best, would appear fuzzy because of diffusion by the [atmospheric] scattering. [Emphasis added.]*

Levin was describing the well-known phenomenon of "Raleigh scattering"—whereby the similar-sized molecules of *all* planetary atmospheres (be it the primary nitrogen of Earth, the carbon dioxide atmosphere of Mars, or even the predominantly hydrogen atmospheres of Jupiter and Saturn) all produce *blue skies* when sunlight passes through them. If you examine the long Martian photographic record—which encompasses hundreds of thousands of images acquired by dozens of observatories even before the Space Age dawned—you can see blatant evidence that Levin is right and JPL is wrong.

In 1997, before the arrival of the Mars Pathfinder spacecraft (the first NASA Lander sent to Mars since Viking), the Hubble Telescope was tasked to acquire a series of "weather forecast images" prior to the landing. This long-distance reconnaissance detected a small dust storm less than a month before the Pathfinder arrival, which, with its potentially high winds, could have posed a serious threat to the Pathfinder entry and landing. The Hubble Investigator on the Pathfinder Landing imagery, Dr. Philip James of the University of Toledo, did note one potential impact on the Pathfinder Mission when the dust storm safely dissipated:

"If dust diffuses to the landing site, the sky could turn out to be pink like that seen by Viking... otherwise [based on the Hubble images], Pathfinder will likely show blue sky with bright clouds."[144]

In other words, based on the Hubble images taken just before the landing, NASA astronomers working with the Hubble data fully expected a surface view to match the blue skies their telescopic images were showing. Instead, when the Pathfinder did arrive, the skies, according to the official JPL Lander image releases, if anything, seemed redder and more "dusty" than on the Viking images from twenty years before (see Color Fig. 22).

In the years subsequent to Viking (and before Pathfinder), a number of other NASA spacecraft had returned a variety of color views of Mars. Remarkably, in *all* those images the tell-tale evidence of major Raleigh scattering is blatantly apparent.

So we have a contradiction. If Mars has a blue sky and looks remarkably like "Arizona" from the ground, why does NASA keep making the images from the surface appear Technicolor red? What possible agenda could it serve to deceive the public about the true colors of Mars? An answer might be found if we go back to *Mars: The Living Planet*.

It turns out that DiGregorio's statement that the NASA administrator was behind the monitor changing incident was based on a confirmation of this from an *official* source—former JPL public affairs officer Jurrie J. Van der Woude—and it had an even stranger and somewhat sinister angle: In a letter to DiGregorio (also reproduced in *Mars: The Living Planet*), Van der Woude wrote:

"Both Ron Wichelman [of JPL's Image Processing laboratory (IPL)] and I were responsible for the color quality control of the Viking Lander photographs, and Dr. Thomas Mutch, the Viking Imaging Team leader, told us that he got a call from the NASA Administrator asking that *we destroy the Mars blue sky negative created from the original digital data.*" [Emphasis added.]

This bizarre sequence of events raises many disturbing questions. For instance, why was the administrator of NASA so determined to conceal the "true" colors of Mars from the American people and the world in 1976? Why would he order the head of the Viking Imaging Team to literally eliminate an important piece of historical evidence from the official mission archive—the original "blue-sky negative"—if the initial release was only an honest technical mistake? Wouldn't that record be an important part of the ultimate, triumphant story of NASA scientists correcting initial scientific errors, in their continued exploration of the frontier and alien environment of another world? And why would a young teenager (the son of one of the key investigators on the Viking mission, no less) be threatened with expulsion by the director of the project for simply tweaking a couple of color monitors around the lab?

In truth, none of Ron Levin's story (or Van der Woude's significant confirmation), makes any scientific sense unless certain individuals at the highest levels in NASA felt compelled—for some arcane reason—to hide at *all costs* the visible appearance of the actual Martian surface.

Beyond that, there is an even bigger "biological problem" for the conventional NASA view of the true colors and environment of Mars. Levin suggested that there were other hues on Mars than just dull browns and reds. This was verified by members of the Viking imaging team, who confirmed there were blue and green patches on rocks that changed seasonally.[145] The only rational explanation for these "changing patches" on the rocks, shifting color with the rising and dropping seasonal temperatures and atmospheric availability of water is *biological entities*, like simple plants or lichens, reacting to changing *biospheric* conditions.

If it is the former, he certainly can point to an extensive, accumulating quantity of startling evidence that calls into serious question the conventional (read "NASA") view of Mars, as "a cold, uninhabitable hell."

As recently as the 1950s, there was a prevailing "Lowellian" view of Mars as a possible "abode of life." This view, so named in homage to Percival Lowell, the nineteenth century astronomer who made the first scientific observations of Mars with a semi-modern telescope, saw Mars as cold, dry and harsh, but not uninhabitable. While his observations of canal networks were pretty much discredited by the 1950s, there were many who maintained that the Lowellian Mars might still be valid, with life functioning at a low level, but surviving. One key clue to support this idea was the so-called seasonal "wave of darkening" that seemed to sweep from each hemisphere's pole toward the equator in their respective spring seasons. This darkening wave rolls across the planet at a rate of thirty-five miles a day, and could be attributed to melting polar caps releasing water into the atmosphere and "awakening" the planet's simple plant life. This was disputed because it was later found that the southern cap was entirely carbon dioxide ice. However, Odyssey's new 2001 observations implied that there were vast quantities of water ice all over the planet that could fuel these darkening waves. While the reality of the "wave" is still disputed, no one argues that certain patches on Mars do darken in the spring and summer.

The countervailing NASA view of Mars (to which most planetary scientists in and outside NASA have quietly acquiesced)—a "dry, dead desert planet" with an atmosphere so thin that not even water could stay in a liquid state on its surface for more than few seconds—was established by the first NASA mission to view the planet close-up. Mariner 4 arrived at the Red Planet in the summer of 1965 and revealed a harsh landscape of barren, cratered deserts. Measurements taken at that time supposedly established that the atmosphere was too thin to support water in a liquid state, and was almost 100% carbon dioxide—leaving the probability for finding life as extremely unlikely.

That view prevailed until the Viking missions of the mid 1970s, when the two Landers were sent to test the soil for signs of microbial life. What most people do not remember is that the Lander tests for life both came back *positive*. NASA, however, quickly moved to suppress this news and present an "alternative" view—that the results were just a "mistake," a chemical reaction and *not* proof of life on Mars. However, Dr. Levin has always insisted that his instrument's results were *positive* for life, and not a result of a mere "chemical interaction." His case was bolstered in 1996 when NASA announced the discovery of microfossils in a meteorite from Mars. Obviously, if there were once micro-organisms living on Mars, there was no reason that they could not be present on Mars today. The only remaining argument against that conclusion was the supposed absence of a "biologically kind" environment, i.e. *liquid* water.

Levin himself had argued for some time that this was not really an issue. He presented a paper describing the circumstances under which water could remain in a liquid state on Mars. He pointed out that the NASA view of Mars as unable to support water at the surface was based on a faulty assumption—that the water was evenly distributed throughout Mars' atmosphere, rather than in "the lower one to three km," as confirmed by Pathfinder.

So, the reality was that there was plenty of evidence that Mars was not only capable of harboring life, but that NASA had in fact already *proven* that to be the case, as far back as 1976. NASA's determined efforts to suppress such a conclusion would seem to fly in the face of the agency's publicly stated mandate. It does not, however, conflict with the agency's probable secret agenda, driven by Brookings, which would almost dictate this behavior as a kind of pre-conditioning over generations—not publicly admitting the truth until the populace was deemed "ready."

Evidently, by the time of the two 2003 Mars Exploration Rovers, we still weren't ready. The two rovers, named Spirit and Opportunity, were scheduled to land in Gusev Crater and Meridiani Planum, respectively, both in the equatorial regions of Mars. The ostensible reason for the landing site selections was officially the search for water, or at least areas where water had once flowed. It did not escape our notice that both of these sites were squarely in the path of outflow channels from our predicted Mars tidal oceans. If NASA was looking to confirm our tidal thesis, there were no two better places to land on Mars. Spirit was the first of the two rovers to land (in Gusev Crater), which it did successfully on January 4, 2004. Almost immediately, the Mars color controversy raged again.

Within a day of the landing, NASA/JPL released the first official full-color image of the landing site, and it once again had the Technicolor red skies and distorted, "red-shifted" color calibration. This made everything, even the rocks, look reddish. Almost immediately, articles appeared on the web noting that the colors in the color calibration dial were all wrong compared to images of them taken on Earth. One reason for this is that NASA chose to use filters in the infrared spectrum instead of visible light on the "Pancam" color imager, and this created the horribly distorted reddish nightmare that was seen in the NASA press release images.

NASA did not respond directly to the new controversy, but instead sent out a number of surrogates, including a Dr. Phil Plait, an astronomer who had a debunking website called "Bad Astronomy." Feeding him and their other surrogates information about the Pancam, Plait claimed that the IR filters were responsible for the reddish hue, and that in any event, it was next to impossible to get the colors correct because of uncertainties in calibrating to the actual conditions on Mars.

However, others, like Keith Laney (who had processed landing-site MOC images for the rover program) had been working with the new color images as well. Using a previously published paper by Dr. Levin as a basis for his color calibration technique, Keith was able to get remarkable results. It turned out that what Plait had said about color imaging with the Pancams was not entirely correct. What makes the process difficult was not that colors are assigned to IR band filters alone, but the fact that NASA used the *wrong* filters for each of the RGB channels, making it almost impossible to correct for. In fact, if you didn't know how to correct for the deliberate misallocation, you could never get close to the true color of a scene at a given moment. Fortunately, Keith *did* know how to make this correction, and he was able to produce color images of Mars that are far more accurate and logical than what NASA has released (see Color Fig. 23).

As you can see from the images (color insert) —contrary to Plait's assertions—it is not only possible to get very accurate, "approximate" true-color images from the Pancam imagers using Laney's techniques, but they can also be produced with consistency. Yet NASA/JPL, with all their resources, can't seem to get even one image close to what Laney produced all alone from his desktop computer. In fact, as you go to the official website, there isn't one color Pancam image from either landing site that looks like anything other than the silly "Technicolor red" skies Plait so readily defends.

The picture of Mars that emerges from Laney's images is not one of some alien, forbidding landscape, but of a familiar, almost comfortable environment that makes sense to the eye. We can easily picture ourselves standing next to the rovers as they circumnavigate the debris fields and rock gardens of the Martian landscape.

So what is a more reasonable picture of the surface of Mars, Laney's carefully crafted work based on the papers of two highly respected scientists, or NASA's absurd, garish, Technicolor rendition of a "Red Planet" that looks like something out of a 1950s B-movie? If Laney can do it, why can't NASA? Well, it turns out, they can, actually.

We found a number of Pancam color images taken on Earth using the same filters as on the color press release images coming from Mars. Miraculously, NASA was somehow able to get the red balance corrected so that the pictures look entirely normal. This capability seems to have eluded them when processing the images from the surface of Mars, but it has not eluded independent workers like Laney.

So is there, really, any reason to suspect that NASA is playing coy with the true colors of Mars? Or is it all, as Plait and others of his ilk have alleged, just a bit of confusion on the part of "pseudo-scientists" like Laney and Dr. Levin? The answer comes, as it always does if one is willing to look, from NASA itself.

What the authors have easily done, but Plait and NASA apparently have not, is to simply do a few color comparisons of objects we know the color of. When we did, we found that the Spirit and Opportunity airbags, far from being the orange-red color in the landing site images, are actually a fairly bright off-white.

Indeed, a simulation video produced by Cornell University depicts them as white.[146] As does JPL's own web page depicting the airbag tests.[147] In fact, all you have to do is look at the skin tone on the arm of the test engineer holding the scale to see that the color in that image is correct (see Color Fig. 24). [148] [149]

By contrast, when you go to the Mars Rovers website and go to the first color images of the Spirit landing site, you see no mention of the IR filters or the fact that the images are false color. In fact, NASA/JPL even *altered* images of the rover airbag tests, distorting the bag color to look as they do in the Technicolor Red polluted press release images.

This is also easily verified. In the background of the fake-color air bag image (from the press release gallery of the MER website)[150] is a snap-on tool box, a site familiar to anyone in the aerospace industry. In scrutinizing the tool box, you'll see that all of the normally silver pieces on the box have a distinct reddish-orange tint to them. This is completely at odds with the way the boxes are painted when they leave the snap-on factory.[151] In fact, the press release pages are littered with images of the Rovers in testing with the color altered to make them appear redder.

So if—as Plait and others flatly insist—there is no intent to deceive on NASA's part, why are the images of the airbag tests done on Earth (and flatly *not* photographed with an IR-filtered Pancam) depicted on the Rover press release pages as *red-orange*? And why is there no mention of false color or IR filters in the initial NASA press releases?

The answer is obvious. NASA has deliberately altered the color balance of the press release images to make the bags appear the same color as they do in the Technicolor images of the Martian surface. In fact, they went out of their way to change the coloring on press release images of the tests to make the bags appear reddish. And if they were to mention the IR filters or false color processes, then everyone would know the sky on Mars isn't really Technicolor red. Fortunately for us, they missed a couple of images of the airbags shown in their actual off-white color.

This pattern goes on in the press release pages. The first mention of "True Color" on the press release pages comes on January 10, 2004 when an image of the color calibration dial was posted. It made no mention of the IR filters or the false color of the Martian sky, or the fact that the "color" images released up to that point were *false* color. The first mention of the term "approximate true color" does not appear in press releases until January 19, *after* the color controversy had erupted and NASA had started to receive a lot of e-mails on the subject.[152] Even so, the term "approximate true color" to describe the Technicolor red skies is at the least misleading. As Laney has shown, NASA's concept of "approximate true color" is laughably wrong.

Amazingly, it's also clear, going though the press release pages, that NASA is quite capable of getting the color right—when they want to. Here is a press release version of the American flag logo seen above, with the color correctly rendered and very close to Laney's results. Evidently, it was safe to make the color right in this image because the sky and horizon were not visible.

Clearly, beyond a reasonable doubt we have shown that NASA has had a decades long desire to give the public the wrong impression of Mars's true colors. And that ruse has continued to this day. But it was only when ESA's Mars Express returned one of its first color images of Mars that we understood exactly why.

Spirit and Opportunity

Just about two weeks after Spirit arrived on Mars in Gusev Crater, ESA's Mars Express took advantage of their first opportunity to image the crater in high-res color. Our hopes that Dr. Gehard Neukum, the principal investigator for the HRSC camera might give us a more honest look at Mars were almost immediately fulfilled. Mars Express got an excellent view of Gusev on January 16, 2004, and it became one of the first images released by the HRSC team (on January 24, the same day Opportunity landed in Meridiani) in full color. What we saw not only surprised us, it stunned us.

Both Viking and Global Surveyor had imaged Gusev in grayscale. In both sets of images, the crater could be seen to have very large dark splotches roughly in the middle. The shape of these splotches had changed from Viking in 1976 to MGS in the late 1990s, indicating they were at least somewhat transient and not permanent markings. What made the Mars Express Gusev image so immediately interesting was the fact that those dark markings were revealed by Mars Express to be various amazing shades of *green* (see Color Fig. 25).

Reaction to this startling European Gusev image was immediate, and highly controversial. The blatant "green" indicated to many the distinct possibility of current plant life on the floor of Gusev. Linda Moulton Howe, a regular contributor to *Coast to Coast AM*, managed something of a scoop when, shortly after the above Mars Express image was published, she managed to get an on-the-record statement[153] from Michael McKay, Flight Operations Director of the European Space Agency:

> "Like the green in the Gusev Crater picture... *it certainly gives rise to the speculation that there* could be algae [there]... *It certainly gives much more weight to such speculation, particularly since here on the Earth's glaciers and [in] the Alps and [at] the North Pole, you can see algae in the ice itself that turns rather a pink color or green-grey color. Just tying that observation on the Earth together with things we are starting to see on Mars, certainly adds a bit more weight and people will seriously be thinking about these questions and trying to put some definite answers to them..."* [Emphasis added.]

Remarkably, right after this extremely leading, extremely provocative statement, the color of the official Mars Express Gusev image on the German Space Agency website was curiously "recalibrated,"[154] while simultaneously the caption on the official ESA site carrying the Gusev "green" image was also altered—with a key line added: "Note the green coloring is an effect of image processing..."

However, inexplicably, the image on the site remained unchanged. Given that a "recalibrated" version of this same image had just replaced the original on the official German website, this is completely baffling. Unless it was designed as a message that the original bright green version of the data was valid.

Our reaction was a bit more direct. We published, on the Enterprise Mission website, a side-by-side comparison [Color Fig. 25] of the provocative old/new Mars Express image and a "colorized" comparison of the same Gusev region.[155] The latter was unofficially created from official NASA THEMIS data and colorized (from the same data) by space artist and NASA contractor Don Davis.

It was certainly obvious from this particular comparison that something indeed was/is very wrong with NASA's Martian colors. Even after the officially attempted correction on the German website, on enlargement of the revised Mars Express image, the wispy streaks were *still* green—albeit a darker bluish-green, with maybe some purple thrown in. What was truly fascinating was that, strikingly obvious in the new color image, the "streaks" emanated directly from the dark crater floors. This visible preference for the wisps to somehow want to interact with craters was not easily explainable in terms of the prevailing NASA model, which still maintains that the sinuous dark features on Gusev's floor are simply random wind streaks, caused by lighter dust being removed by local dust devils from the darker, underlying surface.

In fact, the imaging comparison revealed the opposite. The Martian winds are preferentially removing something dark from the floors of the even darker craters, and depositing it on the plains *between* these craters—as the wispy, blue-green, purple streaks [Fig. 11-5]. This is plainly evident in the recalibrated German image. This is also where two completely independent Mars observations suddenly came together.

When Spirit landed on the floor of Gusev on January 3, 2004, one of its first high-resolution surface color images showed a mysterious patch of "something" lying a few feet from the Lander [Fig. 11-6]. The nickname the rover science team eventually gave this curious surface feature was the "magic carpet." When Spirit descended from its Lander a few days later, instead of investigating the "magic carpet" close-up with its unique array of instruments, the rover was commanded to drive as fast as possible several hundred feet away to Bonneville Crater. The mystery of the magic carpet was literally left behind, never to be solved.

But what if these two issues are connected? What if Mars Express' new color image of the mysterious dark streaks covering sections of the floor of Gusev Crater is somehow connected to Spirit's equally provocative observations on that crater floor, of the mysterious "magic carpet" area. Suppose that the Spirit images of a "mud-like surface feature" were *exactly* that—images of Martian mud? Suppose that a highly concentrated *brine solution* lies just under the surface rocks and dust, beneath major sections of this ancient crater floor? After all, this was supposed to be an ancient crater lake at one time.

Then suppose that, since it was summer at the Gusev site when Spirit landed, this subsurface brine solution had once again seasonally melted (surface Martian temperatures can be as high as 70º F), creating a layer of mud just beneath the surface rocks and dust. Spirit lands... the airbags drag across this partially wet, very sticky surface, and—viola!—Spirit captures the first image of a genuine "mud puddle" on the planet Mars.

So, what has this to do with the "greenish" color and sinuous nature of the streak, and their obvious preference for craters? If the "Magic Carpet" was indeed caused by a briny "water table" lying beneath the ancient, dry lake Gusev surface, then every crater in the area, having punched through this surface crust to varying depths, should extend well *below* this dry and dusty surface, well down into the brine layer. On Earth, such a situation would be tailor-made for all varieties of simple (and even complex) plant life to begin to grow—particularly certain kinds of algae. Some species of terrestrial algae are extremely adapted to highly saline conditions,[156] and often reproduce by creating spores, which are then redistributed by local winds, forming other colonies.[157]

At Gusev, if the craters in the area were indeed harboring conditions conducive to some special algae growth—primarily by extending below the local water table—then one could easily speculate that the algae mats within some craters grow in the Martian spring and summer, and ultimately reproduce. Their spores are then carried by the winds out of the craters to form the long, sinuous streaks across the intercrater surfaces observed from orbit. The "streaks," then, would be more colonies of algae from the craters, spread by algae spores surviving for a time between the crater floors.

And *that* would explain why NASA chose to land Spirit right smack in the middle of these dark, organic streaks.

It was obvious that this was indeed the unstated reason why Gusev had been targeted for one of the Rovers. NASA wanted to check first-hand on the dark streaks and find out if they were indeed simple plant life. The Mars Express team had then effectively "outed" NASA's real objective by publishing the bright green image of Gusev. They probably got a lot of static from the Rover team through the back channels (remember, NASA was already embroiled in a bit of a color controversy over the Spirit surface images), so they issued a "corrected" version of the data, but *never withdrew their initial bright green version.*

To us, this confirmed that ESA was going to take a fairly neutral stance on hot political issues. It might mean that we would never get a press release calling Cydonia artificial, but we would get honest data, at least. NASA, however, was not so committed to disseminating truth.

At the same time as all this was going on, the second Martian rover, Opportunity, had made a successful landing in Meridiani Planum. On Sol 33, its thirty-third day on Mars, the rover was commanded to roll forward to begin intensive investigation of a small section of rocky outcrop rimming the small crater that it had landed in on January 25, 2004. The outcrop, only a few inches high but which spans approximately 180º of the crater's interior, had been dubbed by the JPL Rover Team "Opportunity Ledge." The specific section that Opportunity was ordered to investigate was about in the middle of this outcrop, is approximately ten inches high and was named by the team "El Capitan."

Preparatory to actually drilling into El Capitan and making detailed composition measurements with the array of sophisticated instruments on the Rover's arm, Opportunity was commanded to take a series of close-up images of the untouched surface of the rock with the B&W microscopic CCD camera attached to the arm. One of those images revealed an amazing sight: An apparent Martian *fossil* [Fig. 11-7].

A close-up enlargement revealed apparently snapped-off body geometry, at least five visible cylindrical segments, and a hint of other fossil-like features buried in the surrounding rock itself—all classic hallmarks of a former *living* organism. After we discovered and posted an initial version of this claim on the Enterprise site on March 2, 2004, as might be expected we began receiving e-mails from around the world. They came from amateur and professionals alike, and all pointing out an almost unbelievable resemblance between our Martian fossil and a well-known terrestrial counterpart. Quoting from one correspondent, James Calhoun:

"I have been a collector of marine fossils for thirty-four years, an amateur to be sure, but with years of field experience. When I saw the 'Fossil' pic [on the Enterprise site], it was clear to me that it met a number of the basic criteria of fossilization. RCH was correct in that "scale does not matter," as the physical characteristics of the item are immediately apparent, and it is sad that the MER team did not present a professional paleontologist to comment. In that light, I have heard a varied number of explanations as to what type of fossil this could be, everything from a segmented worm (annelid) to a shrimp (crustacean). I would like you guys to consider that based on the symmetry of the object, that it could be in fact an early Crinoid, a filter feeding marine plant-like animal, a type with a calcium carbonate exoskeleton (this is Earth-based of course, the Martian exoskeleton [could] have been of a differing mineral composition). I have included a couple of pictures for symmetry and scale reference. Notice the triangular symmetry in the 'branch areas,' not to mention the segments, and also that the scale is inline with the 'size of the blueberries.' Your opinion would be most appreciated. Thanks for your time and I appreciate the work the team does."

The images Jim included with his e-mail knocked our socks off, as the saying goes. His crinoids were a perfect match for our Martian "fossil." A crinoid [sometimes called a "sea lily," because of its superficial appearance to a spreading flower] is, as Calhoun described, "a filter feeding, marine, plant-like *animal*." Crinoids first appeared in Earth's primeval seas over 500 million years ago, in the so-called "Cambrian Era," climbing to dominance over the next 150 million years, before receding once again in the terrestrial fossil record.

Crinoids lived in ocean water—ranging from a few feet deep to several miles—anchoring their stems on the ocean floor and feeding on whatever nutrients drifted by. If you look at a combined map of where JPL landed its two rovers and the Odyssey Gamma Ray Spectrometer orbital determination of water abundance in the upper one meter of Martian soil, a glance will suffice to show the rovers are indeed exploring none other than the shallows of our two proposed equatorial Martian tidal oceans, almost as if that had been planned.

It takes almost no imagination to picture this site several million years ago as a quiet tidal pool, filled with gently waving creatures of the sea, until one day something extraordinary happened, and this pool and all of Mars was forever changed.

So, upon making this extraordinary discovery, what did NASA and the rover team promptly do? Did they call a press conference and hail their discovery to the world? Did they head to the White House to brief the president on the most momentous scientific discovery in all of human history? No.

They immediately took the grinder (technically called the "Rock Abrasion Tool"—RAT), and *ground it into powder*. Instead of moving the grinder a couple inches to the left or right, they simply bored down on the fossil, totally obliterating it.

Before calling this a scientific crime against humanity, perhaps we should consider that this was simply a mistake. Theoretically, it is possible they simply didn't see it for what it was, that there was no one on the team that could recognize the tell-take signature of fossilized plant life. That scenario is possible, but not likely, given the "fictional predisposition" of at least one of the rover science team member.

In 2000, Geoffrey Landis—a NASA scientist attached to the NASA-Glenn Research Center, in Ohio—wrote his first Mars science fiction novel, "Mars Crossing." Landis had been a principal investigator for one experiment on the Mars Pathfinder Sojourner rover mission to the outflow of an ancient water channel, Ares Vallis. As a result, his "Mars" in the new novel was highly praised, by both veteran science fiction writers and planetary scientists alike, as "totally authentic." One wrote:

"High-quality hard SF written with the authenticity of a NASA insider... Landis has given us a legend of our own near future..."

Geoffrey Landis is also now a member of the current JPL Rover Science Team, a member of the "atmospheric group." This makes what he did in his novel six years ago very interesting, to say the least. Three quarters through his novel, Landis has one of his characters, Brandon Weber, get lost in the arid Martian desolation of endless dunes and dust. Tired and scared, the astronaut finally climbs a small butte to get his bearings, and makes a startling, serendipitous discovery.

"There was a fracture line running down the middle of the butte; one half of it was two feet higher than the other. It made a natural seat. Without any sense of wonder, without even a sense of irony, he reached out and touched it. Embedded in the layered sandstone exposed by the crack, it held a perfectly preserved fossil. It looked like *a cluster of shiny black hoses, clumped together at the bottom, branching out into a dozen tentacles at the top.* In the same section of rock, he could see others, of every size from tiny ones to one three feet long. There were other fossils too, smaller ones in different shapes, a bewildering variety.

"'I name you Mars Life Brandonii,' he said."

How did Landis—a Pathfinder and MER Rover Mission NASA scientist—somehow *know,* four years before the Opportunity mission even landed? And why didn't he, or anyone else on the Lander science team, say anything about it, or even raise an objection to its destruction?

There is no question that the rover team saw the crinoid fossil. Not only did they use it as the base target for the drill, they gave it a target name: "Guadalupe." Our Lady of Guadalupe is a Roman Catholic religious icon, functionally equivalent to the Egyptian *Isis*, the goddess of *life*. Yet after discovering a fossil that could confirm the existence of complex life on Mars sometime in the past, and even naming it after a "goddess of life," they destroyed it. Since Opportunity's instrument suite was not designed to look for signs of life (it was strictly set up to be a roving geologist), Guadalupe's destruction served no real scientific purpose. It had long ago taken on the properties of its surrounding rock. The only testament to the fact that it once had lived was that unmistakable segmented shape in the rocks—and NASA destroyed it.

So the game seems to be to discover evidence of life beyond Earth, but do everything possible to publicly denounce it. The agency seemed to be following a carefully directed script, one that at this point did not include the finding of life on Mars. Spirit and Opportunity had been sent to do one thing—confirm the existence of ancient water (and current water-ice) on Mars, at the most. Like Viking, a finding of actual life would have been politically unacceptable. Clearly, the agency was on a timetable, a Brookings-like pattern of preconditioning to avoid the "unpredictable" shock that might come with an announcement of "life."

In Mars exploration since 1976, NASA has discovered evidence of microbial life in two Viking-era experiments, discovered that the water table of the atmosphere is conducive to liquid water on the planet's surface today (Pathfinder), detected evidence of a muddy-briny surface in the bottom of a shallow crater (Spirit), detected visual evidence of algae and spores from orbit (Mars Express), detected large amounts of (probably) organically produced methane in the atmosphere (Mars Express), and found an unmistakable fossil analog to a form of life that once flourished on Earth (Opportunity). In each case, at each crossroad, they have chosen to debunk, ignore, suppress and even destroy the evidence that could support a finding of current-day life on Mars. Along the way, the only time they even lukewarmly supported such a possible discovery was when they announced the finding of four-billion-year-old fossilized bacteria in a meteorite on Earth (which raised the likelihood of contamination of the sample), and then promptly backed off from even that timid claim.

Clearly, each successive Mars mission was supposed to move the pieces forward ever-so-slightly—but, as with Apollo 10, NASA would not allow any undesirable findings to alter the game plan, simply because "it wasn't time yet." However, soon after the great crinoid cover-up, we learned that the time might indeed be at hand.

Chapter Eleven Images

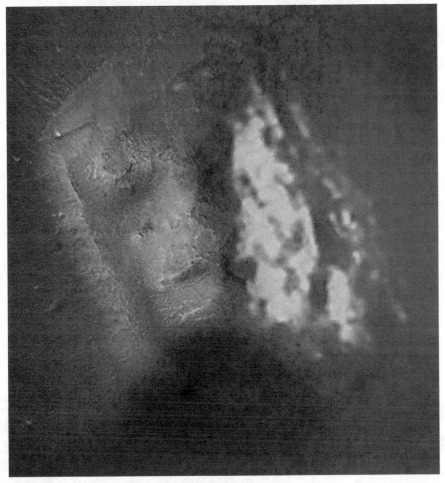

Fig. 11-1 - Composite image made from May 2001 MGS B&W image, and July 2003 Mars Odyssey 2001 five-band color image. Note anomalously bright, geometrically aligned, *structural* elements on eastern side (R) (ASU/Laney/Hoagland)

Fig. 11-2 – Grayscale version of five-band color IR image V0 3814003. Note highly anomalous lighting of the Face and D&M, even in this "pre-dawn" image. (ASU)

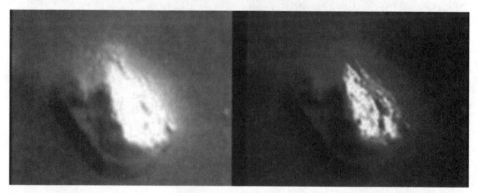

Fig. 11-3 – Side by side comparison of the Face from V0 3814003. Version on left is unprocessed. Version on right has had brightness reduced, revealing mysterious, overtly geometric panels making up the Face's eastern (Cliff) side. No other processing has been applied to version on right. (ASU)

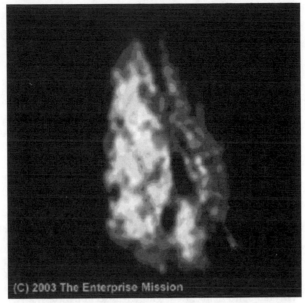

Fig. 11-4 – Close-up of highly reflective, geometric panels on eastern side of the Face. Panels align with central symmetry axis, *not* with the image scan. Unique lighting geometry of this *pre-dawn* image acts like an x-ray, optically revealing the same underlying architectural substructure of the Face seen at much lower resolution (10 X) detected in nighttime THEMIS IR image (ASU).

Fig. 11-5 – Mars Express close-up of wisps of dark green material from the inside of small craters in Gusev. Was this organic looking material what the MER rover Spirit was really sent to investigate? (ESA\Mars Express)

Fig. 11-6 – The "Magic carpet" from the Spirit landing site. Why did Spirit deliberately ignore this seemingly obvious "mud puddle" on a planet on which all water is supposedly frozen solid? (JPL)

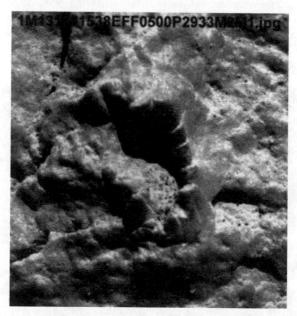

Fig. 11-7- Segmented, chrinoid-like Martian fossil on "El Capitan" rock. Instead of studying this amazing find, NASA ground it in to dust. (JPL)

Chapter Twelve

Where the Titans Slept...

On January 14, 2004, right in the midst of the Spirit and Opportunity Rover Missions, President George W. Bush traveled the few blocks from the White House to NASA headquarters, in Washington, D.C. to make a momentous announcement. More than thirty years after Mankind had last set foot upon the surface of another world, Bush had come to NASA HQ to announce a new direction for the Space Agency. We were, finally, going back to the Moon and then on to Mars.

At the NASA Announcement Ceremony for his "Vision for Space Exploration," Bush outlined three specific goals.

1. Complete the International Space Station (ISS) by 2010, and then retire the Space Shuttle.

2. Develop a new Crew Exploration Vehicle, or CEV, to carry astronauts to the Space Station and then on to the Moon and Mars by 2008, with manned missions to start no later than 2014.

3. Return to the Moon with a manned mission by 2015, if possible, but no later than 2020.

In order to accomplish this, the President siphoned off $11 billion over a five-year period from the existing NASA budget, and then went to Congress to ask for an additional billion dollars to complete the funding. This was a considerably more modest proposal than the so-called "Space Exploration Initiative" proposed by his father, President George H. W. Bush, in 1989. That program had met with resistance over its large budgetary requirements and was subsequently killed by his successor, President Clinton. "W," it seemed, had learned from the mistakes of Bush the Elder and made a proposal that would ultimately receive almost unanimous support in political circles.

Bush 43, of course, had made something of a habit of paying back Bush 41's political debts, and finishing his unfinished business.

As readers of our web-based "Millennium" series will note,[158] Bush 43 seems to be playing the role of the avenging son Horus to his father's "Osiris." Almost from the day of his fathers loss to Bill Clinton in the 1992 presidential election, George W. Bush began plotting his own path to the White House, vanquishing along the way most of his fathers' most prominent political enemies.

He started in his home state of Texas in 1994, running against the one-term incumbent Democratic Governor, Ann Richards. Richards had attained national prominence at the 1988 Democratic National Convention by mocking his father, the first President Bush. Further, she had been equally disrespectful when the elder Bush left office. However, two years later, it was Richards whose political career was over. Bush the younger won a surprisingly easy victory over Richards, who was considered popular and was heavily favored to win re-election. After a landslide victory in his first attempt at re-election in 1998, Bush set his sights on his father's old job at the White House.

Although the 22nd amendment to the U.S. constitution prevented him from running against President Clinton himself, Bush could run against Clinton's hand picked successor and political protégé', Vice President Al Gore. In the summer of 2000, both men swept easily to their respective parties nominations for president, and then faced off in the fall campaign. Gore, like Richards before him, was considered a heavy favorite over the supposedly "light-weight" Bush, but soon found he had all he could handle in his opponent. Bush deftly exposed some of Gore's less pleasing personality traits by goading him in the three presidential debates, and Gore's slim lead had faded away by the weekend before the election. Then, Fox News broke a story that Saturday night about Bush's arrest 25 years earlier for drunk driving. Bush's momentum stalled and the election became a toss-up.

Election night 2000 was certainly one of the most memorable for Americans in their modern history. Using exit polls gathered by a consortium called the "Voter News Service" rather than actual returns, the various news networks called a number of states early for Gore, and he seemed to have a solid lead. When they called the State of Florida for Gore before the polls had even closed in the panhandle section of the state, it seemed to put him over the top. Bush however quickly went on national television and stated that he did not agree with the projections giving the state to Gore, and that he fully expected to win Florida. As the evening wore on and the returns flowed in, Gore's lead in the State was not nearly as large as the exit polls had predicted. When the polls closed in the panhandle, the networks retracted their calls for Gore, and as the returns came in more fully, called the State for Bush. As the final returns came in, Bush's lead shrank to about 1,700 votes, making Florida too close to call and forcing the networks to retract their earlier calls for Bush. Gore, who had already conceded the election to Bush, immediately retracted his concession and proceeded to initiate legal action to force a hand recount. A machine recount also showed Bush the winner by a slim margin, but Gore requested a hand recount only in selected counties of the State. After a month of litigation back and forth, the Supreme Court of the United States declared on a 7-2 vote that the hand recounts were unconstitutional, and ordered the recounts stopped. Gore then conceded and Bush was inaugurated in January 2001.

Having vanquished his father's domestic political foes, it would not be long before Bush 43 would have to face off against his fathers old international opponents as well. After the devastating terrorist attacks of 9-11, 2001, Bush made a speech to Congress declaring a "war on terror" and promising to take the battle not just to the terror groups themselves—but also to the nation-states that supported them. Iraq and its dictator, Saddam Hussein, were first on this list. After being left in power by Bush 41 at the conclusion of the first Gulf War, Saddam had sponsored and financed several attacks against the United States, including the first World Trade Center bombing. He was also strongly implicated in the Oklahoma City bombing of the Murrah Federal Building. He had attempted to assassinate Bush 41 in 1993, and George W. Bush was determined to take him down. The U.S. and a loose coalition of other nations invaded Iraq in 2003 and captured Hussein in 2004. He was tried for crimes against humanity in Iraq and executed in late 2006.

With all of his father's geo-political foes swept aside, all that was left was for George W. Bush to complete his father's one other unfinished task— a workable Mars Program. While the Space Exploration Initiative might not be the most significant memory most Americans have of the first Bush Presidency, there is considerable evidence it was of paramount importance to the President himself.

In fact, it was the *first* major policy initiative of his Presidency.

George Herbert Walker Bush announced the Space Exploration Initiative on the occasion of the Twentieth Anniversary of the Apollo 11 Manned Moon Landing. "The U.S. is the richest nation on Earth, with the most powerful economy in the world. And our goal is nothing less than to establish the United States as the pre-eminent spacefaring Nation," he stated. "And next, for the New Century, back to the Moon, back to the future, and this time back to stay. And then, a journey into tomorrow, a journey to another planet, a manned mission to Mars."[159]

His announcement at a Smithsonian Air and Space Museum ceremony honoring the Apollo 11 crew caught many by surprise, since as late as a few days before the ceremony reporters were convinced that Bush would not be announcing any new space programs. The secrecy enabled Bush to gain a brief political advantage, but that momentum quickly faded as the President made several tactical mistakes along the way.

First, he re-created the National Space Council, an Agency that had once been led by Lyndon Johnson when he was Vice President. Bush appointed his own Vice President, the much-maligned Dan Quayle, to head the Council and make recommendations on executing the President's vision. He gave the Council ninety days to establish "realistic" goals and milestones for the new initiative.

However, both NASA and the Congress viewed the Council as an Executive Branch usurpation of both entities' authority. The lack of communication between the Council and the Space Agency led to the ninety-day study becoming a bloated, incoherent wish list of NASA whims rather than a minimalist, cost-sensitive blueprint for practically achieving the President's goals (one over-the-top-proposal even called for a crane to be built on the Moon, at a cost equivalent to *an entire aircraft carrier group*).

When all of the items on the NASA "wish list" were added up, the cost came to a staggering $541 billion over several decades.

With a price tag like that, the President's space vision was dead on arrival in a hostile Congress controlled by the Democrats. To many, it seemed that NASA had gone out of its way to kill the President's Space Initiative, when they ought to have been embracing it. However, this seeming inconsistency takes on a very different *political* light when we consider the elder Bush's stated motivations for creating the Program in the first place.

As we showed in Chapter Three, the first President Bush overtly stated on at least one occasion his view that the work of Richard C. Hoagland (and the then-"Mars Mission") was a primary impetus behind his SEI program. In fact, on the very day of Bush's announcement (July 20, 1989) of his dramatic new thirty-year plan, CNN called Hoagland for an interview. The producers of the political program *Crossfire* specifically wanted Hoagland, of all the possible NASA Mars experts they could have called upon, to present "the case for Mars," following the President's surprise announcement on the steps of the National Air and Space Museum earlier that day.

This raises two immediate questions: Who recommended Hoagland for this gig; and, what do you imagine they thought he'd talk about on CNN ...?

On the first question, the answer seems obvious—the White House. This was, after all, the President's first major policy initiative of his first term, and *Crossfire* was the pre-eminent political talk show of the day. To think that the White House would allow just "anyone"—anyone other than their own hand-picked "talking head"—to make the case for the new policy strains credulity. Certainly, the White House did not abdicate when asked to provide its own expert.

Hoagland, at that moment, was in the middle of Yosemite National Park in Northern California—about as far from a television studio as you can get in North America. The proposed CNN Mars *Crossfire* for that evening simply couldn't happen for reasons of practical logistics.

On the second question, there can also be little doubt. Obviously Hoagland would have discussed Cydonia and the Face on CNN that night.

So here we have the specter of the President of the United States sending out the number-one advocate of the idea of "artificial ruins on Mars" as his *primary* spokesperson on a new Space Policy.

No wonder NASA moved to kill the SEI so quickly.

A few months later Hoagland was invited to make his presentation at NASA Lewis. Apparently in response to the President's wishes, when Dr. John Klineberg, Director of NASA/Lewis, introduced Hoagland for his presentation at that facility on March 20, 1990, Klineberg told the assembled NASA workers that it was Hoagland's work that had inspired the SEI:

"Richard Hoagland is [also] the man who managed to convince the President to state that a return to Mars is one of our goals..."[160]

Strangely, even though these comments were made in front of literally thousands of NASA scientists and engineers in the main NASA-Lewis Auditorium, and simultaneously broadcast via close-circuit television throughout the rest of the ~4000-acre NASA facility, NASA said it somehow, subsequently *lost* the entire video recording of Klineberg's introductory statement.

Over a week later, a NASA public affairs representative finally "explained" to "Nightline" producers at ABC News (who were eagerly awaiting delivery of Klineberg's extraordinary remarks related to the White House on the tape) that "they'd had a simultaneous failure in *all three cameras* recording the event." Miraculously, however, the three cameras came back to life at the same moment, just as Hoagland began speaking.

Fortunately, a friend of Hoagland's, Mark Dwane (whose father worked for NASA-Lewis), was in the audience with an audio cassette tape recorder, and captured Klineberg's entire "official" introduction.

Naturally, skeptics and critics scoff at the idea that the Cydonia anomalies had inspired the entire SEI. But in order to do so, they must ignore the CNN incident *and* Klineberg's extraordinary remarks at NASA-Lewis. Only minutes before his public statement, Klineberg had informed Hoagland and several other attendees in a private meeting in the Director's office that NASA was "under intense scrutiny from Congress, because of SEI." Why would he then go out and make an untrue and irresponsible statement about the President (and over such an intensely controversial issue)—with multiple television cameras and recorders running—when the President's support would be desperately needed in the Agency's ongoing political battles?

The answer is obvious. He *wouldn't*. And he didn't.

What Klineberg was doing was clearly what he believed the White House *wanted* him to do. His introductory comments only make sense in this context. In any event, there's still more to the story.

In 1997, the brother of one of the authors, Dave Bara, was walking through an office area of the Boeing Company in Renton, Washington, which had just been vacated by a relocated organization. On the wall of an abandoned cubicle, he spotted a poster that he instantly knew to be highly significant. It was from the 1989 White House Mars Exploration Program proposed by then-President Bush. The poster had been commissioned by the Boeing Company and designed, as all such materials are, to raise awareness and inspire enthusiasm for a given program. In this case, the artists saw fit to inspire their workers and the public by creating a depiction of NASA astronauts ascending a cliff (perhaps *the* Cliff, as there is a suspiciously "Face-like" edifice in the background) and encountering nothing less than clearly *artificial ruins* [Fig. 12-1]. To ensure that there is no doubt that this was connected to the President's "space vision," the poster also includes a quote from President Bush explaining "Why Mars?"

In actuality, the illustration says it better than any words. The ruins are a series of partially buried, stacked stones with a variety of Egyptian or Sumerian-appearing symbols and glyphs on them, but the image is dominated by the sculpted face of what appears to be a black man in an Egyptian-styled headdress. It is obvious from this that Bush was as interested in Hoagland's work as Klineberg suggested that day at NASA/Lewis, and as his invitation for *Crossfire* implied.

It was clear to us that these three separate lines of evidence—the *Crossfire* invitation, the Klineberg statement quoting the President and the Program poster—all pointed to Bush 41 being not only familiar with Hoagland's work, but *inspired* by it. Inspired to the point that he would propose a $500 billion program to test its core thesis—that there are ancient ruins on Mars. What was not as certain, as the younger Bush made his own Announcement 15 years later, was whether he shared his father's views on this aspect of the Mars exploration project he was proposing.

Those doubts were quickly allayed.

Once the new Programs got going, under the umbrella title of "Project Constellation," it was soon obvious that they would follow the previously established NASA ritual patterns. The new light and heavy lift rockets to support the program were quickly named "Ares," after the Roman god of war. "Ares" was also commonly associated by the Romans with the planet Mars, so in the context of the new "Moon-Mars" program, this made sense. On July 20, 2006, the Thirty-Seventh Anniversary of the first Moon landing and the (simultaneous) Thirtieth Anniversary of the Viking 1 landing on Mars, word leaked of an even more important ritual linkage—

NASA had decided to name the new Crew Exploration Vehicle (CEV) "Project Orion."

As we discussed in Chapter Five, the symbol of the constellation Orion appeared on the early Apollo program patch, only to "disappear" *after* Apollo 13 [Fig. 12-2].

The three belt stars of Orion, so significant on the Apollo patch, only magically "reappeared" on the Project Orion logo (as we might expect, if "rituals" were still crucial to the inside workings of the Agency ...). Beyond that, the Orion patch is significantly overlaid on a background shaped like an *equilateral triangle* (remember: a two-dimensional stand-in for a three-D tetrahedron.

The true significance of this overwhelming, redundant ritual pattern cannot be overstated.

We have a new manned Moon-Mars program, named "Orion"—which we have previously established is interchangeable with the god "Osiris" in the Egyptian tradition. In the color version of the new patch, this crucial constellation is overlaid on a blue equilateral triangle, which also matches the Egyptian hieroglyph for "Sothis," the star Sirius, which represents "Isis," Osiris' sister and consort in that same tradition.

The name of this new program appeared in the press for the first time on July 20, a recurring NASA "ritual date" which marks the heliacal rising of Sirius/Isis in the ancient Egyptian stellar calendar—and all of this for a Program proposed by a President who sees himself as "an avenging Horus figure." And this new Program not only "resurrects" the first President Bush's grand space vision, but also resurrects NASA's entire listless manned space program, which has been stalled in low Earth-orbit for nearly four decades. And, "Osiris" just happens to be the Egyptian god of resurrection.

They may as well have just named it "Project Osiris."

What's more, apart from this Egyptian connection, "Orion" makes no sense as an official Project name; it bears no resemblance to the Greek "Apollo" tradition of the previous Program, and is not commonly associated with either the Moon or the planet Mars, as "Ares" is. The fact is, "Orion" connects to a space program destined for the Moon and Mars only *one way*—through the original Apollo Program patch!

NASA Director Mike Griffin apparently truly meant what he said, when he flippantly termed it "Apollo on steroids." For that's *exactly* what it is—no more, no less. The blatant symbolism cinches it.

So, is it actually designed to finish—either secretly or (finally!) in the open—what "Apollo" started ...? That's the overriding question, at this point.

* * *

As the summer of 2006 unfolded, it became clear that "Project Osiris" was of the highest priority. A Space.com story[161] explained that certain forces on Capitol Hill, including the powerful General Accounting Office, wanted NASA to slow down their decision on picking a builder for the new Project Orion spacecraft. NASA quickly blew off the GAO's recommendations, an unusual maneuver to say the least. The GAO is highly respected in Washington, and rarely do their recommendations go unheeded, much less get shoved back in their faces. NASA's swift outright rejection of the GAO's recommended timeline was an indication that they were in a real hurry to start work. In fact, word later leaked that the competing contractors were told that they would be expected to be putting out production designs *within one week* of being awarded the contract.

Which begs the question: What's the hurry?

Given that we have already been to the Moon and back, and that there was no "space race" with the Soviet Union as there had been in the 1960s, what could be the true reason for rushing this new Initiative? What could be driving these sudden decisions to not only go back to the Moon, Mars (and perhaps beyond ...), but to also fast-track the Program with a tight timetable and unheard-of demands from contractors? Was there perhaps something about the times we lived in that could be driving this aggressive schedule? Or, had NASA found "something" on their first voyages to the Moon that compelled *this* President to go back?

To be fair, the tight schedules may have just been driven by a desire to set the program in motion so that it reached a "point of no return" politically. As columnist George Will once put it, "government programs that are already in motion tend to stay in motion." By getting the Orion capsule a significant way down the road by 2008 (the year of the next presidential election), Bush would make it harder for his successor to cancel the Programs without a considerable financial (thus, political) penalty.

We, of course, suspected there was something more to the story. We went back to the President's initial January 2004 announcement speech at NASA headquarters. There, we found clues that got us thinking there were bigger— much bigger—considerations, than mere concern for political legacy, driving "Project Constellation."

Signs and Wonders

The President's announcement ceremony at NASA Headquarters in Washington DC in January 2004 was actually a fairly quiet and somber affair. Bush's remarks were short, almost terse. He addressed first the decline of the manned space program over the preceding three decades:

"In the past thirty years, no human being has set foot on another world, or ventured farther upward into space than 386 miles—roughly the distance from Washington, D.C. to Boston, Massachusetts. America has not developed a new vehicle to advance human exploration in space in nearly a quarter century. It is time for America to take the next steps... along this journey we'll make many technological breakthroughs. We don't know yet what those breakthroughs will be, but we can be certain they'll come, and that our efforts will be repaid many times over. We may discover resources on the Moon or Mars that will boggle the imagination, that will test our limits to dream. And the fascination generated by further exploration will inspire our young people to study math, and science, and engineering, and create a new generation of innovators and pioneers.

"Mankind is drawn to the heavens for the same reason we were once drawn into unknown lands and across the open sea. We choose to explore space because doing so improves our lives, and lifts our national spirit ...."

About midway through the speech, Bush turned to and acknowledged the last man to have set foot upon another world, Gene Cernan, the Commander of Apollo 17:

"Eugene Cernan, who is with us today—the last man to set foot on the lunar surface—said this as he left: 'We leave as we came and, God willing, as we shall return, with peace and hope for all Mankind.' America will make those words come true."

And how did Cernan react as the room filled with applause? By refusing to even acknowledge the gesture on the President's part, and even scowling as the words were spoken. The astronaut next to him, who remains unidentified, even cast Cernan a stern look [Fig. 12-3].

Frankly, this reaction, or lack thereof, stunned us. Not only was it rude of Cernan, it was totally inappropriate etiquette, given that he was invited (presumably) at NASA and the White House's behest. How often does one refuse to even respond to an acknowledgement by the President of the United States—especially with the cameras rolling, and surrounded by the very people who invited you to the event?

We considered the possibility that Cernan was simply a dyed-in-the-wool Democrat, but even that would hardly be an explanation given that he was at NASA HQ on a politically neutral occasion, the announcement of *the very thing* he'd asked for in his final remarks on the surface of the Moon.

We began to suspect that there was something more to Cernan's silent protest than a mere political disagreement with President Bush. Perhaps there was something more, something about the Apollo 17 Mission ... and Cernan's role in it at work here.

In reviewing the Bush Announcement, the choice of Cernan to represent the Apollo past had seemed strange all along. Apollo 11's Buzz Aldrin and Neil Armstrong would seem much more logical choices, since they had already participated in numerous NASA events like this one, were far better known to the general public than Cernan, and had always delivered generous and memorable remarks.

However, on the occasion of the Twenty-Fifth Anniversary of Apollo 11, at a White House ceremony on July 20, 1994, Armstrong himself had seemed frustrated. He started his highly emotional address by first comparing himself to a parrot —saying only what he had been told to say—and then let slip his provocative remark at the end, about "truth's protective layers." What exactly had he meant by *that?* Was this a "coded" admission of the reality of a Brookings-like control of the astronauts? Was Cernan's steel-cold silence ten years later, at the 2004 announcement of, essentially, a rebirth of *the entire Apollo Program,* actually a sign of solidarity among the astronauts, of acknowledging that "something" has been hidden for all these years ...? After the President's 2004 announcement Cernan had only this to say—"I've been waiting for this day for thirty-one years."

Not exactly a ringing endorsement of the President's plan.

The astronauts have all seemed abnormally uncomfortable at these events representing NASA, as if they were being trotted out for show but not really being allowed to discuss what they wanted to discuss. Beyond the silent confirmations of our Lunar dome thesis in Alan Bean's artwork, they seemed to be, at times, speaking out in the only way they could, through subtle "messages" and actions, like Armstrong's double meaning "code," or Cernan's definitely out-of-place non-reaction

One of Cernan and Armstrong's fellow astronauts had made a similar protest in even stranger circumstances a few years before. In March 2001, former astronaut and Senator John Glenn made a March appearance on the NBC comedy *Frasier*, in which he made some not-so-subtle comments:

"Back in those glory days, I was very uncomfortable when they asked us to say things we didn't want to say and deny other things. Some people asked, you know, were you alone out there? We never gave the real answer, and yet we see things out there, strange things, but we know what we saw out there. And we couldn't really say anything. The bosses were really afraid of this, they were afraid of the 'War of the Worlds'-type stuff, and about panic in the streets. So we had to keep quiet. And now we only see these things in our nightmares, or maybe in the movies, and some of them are pretty close to being the truth."

There was, of course, something of an uproar when this episode of *Frasier* aired. NASA sycophants, like James Oberg, immediately dismissed these comments as simple humor because of their context—one of the most popular situation comedies on television. But when you actually watch the program, listen to the words and observe Glenn's demeanor—especially when you know what we now know—it becomes much harder to dismiss his comments as simple farce.

In fact, this bizarre appearance has all the earmarks of a straightforward "Brookings" revelation. It is so direct, so fraught with implications, that had it happened in any other forum, it would have sent shockwaves through the press. Instead, because it happened in the context of a network comedy, most observers, including some in the UFO community who should know better, dismissed it and even made light of it as an early "April Fool's" joke on Glenn's part. Perhaps that was precisely why this particular forum was chosen in the first place.

Those dismissals are missing the key point. Regardless of where they were spoken, Glenn's words are a scathing indictment of NASA and its integrity. Let's consider this a moment: Assume that everything the doubters say is true, that Glenn was "just invited" to be on this show, and for whatever reason decided to accept. As a true American hero, a distinguished senator from Ohio and former presidential candidate, Glenn would obviously have some leverage on the script. So imagine his reaction when he is handed his copy of the words he is to speak. In his most crucial scene, he's asked to:

- Admit he is a liar;
- Admit that the Agency which sent him into space twice—making him everything he is today—astronaut, senator and American hero, not to mention a man of some personal wealth—has also lied;
- Admit that the Brookings recommendations to cover up any discovery of extraterrestrial ruins (or life) are true;
- Admit that he and his fellow astronauts were so shattered by what they saw that they have had "nightmares" ever since;
- Admit that some film portrayals about the UFO subject are accurate.

And John Glenn—war hero, senator, statesman, astronaut and American icon says: "Sure. No problem. Anything for a laugh." Can't you just see the belly laughs and knee slapping back at NASA Headquarters when this episode aired?

The simple and obvious truth is that no man with any sense of integrity, loyalty or gratitude would stoop to such a level for a mere laugh. The idea that Glenn would agree to say these things about himself and NASA—even in such a context—is, well ... laughable. Anybody who quickly dismisses this without considering these deeper issues is just whistling past the graveyard. This was a deliberate broadside, directed right at NASA.

When you watch the show itself, this really becomes obvious. The plot, such as it is, deals with Frasier's producer Roz, who wants to do a show on the space program "because it's 2001." Glenn is brought in to narrate, and through a series of circumstances ends up lying to Roz to protect Frasier. Once he is caught in this lie, he apologizes and adds the statement "I was misled. It's not like me to be that underhanded."

An argument ensues between Frasier and Roz, and they retire to the control room to have it out in private. As they do, Glenn—in the studio all by himself—begins to recite the words above over the air, completely out of context with anything else that has happened in the show, or with the action now taking place behind him in the control room. As he does so, he does not address *any* of the characters around him. In fact, he looks directly into the camera, solely addressing *the audience at home* watching throughout America.

When he's done, and realizes that his words have been recorded, he rushes into the control room and he asks for the tape—implying that he is still under some sort of duress or pressure to keep quiet. Roz and Frasier, who—because of their total preoccupation—*haven't heard a word* of Glenn's from the heart "confession," blithely give it to him, completely missing what's just happened.

And the biggest problem with the "simple comedy" model? The whole thing, from start to finish, isn't funny. In fact, the laugh track used at different points while Glenn is making his statement is wildly inappropriate. Beyond that, Glenn's use of the camera as his audience blows the illusion of reality. Glenn actually seems to be making his statement completely outside the universe of the show.

The only real joke in this whole episode appears to be on *us*. Frasier and Roz represent completely self-involved stand-ins for all Americans, arguing over totally petty concerns while right in front of them a genuine American hero is literally baring his soul about what he has actually seen "out there." He goes on to recite the unconscionable behavior of the key American institution, NASA, charged with its exploration. The "joke" is that Frasier and Roz—as "us" for the last fifty years—once again *completely miss it*. Of course, the joke only works if Glenn's now is finally telling us the truth.

And, if we were right, if the Brookings study—which pointed to the "War of the Worlds" scenario specifically cited by Glenn in his speech as a justification for suppression of evidence of extraterrestrial activity—were not merely a forty-year-old "recommendation," but current *policy*, then this is exactly the context in which Glenn would "tell all." Doing it on a comedy show would give him exactly the kind of political cover he would need to come clean.

Perhaps he was hoping somebody in the mainstream press would see the contradiction in his actions, and ask him directly if it were true. Or, perhaps he wanted his "confession" on the record in advance of any official "disclosure" efforts that might one day surface. Otherwise, why would such an individual engage in such an overt act of "NASA bashing?"

There is one last point that puts this whole weird affair into its appropriate context. According to an NBC promotional video, it was *Glenn* who approached the "Frasier" people ... to do this *specific* show.

Now, isn't *that* funny?

There are other indications that all is not well in the astronaut corps regarding the "official story" of how NASA went to the Moon, and what it really found.

Over the last several decades, scores of astronauts have taken to writing books about their historic missions, seeking, understandably, to immortalize the obvious high points of both their lives and their careers. Some of these biographies are—to say the least—inconsistent with their authors' obvious intent: to leave a *positive*, personal record of events.

Some of them, in fact, contain outright, blatant *lies*.

Take one of the original "Mercury Seven" astronauts, Alan Shepard. Shepard's memoir, "Moon Shot" (co-written with another member of the original Mercury Seven corps, Deke Slayton, and two veteran space reporters, Jay Barbaree and Howard Benedict), is the usual tale of The Right Stuff, complicated by the *unusual* medical histories of Shepard and Slayton.

For Shepard, it was a curious inner ear problem, which took him off flight status for six years, from 1964 (following his historic sub-orbital Freedom 7 "hop" in 1961)until he was miraculously "cured" and reinstated in 1969. He had a secret and then-innovative inner ear surgical procedure, which allowed him to ultimately command Apollo 14 to the Moon in 1971.

It was during the last EVA of Apollo 14, you might remember, that Alan Shepard surprised NASA and the Nation on live television—and achieved a unique claim to history as well—with his famous "... first golf shot on the Moon."

For Slayton, it was a chronic erratic heart rate—which effectively barred him, just after his Mercury selection, from *all* future spaceflights (he was given the title of "Chief Astronaut" in Houston as a symbolic consolation, and the actual job of picking other astronauts to go "where he could never go." But, he was obviously not happy with the bitter twist of fate that grounded him. Then—at the very end of the Apollo Program—NASA flight surgeons suddenly, miraculously, cleared him to fly as well ... just in time to catch the last seat aboard the historic first joint U.S./USSR Mission in Earth orbit, the Apollo Soyuz Test Project (ASTP), in 1975.

So far, so good.

It's when you begin to examine the *photographic* section of their joint authorship, "Moon Shot," that the real problems begin. There, right in the middle of Shepard's own account of his amazing against-all-odds personal journey to the Moon...

... Is a *deliberately* faked record of the trip [Fig. 12-4].

If you examine the second, annotated image carefully, you will ultimately catalogue an amazing number of "discrepancies"—including, the MAJOR discrepancy of the image itself; this is obviously supposed to be one of thousands of Hasselblad photographs taken on the lunar surface by the Apollo astronauts, specifically, during Apollo 14 ... of Shepard's well-known lunar "golf shot," right?

However, since there were only *two* Apollo 14 astronauts who landed on the Moon—Shepard and Mitchell—and just two Hasselblad cameras (one for each of them) ... who took THIS Hasselblad shot—published as a two-page spread (note the central binding in the reproduction), in Shepard's own "space biography?" Since—

Both astronauts—Shepard and Mitchell—are seen simultaneously *in the same photograph?*

The only other camera which could have taken this particular image was the color television camera, erected by Mitchell at their landing site soon after first emerging from the lunar module "Antares" two days earlier. In fact, this camera did send back to Earth a running sequence of television images of this event, in parallel with Shepard's live radio transmission as he took his swing, and boasted that the ball "went miles and miles." So, is this a "frame grab" from *that* camera? It was quickly obvious that was not the case. Besides the terrible quality of the TV images—which, obviously, can in no way compare to the B&W "photograph" in *Moon Shot,* we noted, among other things, that the flag was pointing the opposite direction in the photo, Shepard's body position was different, the image of Ed Mitchell in the golf shot fake is simply cut out (and reversed) from a *real* Hasselblad frame, AS14-66-9301, the now-infamous "Mitchell under glass" photo [Fig. 12-5].

This, and a dozen other discrepancies, will more than convince most readers that the "photograph" reproduced as the literal centerpiece of Shepard's ultimate claim to fame during his spacewalk on the Moon—the golf shot—*is an outright fake.*

The truly fascinating question then becomes: for God's sake, *why?*

Why—if you were going to make a big deal about taking the first golf shot on the Moon, why *fake* the photograph of that event? And, why ... right in the middle of your own personal success story of the most amazing achievement of your life?

Unless ... it's another "message?"

When we first saw this image, reproduced as the centerpiece in Shepard's own *official* story ... we couldn't believe how bad it was. After all, if you're going to shade the truth a bit, come up with a decent image of your quirky moment of non-NASA lunar creativity, why not a *good* fake?

Unless ....

You *wanted* to get caught. You *wanted* folks to notice ... and ask "why?"

Not just about this obviously fake "photograph" ... but, maybe, about the overall Apollo 14 Mission ... maybe, the entire *Apollo Program itself.*

Some, in looking at this (and other "anomalous" Apollo lunar surface photographs), and seeing all kinds of major discrepancies on those official images, have eagerly seized on their existence, claiming, "See, that *proves* NASA never went to the Moon!"

Well, we've effectively blown that insipid Moon hoax theory out the window, starting in the very Introduction to this book. So, what other "message" from Shepard does that leave us with?

Maybe something simple, like: "There *are* other fake images from Apollo out there ... and they've been deliberately created *by NASA*—to cover up what we all *really* saw and photographed during *all* our lunar Missions ... which I *can't* put in this book."

Remember, this (below—also see Color Fig. 5) is a *real* Apollo 14 photograph, taken by Alan Shepard of his fellow astronaut, Ed Mitchell—as Mitchell was setting up the TV camera at the beginning of their first lunar EVA. Look what Shepard *really* photographed—above and all around *both* astronauts—on Shepard's amazing Apollo 14 Mission to the Moon.

Wouldn't *you* want to—somehow—tell the world what they were *missing*?

Recently, yet another former astronaut has added his two cents to this growing list of "dissenters." A BBC documentary, entitled *First to the Moon,* was shown in England and then repeated once in July 2006 on the Science Channel in America. In the program, Buzz Aldrin, the second man on the Moon, admits to and describes the Apollo 11 crew's encounter with "a UFO."

He described the object, which seemed to be pacing the Apollo 11 spacecraft for hours, as a "double ellipse" or "bell-shaped" object. The crew was so concerned about it that they made extensive observations (and presumably took photographs, although this is not stated in the show) with the onboard telescope in the Command Module. Failing to identify the object, they cautiously asked NASA to relay the location of the discarded third stage rocket, the S-IVB, relative to their current position. NASA, not understanding why the information was requested, nevertheless told the crew that S-IVB was some 6,000 miles distant and well behind them. Aldrin on the show stated that none of the other two crewmen of Apollo 11—Armstrong or Collins—believed that the object in question was anywhere near that far away. Other footage, of a similar object filmed on a later Apollo mission, was shown as Aldrin spoke about "the mystery UFO."

Eventually, the crew decided that discussing the object on an open broadcast channel to Houston could disrupt the Mission (Aldrin stated that he "feared we would be ordered home, *before* the landing), so they agreed to say nothing more about it until their debriefing upon return to Earth. Aldrin did not state if that debriefing actually took place.

So with all this, we began to really wonder: was there something *specific* to Cernan, and to Apollo 17, "something" that had required *his* presence at the 2004 "Return to the Moon" announcement? Was there something known—but publicly off limits—between him and the President that had led to Cernan's odd and even disrespectful behavior? Had we missed something about Apollo 17 that was meaningful, and that required Cernan to be at the Announcement ceremony ... for "ceremonial" reasons?

Was Cernan—like Bean, Armstrong, Aldrin, Glenn and Shepard before him—trying to tell us "something?"

Where the Titans Slept ....

Fortunately for us, two of our associates, Keith Laney and Steve Troy, had been looking at Apollo 17 data for some time. Steve is an amateur geologist and artist from South Dakota and Keith (as you know) is an imaging specialist working with NASA/Ames. Between them, they had done extensive photographic studies of the Apollo 17 mission to Taurus-Littrow, obtaining many early generation negatives. Keith had even posted an extensive analysis of all of this on his website.[162]

The first thing that struck Hoagland in his original assessment of the Apollo 17 Mission, back in 1998, was the incredibly dangerous look of the landing site. Positioned at 19.5º N by 31º E, the target landing ellipse was in a narrow valley amongst the Taurus-Littrow highlands. This was by far the riskiest Apollo landing of them all, as Gene Cernan would be required to set the lunar module *Challenger* down among gigantic (6,500 to 8,200 feet tall) mountains on a valley floor littered with large craters. The area is in fact so dangerous that it required a unique navigational approach that was completely different than any other Apollo landing. In order to even reach the Taurus-Littrow site, NASA had to abandon long-standing mission rules requiring "free-return" trajectories (a lunar orbit insertion trajectory which would allow the spacecraft to loop around the Moon and return to Earth in the event the CSM engine didn't fire), as well as prohibitions against launching at night. They even abandoned mission rules covering the roughness of landing sites in order to accommodate the desire to land at Taurus-Littrow.[163]

In originally looking at the Apollo 17 landing site images from Lunar Orbiter, Hoagland began to wonder: what could possibly be so significant about this site that NASA would attempt such a hair-raising landing?

The most obvious clue was the ritual aspect of the 19.5º landing ellipse. Just like the *Pathfinder* landing site selection on Mars years later, it was very close to a 19.5º x 33º tetrahedral location—but, 2º on the Moon equates to nearly forty miles away from the Taurus-Littrow landing site, and that site rests at 31º longitude, not 33º. What was so compelling about this particular area?

In more closely examining the proposed landing site years later, it became obvious to Keith Laney what the attraction was. There, in almost the center of the ellipse, was a massive, hexagonal mountain. Officially dubbed the "South Massif" for purposes of navigation, the mountain has at least four clearly visible and near-equal length sides, and the implications of two more sides that were obscured in the collapse of the main structure [Fig. 12-6].

In looking closer at the South Massif, it became evident that the south side of the structure had collapsed inward, perhaps forcing the bright material visible on the north face out from under the structure. It is highly unusual for a solid rock mountain (presumed by geologists to be a result of a volcanic uplift) to collapse inward like this. Cinder cones on Earth frequently show evidence of some internal collapse, but these deformations are uniformly circular. The South Massif would therefore have to be one of the most unusual cinder cones ever discovered. For one thing, cinder cones are exactly that, conical-shaped volcanic uplifts with distinctive rounded crater-like depressions ("vents") at their peaks. Rarely, if ever, do cinder cones take on geometrical shapes, and the authors know of no cinder cone ever identified that took on a hexagonal configuration. Further, the "vent" in the South Massif—if that's what it is—is square. This is also virtually unprecedented.

The ostensible geologic reason for selecting this landing site was the opportunity to sample the "dark mantle" material that covered the valley floor. Supposedly, this would be from the earliest impact that formed the Sea of Serenity basin that the Taurus-Littrow highlands bordered. Lunar Orbiter images of the region also showed large boulders deposited along the bases of the mountains (particularly the South Massif). There is also a bright patch of material overlaying the dark mantled valley floor extending from the north end of the massif all the way to the sinuous Lincoln Scarp. The bright material was presumably deposited from the event that led to the collapse of the mountainous structure. There were also several dark rimmed "halo" craters that were thought to be volcanic.

Using a landing approach that required him to drop the LM sharply down among the mountains at a steep angle and with little margin for error, Cernan guided the lunar module Challenger to a landing at the outer edge of the landing ellipse on December 11, 1972 at 19:54:57 UTC. The precise location was at 20° 11' 26.88" N x 30° 46' 18.05", or just beyond the magical 19.5° location [Fig. 12-7].

Cernan managed to land the *Challenger* near Camelot crater (wistfully echoing the theme that Jackie Kennedy had given President Kennedy's entire, brief Administration after his untimely death ... the President who had made it possible for the Gene Cernan to even land here at all). "Camelot" was just one of several large and dangerous craters threatening a safe landing, located in the Taurus-Littrow Valley between the giant mountains known as the North and South Massifs, the "Sculptured Hills" and "Mons Vitruvius." The crater "Isis," named by Mission Commander Cernan, was also was nearby.

The EVA (Extra Vehicular Activity) plans for the Mission were the most extensive ever attempted for an Apollo mission. The science lobby inside NASA had pulled strings to get Harrison "Jack" Schmitt, a geologist, reassigned to this Mission after his previously assigned mission, Apollo 18, was abruptly canceled. Schmitt and Cernan had an aggressive schedule that called for them to unpack the Lunar Rover, emplace a number of seismometers and explosive charges at key points within the Valley, while traversing and exploring nearly 39 kilometers in total. They were also tasked to deploy a mysterious and classified experiment called "Chapel Bell," about which virtually nothing (to this day) has been revealed.

After unpacking and deploying the Rover and ALSEP instrument packages (along with the American flag), Schmitt and Cernan headed to their first "geology station," the nearby crater Steno. After taking various samples there, the crew returned to the Lunar Module Challenger and rested up for day two, which would turn out to be the biggest day of the Mission.

EVA-2 called for Cernan and Schmitt to head directly for the South Massif and the odd "crater" named Nansen. After that, they were scheduled to visit locations on the Lincoln Scarp and then stop at Shorty Crater, one of the key targets for the Mission.

In looking at the orbital images of the landing site, it becomes quite obvious why these locations (and EVA-3's planned visit to the "Sculptured Hills") were so intriguing: recent enhancements done by Laney and Hoagland of the Apollo landing site orbital images show "boxy-looking patterns" on the faces of virtually *all* the "hills" and" mountains" in the landing vicinity.

Other views of the landing site, taken by the high resolution panoramic camera on the CSM America, show the entire South Massif is criss-crossed with highly unusual geometric patterns, both on the visible face of the massif and in the hollowed-out interior.

New enhancements done by Hoagland show that the still-standing sidewalls of the "vented" massif are actually *overhanging the empty space inside—* as if they were supported by some internal truss structure that is still holding them up!

In fact, in looking at the "vent hole," there is a curious lack of material inside the walls of this formerly hexagonal mountain. It is almost as if the mountain was a hollowed arcology-like structure all along—but if this were truly once a whole mountain (or arcology), where did all this "missing material" go?

Laney is of the opinion that the collapse pushed a great deal of the bright material out from under the South Massif through Nansen, creating the bright blanket of debris on the valley floor to the north of the massif. This may be partially correct, and some of the material that makes up the bright blanket may have come off the top of the massif as well [Fig. 12-8].

EVA-2–a "Valley of Surprises"

The mission plan for EVA-2 of Apollo 17 called for one quick stop to take geological samples and gravimeter readings, and then the astronauts were to make a B-line for a bizarre feature at the base of the South Massif called "Nansen."

Officially listed as a crater, Keith Laney has shown categorically that Nansen is nothing of the kind. Recon photos of the South Massif show Nansen as a V-shaped depression at the base of the massif, over which the "rim" of Nansen seems to be an overhanging shelf. Views of Nansen strongly imply that it is a hole in the base of the South Massif, possibly an entrance point (or exit wound) *into* the massif [Fig. 12-9]. Certainly, if there were anything unusual about the South Massif, Cernan and Schmitt would be able to spot it either from "Geology Station 2," which was at the base of the South Massif atop Nansen, or on their way up (or back from) the station.

Everybody seemed very excited about prospects for this second EVA. Excerpts from the Apollo Lunar Surface Journal[164] show that as the astronauts prepared the rover for the second EVA, Mission Control and Schmitt had this quick exchange:

> 141:02:06, Parker: "And, Jack, if Gene's working there on unstowing SCB whatever-it-is—five, yeah, five—maybe when you put the camera down, you might want to shoot off a few five-hundred-millimeter frames of the North and South Massifs, if they look interesting. I can't tell from the TV. That might be an opportune time to grab a couple."
>
> 141:02:27, Schmitt: (Incredulous) "If they look interesting!? If they look interesting!? Now, what kind of thing is that to say?"

Their first waypoint was an area on the bright blanket of material north of the South Massif called "Hole in the Wall." This was a shallow rise up the eighty-meter-high Lincoln Scarp between the craters Lara and Candide, which would enable them to approach Nansen (and the base of the South Massif) from the east. Other traverse maps had this area near Nansen curiously labeled "Access Region." Were they headed to Nansen hoping to obtain "access" to the interior of the massive hexagonal structure?

As they neared the South Massif, Cernan continued to make observations and relay them over the radio. One of the first things he noted was the curious linear geometry on the surface of the massif:

> 141:52:03, Cernan: *"Jack, can you see over there to the left—I'll turn a little bit (for a better view)—on the dark area of the South Massif where you get those impressed lineations? See them going from left upward to the right?"*
>
> 141:52:11, Schmitt: *"Yeah. I see what you mean; right."*
>
> 141:52:14, Cernan: *"That's what I saw out my window."*
>
> 141:52:15, Schmitt: *"Yeah, they go obliquely up the slope."*
>
> 141:52:20, Cernan: *"They're more like wrinkles, they're linear wrinkles."*
>
> 141:52:22, Schmitt: *"Yeah. Crenulations, you might say, in the slope that looks something like those I saw from orbit, looking in the shadowed area... or, at the edge of the shadows."*

Schmitt would later downplay these comments while debriefing on the EVA:

"It was a puzzle, seeing apparent lineations on the slopes of mountains. Some people, as I recall, did some simulations, building models, putting random roughness on the surface, and then dusting them and moving the light around, and they were able to create apparent lineations just with light position. Generally, I think, people don't feel that they represent any underlying structure—it's just an accident of dusting and lighting. The massifs do have layers in them—layers of debris—and I think the fact that you see what appear to be zones of blocks at the top is probably a layer of relatively hard material. But they really are gross layers."[165]

Pretty clearly here, Schmitt is using the old "trick of light and shadow" argument to explain away the totally anomalous nature of the South Massif. As we will see, it will become harder and harder for both of the astronauts to disguise their shock at what they see:

> 142:12:30, Cernan: *"Jack, look at the wrinkles over there on the North Massif."*
>
> 142:12:34, Schmitt: *"Yeah. There's no question that there are apparent lineations all over these Massifs, in a variety of directions. Hey, look at how that Scarp [sic] goes up the side (of the North Massif) there. There's a distinct change in texture."*

142:12:46, Schmitt: "As a matter of fact, the lineations are not present on the Scarp [sic], that we can see, where it crosses the North Massif. There is no sign of those lineations on there."
142:12:58, Cernan: "Oh, man; yeah. I can see what you're talking about now."

This implies that the scarp and the South Massif are not of the same origin, since the "lineations" they keep talking about begin specifically at the base of the massif. Later images, along with the orbital photography, have confirmed that the "lineations" are not tricks of light and shadow, but undeniably real features of the mountains in this Taurus-Littrow Valley. Schmitt's inability to explain them stems from the fact that such "geologic layering" is almost always associated with sedimentary deposition, caused by standing water (the other, more infrequent cause is "successive lava flows" ...). Since no water has ever flowed (let alone pooled!) on the Moon, and the lineations on the massifs were literally *thousands* of feet above the (presumed) lava Valley floor, such geological explanations were totally untenable.

The only viable alternative is that this mysterious "layering"— repeatedly seen (and photographed) by the Apollo 17 crew—is only a surface manifestation of some kind of now heavily eroded, repeating, artificial, *3-D* "cell-like" *constructions*. It is these former highly geometric rooms which are constantly being exposed (and then eroded back ...) by an incessant meteorite bombardment of these massive lunar "mountains"—an "explanation" which certainly isn't going to appear in any official NASA documents.

Strikingly *identical* (and equally mysterious) "rectilinear" formations had been seen and photographed by the crew of Apollo 15 (which landed far to the west of Apollo 17, on the "shores" of Mare Imbrium), as well as in "the highlands of the Moon," half way between the 15 and 17 sites, by the Apollo 16 astronauts, John Young and Charlie Duke [Fig. 12-10].

As they cleared the top of the "hole in the wall" and headed for the east slope of Nansen, the Apollo 17 astronauts at first couldn't see down into Nansen. Cernan had to drive to his left (east) and approach the overhanging shelf from that side.

At this point, the astronauts are about 1.5 km from their destination atop the shelf that overhangs Nansen. Schmitt, who has the task of taking a photograph every thirty yards or so during the traverse, inexplicably stops taking pictures at this time. Or, at least there are no more pictures showing the traverse to geology station two available in the official records. A few moments later, the astronauts get their first look down into Nansen, at which point Cernan decides to park the rover up on the shelf, facing north, so the TV cameras cannot see directly into the opening at the base of the massif. Schmitt and Cernan, however, *could* see directly into Nansen.

Unfortunately, when we did get the television picture, it showed us little to nothing. They had parked the rover on the shelf above the entrance to the South Massif, and all we had was a view looking back toward the light mantle avalanche runoff [Fig. 12-11]. The astronauts were also completely out of sight of the camera for most of the next twenty minutes. In fact, they are out of sight of the camera for fully 85% of the entire sixty-four minute visit to the upper shelf of Nansen. The camera just pans around aimlessly. This would have given them plenty of time to descend the hill and investigate the interior of Nansen, including examining the opening below the overhang.

As they are nearing the end of the station, Cernan stops (off camera) to take some pans of the view from the base of the massif:

> 143:22:08, Cernan: "Well, I have some good pictures of Nansen, anyway, and... (long pause)... You know, I look out there, I'm not sure I really believe it all."

A bit later, completely out of context, Schmitt seems to address their "off-camera" time to mission control:

> 143:27:11, Schmitt: "We haven't had a chance to look around any more than you've heard."
> 143:27:14, Parker: "Okay."

Is this an indication that they *didn't* descend into Nansen during their off-camera time? Or a hint that there was always a secret, second mission to accomplish at Nansen? And what was so unbelievable about Nansen that Cernan had to mention it? Certainly, there is nothing in the released images of the station that implies anything unusual.

Next, Cernan and Schmitt drove to station 2A, just a few hundred meters back down the slope they had ridden up to get to the lip of Nansen. From this perspective, they could have had a perfect vantage point looking directly into the hole leading into the South Massif [Fig. 12-12].

Amazingly, there are officially no pictures taken of Nansen from this perspective, looking directly into the "opening" under the ledge and back up at the previous geology station. At station 2A, Cernan turned the rover in a circle, so Schmitt could take a panoramic view of the area and the valley below them. As he did so, there was this exchange:

> 143:50:20, Cernan: "Wait a minute. Wait a minute. Okay. Let's take one from right here. I want the whole thing. (Pause.) You ready to start?"
> 143:50:26, Schmitt: "Yeah, I got it."
> 143:50:27, Cernan: "Start taking. Take the whole thing."
> 143:50:54, Cernan: "Isn't that something? Man, you talk about a mysterious looking place." (Pause.)
> 143:51:03, Schmitt: "They can cut some frames—some parts of those pictures out—and make a nice photograph. (Laughing.) [With the] TV camera, [and] maps [in the way]."

According to the transcripts, the pan "turned out to be relatively uninteresting because of the sun glare." Published images do not seem to be a full 360° pan, because if they were, the entrance down into Nansen would be visible. In fact, it's nowhere on the published sequences of this area. And how does Cernan's statement "Man, you talk about a mysterious looking place" jibe with the "relatively uninteresting" description given in the transcripts? This whole area is fascinatingly anomalous, so much so that Cernan stopped to make sure Schmitt got a complete photographic record of it. A record, incidentally, which now does not correspond to the descriptions of the astronauts at that time. And what about Schmitt's nervous talk about cutting "some parts of those pictures out?" What does he see that NASA doesn't want the folks back home to see?

As you read back through these sequences, it's clear from the comments of the astronauts that something is amiss. First, Cernan is concerned that they can see directly into Nansen, so much so that he parks the rover up over the lip of Nansen so the TV cameras can't see directly into the opening. Schmitt—the trained scientist—is so astonished as he looks into Nansen for the first time that he exclaims, "Look at Nansen!" Cernan goes on to describe the whole area as a "mysterious place," and "unbelievable," and Schmitt cryptically informs Mission Control that the astronauts "haven't had a chance to look around any more than you've heard," despite their thirty-plus minutes off-camera.

Later, as they take pictures that should show what lies beneath the shadowed "overhang" of Nansen (which is clearly visible from orbit), they joke about cutting out certain parts of the pictures. And in all this time, not once did they take a picture showing Nansen from a vantage point that would reveal the interior of the "crater." They evidently just missed the opportunity, and nobody at Mission Control decided to ask for such a picture.

The reality, however, is that between their off-camera time and the traverse to Station 3, the astronauts would have had plenty of time to descend from the rise and examine and photograph the interior of Nansen. That these photos could have been simply lifted from the photographic catalogs is easy to conceive, especially since some were described as overexposed or "uninteresting,"—eerily reminiscent of the blacked out catalog images from the Apollo orbital cameras. Perhaps they found that getting into the South Massif through Nansen was impractical—there were descriptions of a great deal of debris in the crater. Or perhaps they tried and failed, leading to Schmitt's admission that they didn't get to look around as much as they wanted to.

Whatever the case, they were on a tight timetable to get on with the other stations. What they could not have known, however, was that they were also on a collision course with an even more unbelievable and mysterious destiny.

Data's Head

"Mr. Data, your head is not an artifact." —Commander Riker, from the Star Trek: The Next Generation episode "Time's Arrow."

Along the way to Shorty Crater, a black, halo-rimmed crater on the outskirts of the light mantle material from the South Massif avalanche, Schmitt and Cernan stopped to take some samples along the rim of another small crater. The stop, which was supposed to last twenty minutes, was made to take a double core sample, get a gravimeter reading, and take some 500 millimeter pans of the general area. The station turned out to be a disaster, as the astronauts had numerous equipment problems and Schmitt took a spectacular spill trying to retrieve some sample bags. The crew later named the station "Ballet Crater," in honor of Schmitt's fall, and the astronaut later attempted a few ballet moves in his suit after being kidded that the Houston Ballet Foundation had called to enquire about his services.

It took nearly thirty-seven minutes for the astronauts to complete their tasks at Ballet Crater, and from there it was straight on to Shorty, which was a primary stop for the EVA along with Nansen. Upon arrival at Shorty, the astronauts took care of some housekeeping chores, then got their first look at the crater.

> *145:22:22, Schmitt: "Shorty is a crater, the size of which you know [about 100 meters in diameter]. It's obviously darker rimmed, although the fragment population for most of the blanket does not seem too different than the light mantle. But inside… Whoo, whoo, whoo!"*

Schmitt's description seemed to imply that while Shorty was relatively unspectacular on the outside, the area inside the crater was, at the least, *very* interesting. Unfortunately, when the camera started up, it was pointed at the Rover and the distant South Massif. It stayed positioned there as Schmitt [it's Cernan who moves off to the take the pan—because the "reflections" Schmidt is talking about in a minute come from the Rover's gold myler—which is where we see Schmidt imaged … on *Cernan's* images.] moved away to take a panorama of the crater. Several minutes into this sequence, Cernan oddly states "O-kaay! O-kaay." At this point, Schmitt begins to discuss something odd he has noticed through his visor. Raising his filter, he is suddenly totally absorbed by what he's seeing.

> *145:26:25, Schmitt: "Wait a minute…"*
> *145:26:26, Cernan: "What?"*
> *145:26:27, Schmitt: "Where are the reflections? I've been fooled once. There is orange soil!"*
> *145:26:32, Cernan: "Well, don't move it until I see it."*
> *145:26:35, Schmitt: (Very excited) "It's all over! Orange!"*
> *145:26:38, Cernan: "Don't move it until I see it."*
> *145:26:40, Schmitt: "I stirred it up with my feet."*
> *145:26:42, Cernan: (Excited, too) "Hey, it is! I can see it from here!"*
> *145:26:44, Schmitt: "It's orange!"*
> *145:26:46, Cernan: "Wait a minute, let me put my visor up. It's still orange!"*

145:26:49, Schmitt: "Sure it is! Crazy!"
145:26:53, Cernan: "Orange!"
145:26:54, Schmitt: "I've got to dig a trench, Houston."

The astronauts then began to sample the orange soil, which was later found to be highly oxidized titanium, a discovery which had tremendous implications for later colonization of the Moon. Extracting oxygen and metals from the lunar soil would make the idea of a permanent lunar base much more viable.

Apropos of the "orange soil's" totally melted and oxidized titanium, in some of Cernan's previous Shorty images "strange objects" can be seen which do not resemble the fractured, volcanic rocks that would be expected in and around a major impact crater. Instead, they look like large chunks of broken, metallic *machinery and shattered glass.*

Intrigued by the multiple Hasselblad photographs available of this juxtaposed "mechanical-looking" debris and "orange soil" at Shorty, Hoagland downloaded the highest resolution versions available from the Apollo Lunar Surface Journal website.

In doing so, it became clear that some of the "debris" in Shorty was definitely "unusual"—to say the least. Color enhancement showed that many of the "rocks" had highly unusual spectral qualities, reflecting light more like crystals—or, highly polished metallic "boxes"—than a simple "rock garden" would suggest.

In jointly examining Hoagland's enhancements, both authors independently noticed one very large, definitely strange-looking artifact—that strongly resembled a somewhat mangled "pump mechanism" or "engine housing." Nicknamed "the turkey" (because of its obvious "fan tail" configuration at one end, and dark "beak" at the other), this object appears to have a series of tubes and mechanical features visible inside a geometric metallic (or glass-like) case. There even appear to be "connectors" or "mounting points" on the object [Fig. 12-13].

It was while carefully studying this bizarre, apparently blatant piece of lunar junk that Hoagland spotted an even more unusual artifact, obviously lying some distance beyond it on the crater floor [Fig. 12-14]. Even as he suddenly realized what he was seeing, he couldn't bring himself to admit what it *appeared* to be ....

A *human* head!

In a crater–

On the Moon .... [Color Figs. 27 and 28].

After recovering from his initial shock he swiftly surmised that it couldn't possibly be a *human* skull. After all, it was lying in a debris field from an impact crater which had tossed up all manner of junk and material from just below the regolith of the Taurus-Littrow Valley floor. Something as fragile as fossilized bone could not possibly survive the energies of such an impact. No, this object had to be related to everything else he was seeing in this frame—a lot of which appeared quite *mechanical* in origin.

A robot's head, then?

As his mind grappled with that incredible possibility, Hoagland kept coming back to Schmitt and Cernan's previous statements on what they were witnessing during this entire EVA. As Cernan put it, even though he was seeing it with his own eyes, he still couldn't "quite bring himself to believe it"—and he had dubbed the entire Valley "one mysterious looking place." Had he and Schmitt gazed into the abyss at Nansen, seen chunks of similar-looking mechanical debris, and then stashed the photos away for later breakdown? Was Shorty simply another example of the kind of "unbelievable" things they had seen all along on this second EVA?

Color enhancements (see Color Fig. 28) showed that the "head" had a distinctive red stripe around the area where the upper lip should be, a feature that clearly appeared to be *painted or anodized* on the object. Composites of other frames Cernan had taken from the rim of Shorty showed that the head had two eye-sockets, a forehead, brow ridges, a nose with nostrils, twin cheek bones and the upper half of the jaw; the "lower jaw" seemed to be missing [Fig. 12-16]. Still, it was an astonishing photographic find. And the resemblance to another, even more familiar figure did not escape the authors [Fig. 12-17].

What was most striking about the C-3PO comparison—and most telling— was the *eyes*. Like C-3PO, "our" robot's head had indented, stereoscopic, *rounded* inset eyes.

Camera lenses.

Just like ... C-3PO.

In looking at the context panoramas from which the close-up was taken, Hoagland was able to confirm that the head was approximately the same size as a human's—which meant, among other things ... Cernan and Schmidt *could have brought it back.*

The transcripts for the Shorty EVA show that the astronauts were certainly rushed at this station because of the time they had lost at Ballet Crater. It is very possible that Schmitt and Cernan never saw the object in question, or, that they decided it would be too risky to try and retrieve. However, they certainly had enough off-camera time to descend the crater unobserved and bring it back ... if they wanted to.

In looking at these images, the authors are uncannily reminded of a *Star Trek: The Next Generation* episode called "Time's Arrow." In it, the *Enterprise* is summoned to Twenty-Fourth Century Earth, to an archeological dig below San Francisco. In this dig, Captain Picard and Commander Data (an android) are shown a puzzling artifact: Mr. Data's disassembled robotic head.

In the course of the story, Data's head and the information contained in his "positronic brain" are crucial to unlocking the mystery of Earth's past. By tapping into the memory of this ancient and damaged artifact, the crew of the *Enterprise* is able to stop human history from unraveling and their very existence from being threatened.

Was this perhaps "the great secret" of Apollo 17? Was this the reason for Cernan's decidedly odd behavior at the NASA ceremony, let alone his very presence? Had he waited in vain—literally for *decades*—for NASA to reveal the contents of "Data's head?"

Was he angry that he was being asked to participate in another ruse on the American people, after having already participated once before ... on Apollo 17, only to wait thirty-one years for a chance to go back, and perhaps set the record straight?

In the short-term, it doesn't matter if it was truly "Data's head" or other mechanical artifacts that Cernan and Schmitt brought back from the Taurus-Littrow Valley. As we have seen over the preceding pages, there was plenty of evidence all around these two obviously awe-inspired astronauts, that substantial areas in the vicinity of the Lander were *artificial* ... but now in ruins.

If the Mission of Apollo 17 was to secretly confirm the artificial nature of the Taurus-Littrow Valley, it overwhelmingly accomplished its "hidden mission." However, even ruins inevitably leave haunting clues to the majesty of the original structures and civilizations they once housed.

And—nothing stays buried *forever*....

Land of the Giants

As we continued to download and microscopically examine every available image from the Apollo 17 Mission, we came across one notable fact: none of the photographic surface pans that were available reflected the odd geometry—the "lineations"—that Schmitt and Cernan described on their approach to South Massif. The published surface photos did seem to match the available TV footage, showing the massifs to be just "bland, featureless hills ..." with no distinctive features. Noticing the difference between the astronaut's reports and the photographic record, we began to question everything ... especially the validity of the TV recordings and immediately released NASA prints.

It was soon clear that the orbital photography and ground-based images simply didn't match. Even Ron Evans, the command module pilot for Apollo 17 who made constant observations of the landing site, commented on the "non-monotone" nature of the region and the Moon in general.

"You know, to me the Moon's got a lot more color than I'd been led to believe. I kind of had the impression that everything was the same color. That's far from being true."[166]

As we tried to gather more information on all these related problems, Hoagland took to watching late-night NASA TV in the fall of 2005. Dr. Jim Garvin, who had once promised so much to us when he was NASA's Chief Scientist at NASA Headquarters, had recently been moved from downtown Washington DC out to NASA Goddard, in suburban Maryland, where he had taken over as NASA's chief scientist there, serving the new NASA administrator, Mike Griffin. One of Garvin's first acts at Goddard was to have the Hubble telescope pointed at the Moon to take new images (this was well after NASA stopped falsely claiming that "the Moon is too bright for Hubble to take an image of ..."). This new Hubble lunar campaign was done in concurrence with promoting the President's new Space Initiative.

Garvin's first choice for a target image: the Apollo 17 landing site in Taurus-Littrow.

When the new Hubble image was released to the web in October of 2005, it contained three views in different "panes." The leftmost pane contained the Hubble image of the Apollo 17 landing site with the actual location marked with a red "X." The two rightmost panes showed a derived 3D model of the Taurus-Littrow Valley, complete with the same red X, and ... curious "layering" on the South Massif. The upper right pane contained a portion of one of Gene Cernan's pans at Shorty Crater—one of the same frames that contained images (cropped out) of "Data's head." We took all this as a gigantic "pay attention" signal from inside NASA. We felt they were strongly hinting that when we eventually went back to the Moon, Taurus-Littrow would, in fact, likely be the first place (for some reason ...) to which they would return.

Of course, all that was just an assumption. To that point, we had no proof positive—other than the intense Apollo 17 Mission interest, shown 33 years before in South Massif and Nansen—that NASA saw anything "special" or "unique" about Taurus-Littrow.

Then "NASA Select TV" happened [Fig. 12-18].

Late at night, shortly after the release of the new Hubble image of the Apollo 17 landing site, Hoagland was watching a NASA TV broadcast of a press interview with Garvin on the new images. During the portion of the interview concerning the Hubble lunar shots, NASA showed some old footage of the first Apollo 17 EVA to Steno Crater. In this footage, Hoagland saw for the first time what the "ground truth" at Taurus-Littrow really revealed.

There, in the distance beyond the astronauts, Hoagland could see the astonishingly clear criss-cross patterns that were visible on the orbital photography. These film images were orders of magnitude better than either the still-frame photography or the original TV transmissions from 1972. Clearly, NASA was in possession of far better versions of the EVA films than had ever been broadcast. Now, for the first time, the blatant "truss pattern" construction of the Taurus-Littrow arcologies could be confirmed.

It didn't take long to determine that the arcologies visible in the background were seen against the Mons Vitruvius highlands southeast of the Apollo 17 landing site. These were definitely the same inexplicable "lineations" that Schmitt and Cernan had seen and described on the "air to ground" during their EVAs—and these images simultaneously reinforced a number of other impressions about the Mission [Fig. 12-19].

As noted earlier, NASA had taken great care to get the astronauts to place various seismic monitors and remote-controlled explosive charges and mortars all around the Valley floor. They seemed particularly interested in the seismic properties of the South Massif, as they had monitors placed all around the mountain, and even crashed the used Lunar Module ascent stage of Challenger into it *deliberately* (to create a major seismic signal of known energy input).

And then, there was the curious question of the (still) classified "Chapel Bell" experiment, whose very name implies that they wanted to test the resonant properties of the region. If, in fact, these supposed "massifs" were not that at all, but in fact "hollow, titanium-glass truss structures"—as the new footage tends to strongly imply—then such seismic tests would prove that out beyond any doubt, providing a three-dimensional "ringing map" of the substructure of everything in the Taurus-Littrow Valley.

Unfortunately, the results of "Chapel Bell" are still unknown, and everything about it remains *classified*—on a "civilian" scientific mission to the Moon, after more than thirty years.

At this point, we were as certain as we could be that the Apollo 17 mission had some connection with the President's new Space Initiative. Garvin was pulling out the real data in an effort to do... what? Shock people into asking questions? Test to see if anybody noticed the tremendous difference in quality between the new and old versions of the Rover TV data? Or, was he also simply signaling to the "in crowd" what NASA's *real* intentions were regarding the new Space Initiative, i.e., the *real* reason for the President's sudden "return to the Moon?"

We didn't know—but we were about to find out that all of this would have a very familiar "ring" to it ourselves.

Ares and Orion

By the early portion of 2006, two years after the President's NASA Headquarters' "Vision" Announcement, two major aerospace consortiums were working full speed on their proposals to bring President Bush's 2004 new space vision to life. Once the programs got going, as previously noted, under the curious umbrella name of "Project Constellation," it was soon obvious that they would follow the previously-established NASA ritual patterns. One of the consortiums, Lockheed Martin, had experience building the original Apollo spacecraft that had safely ferried astronauts to the Moon and back (having absorbed many of the original Apollo Command and Service Module contractors). The other, a team made up of Northrop-Grumman and Boeing, had long-since merged with the original 1960's contractors on the Lunar Module and Lunar Rover, respectively.

As the spring of 2006 bloomed, all signs pointed to Northrop-Grumman-Boeing getting the initial contract. Lockheed had botched NASA's initial call for "Apollo on steroids" and, rather than design an evolution of that proven concept, submitted a radical, Space Shuttle-like lifting body design. NASA was unimpressed with their approach, and at a preliminary design review in March told the contractor to go back to the drawing board and produce an Apollo-style capsule more closely resembling Northrop's concept. Unlike Apollo, which only carried three men to the Moon and back, the new CEV would have to accommodate four men to the Moon and up to six to the International Space Station. Designed to replace the aging and dangerous Space Shuttle fleet, the CEV would be NASA's workhorse well into the next two decades.

NASA had also started to look at the launch requirements for the new program, and settled on a derivative design that would use updated Space Shuttle solid rocket boosters (SRBs) to place the CEV and Lunar Surface Access Module into space separately. Larger payloads, consisting of equipment and construction supplies for the ISS and eventual lunar bases, would be lifted using a new heavy lift vehicle that would strap as many as five SRBs onto a new liquid rocket design. NASA quickly dubbed these new rocket concepts Ares 1 (for the CEV and LSAM) and Ares 5 (for the heavy lift vehicle).

Considering that the eventual stated goal of the new initiative was Mars, the name Ares made a certain degree of sense. Ares was the Roman god most closely associated with Mars, although he was also the Greek god of war. It is this association that has led to the planet Mars being most closely associated with war and destruction. Given that NASA had always tended to use the Greek (rather than Roman) mythology in its naming conventions, the association seemed odd. Did NASA really intend to name their new space boosters after a god of war?

As the year wore on, there were more signs that Northrop's design had the edge over Lockheed's. Lockheed had proposed a new suite of avionics that would require extensive technological development, while Northrop's emphasized existing proven electronics. Lockheed also proposed using their satellite division in Colorado as the prime design center, and they had never built a manned spacecraft before.

As the final designs were submitted in May 2006, Northrop officials were confident their proposal had surpassed the Lockheed Martin design. Even though Lockheed's proposal had promised to deliver a working spacecraft a few months earlier, they had a dismal track record with an earlier Shuttle replacement Project, called "the X-33." Lockheed had spent several years and nearly a billion dollars on the X-33, and never produced anything but some quarter scale models to show for it.

As the summer wore on, though, odd signals began to emerge from inside NASA. First, some contractors had protested that the Agency had awarded contracts to competitors simply because of the Mechanical CAD package they were using. NASA's choice for MCAD software (Mechanical Computer Aided Design) was a package called "Pro-Engineer," from PTC Systems. However, virtually all aerospace contractors preferred a French-produced MCAD software called "CATIA, "while a minority still used a product called "Unigraphics." Pro-Engineer was considered inferior by the aerospace industry for a whole variety of reasons, and Pro/E had never won a head-to-head contract against either CATIA or UG at any major aerospace or automotive firm.

A General Accountability Office (GAO) review took NASA to task for its stance, essentially supporting the contractors' position that NASA had tried to force Pro/E down their throats in violation of federal laws. They further found that NASA had violated repeated promises to rectify the situation and give all bidders an equal chance.

Then, the Space.com story[167] broke asserting that certain forces on Capitol

Hill, including the powerful GAO, wanted NASA to slow down their decision on picking a builder for the new CEV spacecraft. With the shuttle fleet scheduled to be retired in 2010, NASA had decided that they wanted as little downtime as possible between the last shuttle mission and the first CEV missions. This pushed the timetable for the first launch up to 2012. Still, the original Apollo Program had been developed in just four years, and that was when NASA also had to build the entire infrastructure to handle the launches and recoveries as well.

What could be driving these sudden decisions to not only go back to the Moon and on to Mars (and perhaps, beyond ...), but to also "fast-track" the program behind the scenes? To, paradoxically, publicly aim for "the first new lunar mission" some *14 years away*—not until 2020—*twice* as long as it took to carry out the original Apollo Program, yet to make almost impossibly tight timetable and unheard-of-schedule demands on contractors out of public view?

Was there, perhaps, something about the times we lived in that could be driving this aggressive, *hidden* schedule? Or, had NASA actually found "something" on their first voyages to the Moon—perhaps, at Shorty Crater—that finally *compelled* this President to go back?

Or—after more than thirty years—is it simply "time?"

It was only when a website called "Collectspace.com" (on July 20, 2006) broke the story—that, the name of the new space capsule was to be "Orion"—we began to fully understand.[168] What NASA once sought to hide behind the symbolism of the infamous original Apollo mission patch, they now openly displayed. Such is their apparent confidence that *no one* in the mainstream press will question either their motives or sincerity around this issue.

The final shock came in late August, when NASA surprisingly announced (well, not that surprisingly; Hoagland had correctly been predicting *for weeks* what was going to take place, which his co-author *steadfastly* refused to believe ...) that *Lockheed* had been awarded the Orion "Apollo on steroids" CEV Program.

NASA's formal reasons were not immediately clear, except that Lockheed had promised to launch sooner and agreed to do final assembly in Florida. There were some technical reasons given, like the fact that Lockheed's design had "circular" solar panels rather than rectangular ones, but these seemed more like excuses than reasons (unless, of course, you quietly noticed the eerie resemblance of the Lockheed "Orion" panels to the look and placement of the actual "belt stars" in the constellation of *Orion*).

The aerospace press was quick to criticize the decision, mostly because the whole thing seemed very rushed and likely to lead to cost overruns. At the official "contractor unveiling" press conference, it was also learned that the final decision for Lockheed's design had been made by new NASA Administrator Mike Griffin himself rather than any "review committee."

But as previously noted, it was when we saw the official production art renderings of the proposed Orion spacecraft and the projected Lunar Lander that we *knew.*

The Orion capsule was shown orbiting above rectilinear lunar ruins—features that eerily just happened to resemble Hoagland's "Los Angeles" region, east of Ukert, at Sinus Medii! And, the LSAM was depicted on a lunar surface that obviously references the memorable Taurus-Littrow Valley—with bits of obviously *broken and twisted machinery* literally spread out all around the Lander's feet [Fig. 12-20]!

The symbolic public message to the "in crowd" seemed abundantly clear:

Project "Osiris"—the literal resurrection of Apollo, if not of NASA itself, after more than *forty years* of "marking time"—is *returning* to a Moon that is (we now know) literally *full* of ancient, waiting ... *artificial* wonders.

Chapter Twelve Images

Fig. 12-1 – Section of NASA/Boeing poster for President Bush's Space Exploration Initiative, circa 1989. It depicts two astronauts coming upon Martian ruins.

Fig. 12-2 – Comparison of pre and post Apollo 13 Apollo program patches, and the new "Orion" program patch.

Fig. 12-3 – Apollo 17 commander Eugene Cernan (and friends) reacting to President George W. Bush's announcement of a new initiative to return to the Moon. What's ultimately behind Cernan's stoic (if not rude) reaction?

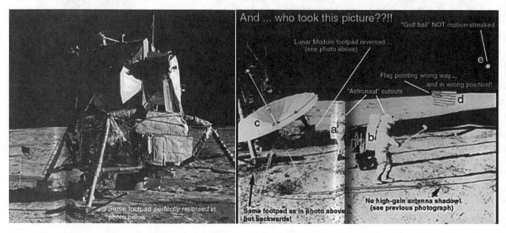

Fig. 12-4 – Faked "centerfold image" from Alan Shepard's book "Moon Shot." Portions of Apollo 14 Lunar Module Antares (L) is copied and reversed in right hand image. In addition, astronauts, "golf ball," TV antenna, shadows and flag are added "cutouts" from other Apollo 14 imagery—with flag pointing in opposite direction from actual Hasselblad photograph (L) and live TV downlink. (NASA/"Moon Shot")

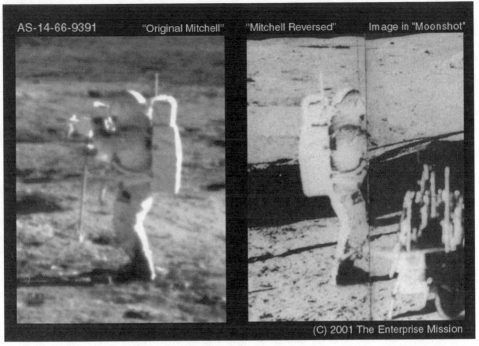

AS-14-66-9391   "Original Mitchell"   "Mitchell Reversed"   Image in "Moonshot"

(C) 2001 The Enterprise Mission

Fig. 12-5 – Side by comparison showing that "Edgar Mitchell" in faked image from Alan Shepard's book "Moonshot" is just a reversed cutout from the AS-14-66-9301 "Mitchell under glass" frame.

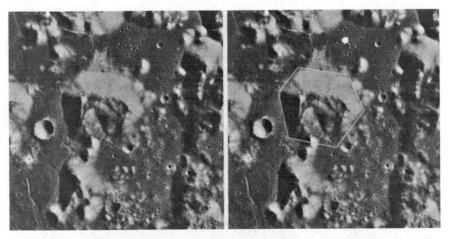

Fig. 12-6 – Side-by-side crops of Apollo 17 landing site. Hexagonal "South Massif" is near the center. Note collapsed backside of the massif and implications of six-sided geometry. Right side image markup shows assumed hexagonal reconstruction (white lines) and actual Apollo 17 landing site (white dot).

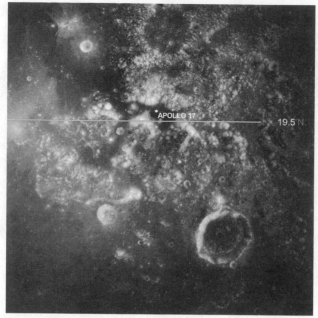

Fig. 12-7 – Apollo 17 landing site—as close to NASA's "magic latitude" of 19.5 degrees N. as astronauts could get, within driving/walking distance of anomalous, hexagonal "mountain" called "South Massif"—amid rugged additional "mountains" of Taurus-Littrow (NASA/ Enterprise)

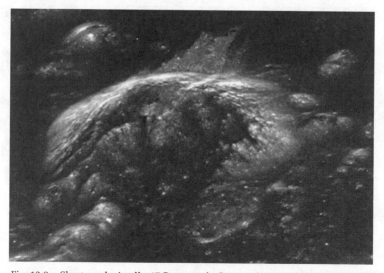

Fig. 12-8 – Slant-angle Apollo 17 Panoramic Camera image of "South Massif," looking North. Note geometric, cell-like pattern of internal structure, and parallel, raised 3D patterns on north face. Interior of "mountain" appears to be hollow, and partially filled with debris from catastrophic internal collapse. Strong evidence supporting "ruined lunar arcology model" (NASA)

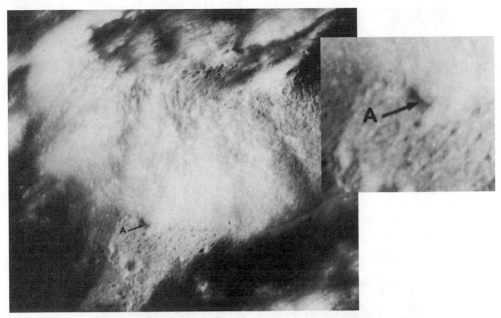

Fig. 12-9 – Two orbital views of Nansen (A), primary target of the Apollo 17 day two EVA.

Fig. 12-10 – Striking example of "geologic layering" (on a waterless Moon...) from Apollo 16 —similar to Apollo 17 landing site, bearing a remarkable resemblance to heavily-eroded, artificial "honeycomb" construction (NASA - AS16—112-18231HR)

Fig. 12-11 - Piece of mangled, reflective mechanical debris located inside Shorty crater (NASA – AS17-137-20996HR/Bara)

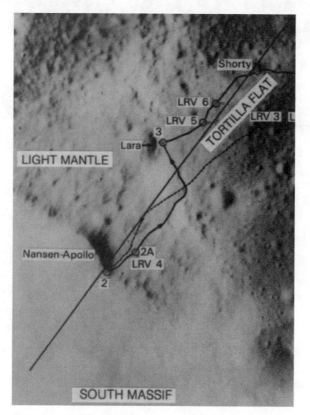

Fig. 12-12 - EVA-2 traverse map showing the locations of geology stations 2 and 2A at the base of the South Massif. Note that station 2A would allow a direct view into the Nansen opening.

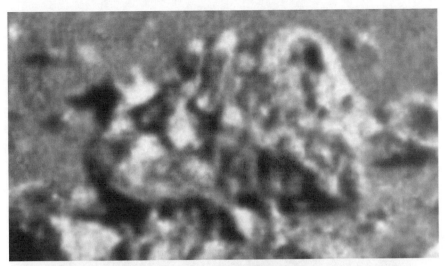

Fig. 12-13 – A piece of mechanical debris from inside Shorty crater.

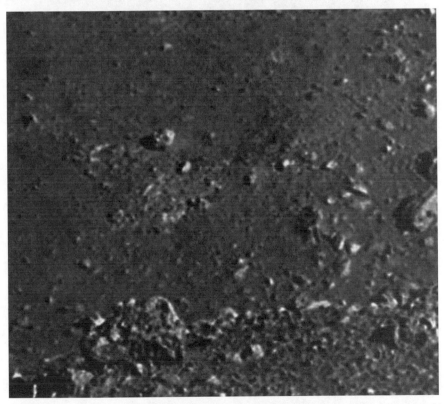

Fig. 12-14 – Context shot, showing previously mentioned mechanical debris (bottom, left) and (above it) "Data's head"—on the floor of ˜300-foot-wide Shorty crater. See also Color Plates 27 & 28 (NASA – AS17-137-20997HR/ Enterprise Mission)

Fig. 12-15 - Grayscale version of "Data's head" (rotated)—on the floor of Shorty Crater. See also Color Plate 28 (NASA-AS17-137-20997HR/ Enterprise Mission)

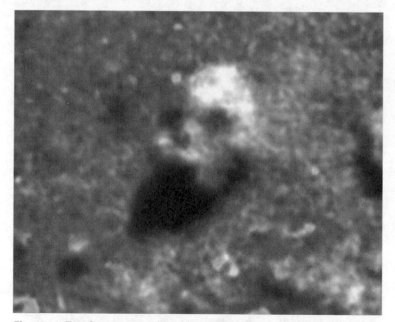

Fig. 12-16 -Two-frame composite of "Data's head"—to improve "signal-to-noise" image detail. Note "iris-type" eyes (strikingly similar to mechanical camera apertures)—evidence of an artificial "life form" (NASA-AS17-137-20996HR and AS17-137-20997HR /Enterprise Mission)

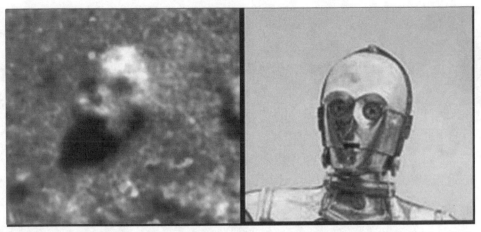

Fig. 12-17 – Comparison of "Data's Head" with another extremely popular "artificial life form"—C-3PO. Note eerie similarity of the eyes .... (NASA/Lucas Films/Enterprise Mission)

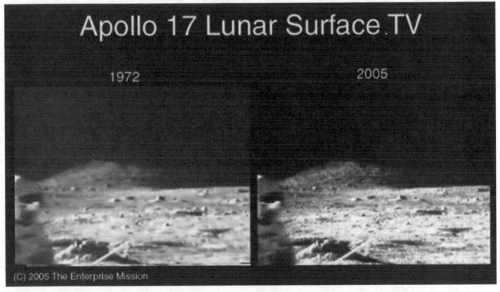

Fig. 12-18 - Apollo 17 lunar surface footage from original 1972 NASA transmission (L), and 2005 re-broadcast (R). Note distinct "honeycombing" on small background "arcology" (feature known on NASA maps as "Mons Vitruvius"), in 2005— in "enhanced" version of original broadcast (NASA/Enterprise Mission)

Fig. 12-19 -- Hasselblad photographic (film) image of Mons Vitruvius arcology, taken by Apollo 17 astronauts from Lunar Rover. Note reflective geometric "waffle" surface pattern on arcology, and "fuzzy" detail on top – both indicative of designed, eroding glass. Metallic tube-like object in foreground is counterweight for Lunar Rover high-gain antenna (NASA/Enterprise Mission)

Fig. 12-20 - NASA artist's concept of the new Lunar Lander on the Moon. Note the similarity of background mountains to "South Massif," and pieces of obvious mechanical debris at Landers feet.

Epilogue
Richard C. Hoagland

The big picture that Mike Bara and I have tried to paint in the preceding pages is imperfect, at best. We know that.

It is, however, an honest effort—based on official NASA data—but data which has been (as you have repeatedly seen throughout this book) both deliberately hidden ... and deliberately corrupted.

It is a picture also based, in part, on correlating what objective evidence we've been able to gather (primarily, the astonishing official NASA photographs ...), combined with the first-person testimonies of a small handful of human beings who have personally witnessed this remarkable deception (a few ex-NASA employees and the Apollo astronauts).

As noted earlier, the latter have been demonstrably compromised—either by having their own memories deliberately altered after seeing first-hand the wonders we have laid out here—or, from their own continuing, misguided "allegiance" to the security oaths they signed when they joined NASA. Remember, according to the official Charter NASA is *not* a civilian research institution, but "a *defense agency* of the United States [emphasis added] ...").

It is an extraordinary picture based, at best, on an investigation also severely hampered because it lacks the force of law to *compel* the truth. As authors and investigators, we do not have the power of the subpoena, the legal process of "discovery" or the simple ability to have witnesses testify *under oath*—all crucial tools in confronting a government-wide cover-up of such a sweeping, overwhelming magnitude and implication.

But, we *do* have evidence—extraordinary *scientific* evidence—which has been leaked to us quietly out of NASA by a few of its truly patriotic former employees over the years. And now suddenly—literally as we go to press—a brand new flood of even more amazing imagery is being posted directly, on official NASA websites available around the world.

A Few "Loose Ends"

As Mike and I were coming to the end of the four years it took to put this book together, our own original focus began to seriously shift—from our decades-long *Enterprise* investigation of the Martian anomalies, and calls for new and better NASA data on the artifacts on Mars—to several major "creeping breakthroughs" in our quiet, ten-year investigation *of the Moon:*

The first of these breakthroughs came literally from within NASA itself, or at least from several former associates or employees of the Agency. In mid 2006, a story broke in the national media concerning the existence of high-quality TV recordings of the Apollo 11 mission having "gone missing."[169] As it turned out, the original TV broadcasts of the historic first lunar landing were far below the quality of what was originally transmitted from the Moon.

The Apollo 11 TV images were sent from an antenna mounted on top of the Lunar Module to three tracking stations on Earth: Goldstone in California and the Honeysuckle Creek and Parkes facilities in Australia. These original signals were in a format called Slow Scan TeleVision, or SSTV. The camera carried to the lunar surface provided a progressive 320 scan lines at 10 frames per second (hence the Slow Scan TV moniker) in black and white. Although still far below the resolution of the standard broadcast quality of the day (525 lines [interlaced] 30 frames per second), the original SSTV signals were orders of magnitude better than what was finally broadcast to the American public on July 20th, 1969.

Because the SSTV signal could not be directly translated to broadcast TV, RCA had to develop a Slow Scan Converter to provide a TV signal to the viewers in "real time." This converter had serious limitations however, and they led to a broadcast signal that was only 262.5 vertical lines of resolution, about half as sharp as a then standard TV signal and on a par with the old "kinescope" recorders of the 1950's.

The result was the dark, ghostly images we all saw on July 20th, 1969 when Neil Armstrong finally set foot upon the lunar surface. Polaroid pictures taken of the SSTV monitors in Australia (the signal from Honeysuckle Creek was the one that was eventually used by NASA to broadcast the historic "one small step" to American audiences) show just how dramatically better the SSTV signals were compared to the scan-converted images the public received [Fig. E-1].

Fortunately, the original SSTV signals were recorded, using a video technology that would eventually find its way into the Betamax and VHS recorders of the 1980's. Unfortunately, when a group of enthusiasts, including some of the original site engineers from Goldstone and the Australian tracking stations tried to find the original SSTV tapes, they found that they were missing. For several years, they have been searching for them, only to discover a very dry hole.

Following the Apollo 11 mission, procedures required that the tapes be shipped to the Goddard Space Flight Center (GSFC) in Greenbelt, Maryland. In 1970, the tapes were moved to the U.S. National Archives in Accession #69A4099. By 1984, all but two of the over 700 boxes of Apollo era magnetic tapes placed in the Accession had been removed and returned to Goddard *at their request* for "permanent retention." There are no records at Goddard showing receipt of the boxes of tapes. In fact, *all* of the SSTV tapes sent to Goddard are now missing, and to this date *not one* of the original Apollo 11 SSTV tapes has been found.

All of this might just be a sad story of incompetence and misfortune, as long as you don't consider the context in which all of these events took place.

Remember, if our thesis is correct, NASA was sending its astronauts to the Moon on an archeological reconnaissance mission, looking to find the lost power of the gods that may have been left lying around the lunar surface in the form of instrumentalities from eons past. As we have also seen from the many images presented in this volume, there were awesome glass structures all around the landing sites on most of the missions. If NASA actually suspected this—that Armstrong and Aldrin may have been descending into a vast, mysterious wasteland of enormous ruins—then a whole lot of things about this "missing tapes" affair begin to make sense.

First, given that color TV was commonplace by the late 1960's, why did NASA send a crappy, low resolution black and white camera to record the most momentous journey man had ever undertaken? Westinghouse had begun development of a color TV camera called "The Westinghouse Field Sequential Color Camera" for use in space (and on Moonwalks) in 1968 and had perfected it by 1969. The color camera had none of the "down converting" problems of the primitive SSTV cameras and could produce a sharp, clear, *color* picture that was comparable to broadcast quality. It was tested on Apollo 10 in 1969 and worked flawlessly, transmitting over three hours of clear, color pictures back to Earth from the Moon.[170] Following this test, you would have thought that NASA would enthusiastically pursue the use of color broadcasts from the lunar surface.

Not so.

According to a NASA paper by Bill Wood, former Apollo MFSN station engineer from Goldstone, there were a great many at NASA who were utterly stunned when it was decided to use the black and white SSTV camera instead of the Sequential Color Camera to cover the first Moonwalk. Max Faget, the designer of the Mercury capsule and considered an "icon" of the glory days of NASA, was beside himself that so many of the pictures (and all of the TV) from the lunar surface would be in black and white. According to "Chariots for Apollo" the official NASA history of the Apollo missions:

"Faget was more than mildly upset when he learned that so much of the television, motion, and still photography planned for Apollo 11 would be in black and white. To him, it was 'almost unbelievable' that the culmination of a $20-billion program 'is to be recorded in such a stingy manner.'"[171]

His objections were based on the fact that the color camera had been so successful on Apollo 10. Now, it might be argued that with only one mission under its belt, the color camera was too risky to take to the lunar surface. However, consider this other point; *neither had the SSTV black and white camera*. Both systems were equally "untried" in the exposed lunar environment.

Unbelievably, the Sequential Color Camera was eventually approved for use aboard Apollo 11, but only *inside the Command Module*. It never was allowed inside the LEM. NASA didn't want that high resolution Color camera anywhere near the lunar surface.

You would think, just based on the high political cost of a TV camera failure (assuming the real objective the Apollo program was to simply "beat the Russians to the Moon"), that they would have at least taken the color camera on board the lander as a backup, in case the SSTV camera failed. In fact, this is exactly what was done on later missions, with the exception that the Sequential Color Camera was the primary TV camera and SSTV camera the back-up. Instead, we had only one chance to record the historic events of that day, and it would all be recorded by a black and white camera that used about as much power as single Christmas light bulb.

Imagine the political embarrassment had the SSTV camera failed to operate.

All of this really only makes sense in the context of our arguments. If NASA had nothing to hide, if the high level puppet masters that ran the program under the auspices of the various secret societies had nothing to hide, then why not send the best possible TV camera to record our historic victory over the Soviets and communism itself?

Unless, of course, they could not be sure just what the audience at home (or the honest engineers at the monitors) might actually *see* if they were allowed to. So they sent the worst TV system they had, which was then downgraded even more by the scan conversion process. Fahrouk El Baz, Ken Kleinknecht and Wernher Von Braun must have breathed a huge sigh of relief when the images of Armstrong came back ghostly, dark, and full of contrast.

Interestingly, the color camera *was* carried on Apollo 12 as the primary camera, without the SSTV as a back-up. A few minutes after deployment of the camera, Alan Bean managed to violate his training and point it directly into the sun, which badly damaged the receiver. This rendered the camera useless.

On subsequent missions, the TV transmissions got better and better. However, the Sequential Color Camera only received a gamma correction capability on the last three missions (Apollo 15, 16, and 17) and all of the transmissions were run through a bandwidth limiting "low-pass" filter which dramatically reduced the image quality.

When you put all this together, we feel it strengthens our arguments considerably. Despite the existence of a vastly superior color technology, NASA chose to send an unproven and very poor quality black and white camera to the Moon on Apollo 11. This is exactly the behavior you would expect if the powerful cabal inside the agency was concerned about how much viewers would be able to see. Further, when they did finally send a color camera, they ran the signal through a "brute force" filter to effectively limit the bandwidth of the color image, and then managed to "stage" an accident once they had enough test footage. On subsequent missions, the cameras got better, but were always limited in both contrast and detail by the gamma correction and low pass filter problems.

So the only existing unfiltered TV images of the lunar landings are the original SSTV images. With today's digital processing capabilities, how much we might glean from those nearly 40 year old tapes? Instead, all we know for certain is that the same agency that ordered Ken Johnston to destroy the only remaining four sets of early generation negatives and prints of the hand held Apollo photography also removed these tapes from the National Archives for "permanent retention" in their own facility.

And then promptly "lost" them.

We doubt, quite honestly, that they will ever be found.

* * *

The second breakthrough of which we spoke was the sudden, public availability of a veritable flood of new Apollo *lunar* images (which would ultimately turn out to be literally *thousands* ...), placed without fanfare on the web.

These high quality "close-to-original" Apollo films—scanned at very high resolution (averaging around 16 MB) by NASA's Houston-based Johnson Space Center (JSC)—had begun to quietly be "leaked" to the general public beginning in 2006, through multiple, *official* NASA websites. This immense amount of data, suddenly "dumped" on the web [on such sites as the Apollo Lunar Surface Journal (ALSJ)], abruptly made possible *Enterprise* analyses never before practical—starting with a one-to-one "calibration" of the validity of the entire database.

By comparing the new NASA "JSC" scans to Ken Johnston's original, 30-year-old Houston data, stored safely away in his private archive for all those years, I realized we could *quantitatively* test the information contained in the newly-released images, by directly comparing them to the pristine details present on Ken's originally-preserved images ....

The immediate results were *outstanding* [Fig. E-2]

When the two versions of AS14-66-9301 ("Mitchell Under Glass" ...) were compared side-by-side, it was obvious that much of the amazing "geometric sky detail" seen on Johnston's original 1971-era Apollo 14 "C-print" (above-left), was *still* visible on the 2006 ALSJ scan (above-right). However, *differences* were also very obvious ... with Ken's 30-year-old print showing, by far, that it was MUCH closer (as it should have been) to the original NASA data acquired on the Moon than the new Apollo Lunar Surface Journal "web scans."

Since this first experiment had proven that a significant percentage of the original geometric information on Ken's print had been successfully passed on—even across the countless generations of analog copies made over those 30+ years between its original Hasselblad source negative and the much later ALSJ scans—I decided to press on with a search for a *second* Apollo 14 ALSJ Archive image, one that might have independently captured the same crucial "shattered dome geometry" at the Apollo 14 landing site.

I quickly found it.

ALSJ Archive frame AS14-66-9279 [Fig. E-3])—a photograph Alan Shepard took while standing on the east side of the LM—contains major elements of *the same scattered light geometry* as seen on Ken's version of "9301" (where Shepard was standing to the west when he took that amazing photograph)..

Though, again, nowhere near as detailed as Ken's pristine 30-year-old version, the correspondence of the major sky features in the two separate Apollo 14 images definitively, scientifically *proves* that the deep blue, ancient lunar glass dome—seen arching over the Edgar Mitchell on frame 9301—is *not* a photographic "fluke."

In fact, after going through the entire newly-released ALSJ Archive for Apollo 14, I ultimately found *four* independent Hasselblad scans—all showing the same general "towering glass geometry" visible on Ken Johnston's original print. You can't get much better scientific validation for a controversial optical phenomenon than *four independent photographic confirmations!*

Having these immediate positive results, I was encouraged to begin downloading as many of the newly added, high-resolution frames—from *all* the Apollo Lunar Missions—as I could over the next several months. My objective: find independent, multi-mission confirmations of anomalous phenomena indicative of lunar glass-like ruins—such as the tell-tale brightening on images *above* sunlit lunar features. Since natural rocks and craters have sharp, solid edges, "fuzziness" of lighting along the lunar limb, for example, would be significant evidence of light-diffusion and scattering by meteor-eroded glass.

If I could identify a handful of such criteria, and then find corroboration for each of these on a succession of separate NASA photographs taken during a particular Apollo mission—such as the preceding confirmation of the diffuse "sky geometry" first seen on Ken Johnston's original Apollo 14 print—and, in addition, find similar confirmations on *multiple* Apollo missions ... then I could consider those collective anomalous phenomena as "proven."

Example:

One striking Archive image was an Apollo 15 frame (AS15-88-12013 [Fig. E4]), taken "post-Trans Earth Injection" (TEI)—after the Service Module engine had placed the combined Command/ Service Module on a lunar escape trajectory, homeward-bound toward Earth after the successful three-day Mission to Hadley Rille.

Taken "looking back" as Apollo 15 was rapidly climbing away from the Moon, this image provides startling evidence that much of the Moon's Earth-facing hemisphere was/is *"domed over"* ... as determined by the scattering of *diffuse* sunlight being reflected off remnants of the surviving "glass-like domes." These domes are visible in this amazing image as a ghostly, bluish, cloud-like semi-circle faithfully following the curving lunar limb (see also Color Fig. 16). This intense backscattering quickly fades with altitude—like a "meteor-bombarded 'prairie fire'"– eventually blending into the expected black background of space *tens of miles* above the lunar surface. It looks, for all the world, exactly like the Earth's bluish "airglow limb." Except the Moon, as we all know, has no atmosphere.

So it has to be something else.

Combining this key Apollo 15 observation of the whole Moon with close-ups of this same phenomenon photographed from close-in lunar orbit (the stunning geometric glass "rebar" photographed over Sinus Medii by the crew of Apollo 10—see Color Figs. 2 and 3) provides two, completely *independent* confirmations of this *same* anomalous light scattering phenomenon.

The only reasonable interpretation of these independent Apollo observations is that both missions were, in fact, photographing the diffuse light-scattering created by *trillions* of surviving fragments from miles-high, glass-like *lunar domes*.

But the next late developments win—hands down—the Disclosure Prize ... for revealing just how *much* NASA has lied to us, all these years, about what's *really* on the Moon.

*   *   *

The story begins with our long-time friend and colleague in this *Enterprise* investigation, Steve Troy. As noted earlier, Steve is our "analog" photographic expert; he has spent a small private fortune over the last ten years ordering the lowest generation (best quality) hard-copy NASA lunar negatives from various NASA photographic archives including the National Space Science Data Service (NSSDC), the Lunar and Planetary Laboratory (LPL), and NASA Headquarters itself. With a NASA negative in hand, Steve then commissions a commercial photo lab to make enlarged, sectional prints—which he then both goes over almost "grain by grain" (literally, with a hand magnifying glass) as well as scans with a computer. The result is a careful tabulation of increasing numbers of "lunar anomalies"—for future close-up imaging when new NASA (or other lunar missions) someday return to photograph the Moon again. Some of these planned new spacecraft will contain amazing optical instruments—such as NASA's new "Lunar Reconnaissance Orbiter," scheduled to begin a "meter-scale" photographic survey of the entire Moon, beginning in late 2008.

Just a few weeks before this book finally went to press, Steve forwarded an e-mail from a correspondent in Italy; the individual in question had his own NASA anomaly website and wanted Steve's opinion on some recent lunar images he'd found and posted.

After looking at the site, I was curious about the *source* for the high-quality NASA lunar images Steve's friend had somehow located.

A quick search revealed *another* Italian site—www.spacearchive.net/— where Steve's correspondent procured the images he'd originally asked about. The "SpaceArchive" site was not only professional and well-organized (with downloadable pages of "contact sheets," arranged by specific NASA missions) it seemed to contain some truly remarkable *low-generation* NASA lunar images.

In fact—the best I'd *ever* seen, outside of Ken Johnston's 30-year-old private NASA stash.

One such image, after downloading [Fig. E-5], revealed a *stunning portrait* of the REAL "glass landscape" arching over Taurus-Littrow, the Apollo 17 landing site.

Blatantly visible is not only an almost blinding lunar sky—FILLED with obviously *battered glass constructions*—but one also riddled with "huge gaping holes" and quasi-vertical, mysterious "dark linear formations."

And, on one of the major background mountains to the SE is a huge, long, brilliant "spar" ... casually leaning on the massive "massif!"

The view fully corroborates Gene Cernan's own emotional reaction to actually seeing all this, first-hand so many years ago:

"Man, you talk about a *mysterious looking* place!"

Here [Fig. E-6] is an annotated version of this astonishing new image.

This astonishing close-up [Fig. E-7], from frame AS17-136-20767, confirms several extraordinary additional details of "lunar dome construction" that we've been proposing over the years—starting with the "prairie fire effect" seen along the left-hand crest (and on the far right ...) of the massive "mountain" that dominates the view (the "mountain" is "Mons Vitruvius"—according to NASA—just another ancient, eroded lunar massif ... over a mile high). This totally anomalous "scattering phenomenon" is revealing confirmation of a key prediction of our overall "dome model"—obviously representing the surviving, battered shards of meteor-smashed glass, *whose remaining density is directly proportional to their ability to scatter sun light.*

The "lunar mountain" beneath this shattered glass is obviously being systematically eroded from the *top* down (which is why the optical density is highest just above the "mountain's" surface); this incessant micro-meteorite "rain" is thus relentlessly whittling away—over millions of years—at the vast mass of what is, in fact, the ruins of another former three-dimensional, *honeycombed, mile-sized* lunar structure ... an ancient lunar *"arcology!"*

Extending far above this ancient eroding arcology—in this brightness-enhanced version of the original NASA image—is not the pure black of space expected on any official photograph taken from the surface of the Moon, but an obviously three-dimensional *matrix,* of more porous, semi-transparent, similar *light-scattering* material, effectively turning the lunar blackness overhead into a glittering, shiny "curtain" ....

Photographic confirmation of Alan Bean's own haunting lunar memories ... "up there, space has a real *shiny* look. It reminded me a little bit of [black] *patent-leather shoes ....*"

In our continuing analysis of "the real Moon," this one image is now total confirmation of the cumulative "optical depth effect" of literally *tens of miles* of an extremely sparse, now almost totally obliterated (otherwise, the Lunar Module *Challenger* couldn't have gotten down through it safely to the surface ... and returned to orbit!), former full-blown *lunar glass dome* ... arching far above the surface of the Taurus-Littrow Valley.

And, if you look closely at this image, you will also see—apparently imprisoned within this all-encompassing light-colored "matrix"—a few surviving, *larger fragments* ... more-resistant objects "still hanging" *above* Mons Vitruvius ... embedded *in the glass!* (These cannot be, by the way, stars or other distant background reflections—the photographic exposure times on the lunar surface were far too short).

But the most amazing thing, by far, is the gigantic, unquestionably artificial-looking *"linear spar,"* leaning up against the side of Mons Vitruvius ... gently sagging—like a "Titan's straw"—under the obvious effects of lunar gravity!

At the upper end, where this amazing artifact is visibly extending out of the shadow of the supporting "mountain," there is some kind of obvious "mechanical linkage." There is also a hint of a corresponding mechanical "fitting" is attached to the lower end, a thin filament partially hidden beyond and below the relatively near-by lunar horizon. This then raises the obvious amazing question:

Did Gene Cernan and Harrison Schmidt, during their first EVA, drive the lunar rover the few miles to the lower end of this amazing artifact ... and retrieve an obvious sample of "lunar dome construction" for return to Earth!

And, would they 'remember" ... if they had?

Stunned by discovering such an obviously pristine NASA frame—and on a public, *international* website—I immediately downloaded three, sequentially numbered similar images of Taurus-Littrow, also listed at the site ... and promptly confirmed even more extraordinary aspects of this amazing scene.

On a wide-angle panorama [Fig. E-8]—assembled by taking four of these Hasselblad frames and fitting them together—the pervasive light-colored "matrix" in the sky is revealed to extend across the *entire Valley*; the "holes" and "dark, vertical formations" are clearly the result of "something" smashing *through* the glass, and removing a significant fraction of the "light-scattering material" in those locations. Exactly what you would expect of an extraordinarily ancient, physically *real* "lunar dome" ....

The "prairie fire effect" of the densest, surviving glass still covering individual features underneath this former dome, can be seen in this panorama to extend *all along* the optical "ridgeline," formed by the silhouettes of the other "massifs" that create the southern boundary of the Taurus-Littrow Valley—from "Mons Vitruvius" (on the far left) ... to "South Massif" (on the right). The glass is apparently densest just over the summit of South Massif, indicating that (maybe), as Keith Laney has proposed, this is the "youngest arcology" built within the Valley ... of course, at "19.5."

[Fig. E-9] is a close-up of the center-section of mysterious "dark streaks"—and the stark "missing slices" in the glass-like material still hanging in the sky.

The ultimate political explanation for the sudden appearance of this astonishing set of *original* NASA images—and on an international website!—would be extremely revealing. But even without knowing the details (like how a webmaster of an obscure Italian NASA archive site suddenly came by *untouched, original NASA images*—and of the *real* Taurus-Littrow—photographed over 30 years ago by Schmidt and Cernan), we can certainly speculate.

In the same timeframe that official NASA websites in this Country are abruptly posting thousands of never-before-seen "best" scanned NASA images of the Apollo Program, I find it hard to believe the appearance of a set of really *best* lunar images (apart from Ken Johnston's, of course)—and on a *foreign* website—"just happened" to occur simultaneously; this really looks like a *deliberate* leak from "someone" inside NASA, with access to some of the astonishing *originally* suppressed Apollo imaging.

And, the "coincidental" e-mail to Steve Troy, which prompted me to do a bit of searching—just as Mike and I were wrapping up this book—is also just a bit too cute in terms of timing ....

Bottom line: the stunning "Italian images" are profoundly revealing—not just for what they show ... but for what they *politically* confirm:

That the *other* recently scanned and released Apollo images, on the Apollo Lunar Surface Journal website (and others)—stated as "coming directly from original JSC film" did *not*. That, like everything else with NASA—going back to its very creation just under half a century ago—they are merely another carefully controlled *version* of the Truth ....

Which "someone" inside NASA apparently decided to expose—by "leaking" (with exquisite timing and "plausible deniability") a smidgeon of stunning *real* Apollo data to this Italian website.

Or, to paraphrase John Erlichmann (of Watergate notoriety):

The ALSI images represent, at best, "just another *limited* 'NASA hangout.'

*   *   *

Speaking of revolutionary NASA space photography ....

Another *major*, late development was the sudden acquisition of the first, long-awaited MRO image of the Face on Mars.

For literally a year, we (and a lot of other folks ...) had been impatiently and intensely looking forward to the first Cydonia image that would be taken (due to overwhelming "popular demand"—according to NASA spokespersons) by MRO's "HiRISE"—the High Resolution Imaging Science Experiment telescope/camera. HiRISE—with a 19.5-inch-wide telescope mirror (I kid you not!), and CCD camera capable of acquiring images over 20,000 pixels wide—represents by far the largest and most powerful imaging system ever sent to Mars."

As we were closing out the book, NASA finally, quietly, acquired and released from HiRISE the first truly spy-camera quality, ultra-high-resolution image of the Face [Fig. E-10]. At slightly less than "11 *inches* per pixel" (compared to previous "high-resolution" images of some ~4 *feet* per pixel) the MRO Face image represents, by far, the best *ever* taken of this still intensely controversial Martian surface feature.

As usual, all the critics immediately proclaimed that the new Face image "finally, overwhelmingly, *proves* that the 'face' is just a pile of rocks!"

Not quite ....

As with any work of art, there is the persistent problem that I have dubbed in previous years "the Gigi Factor" (from the classic 1950's film by the same name, in which Maurice Chevalier plaintively asks in song "have I been *standing up too close ... or back too far?*"); if you're too far way, you won't be able to recognize the art; if you're too close, all you will see are brush strokes ....

In all the typical presentations of the Face on Mars, the image above is how it is inevitably published: a small-scale reproduction, where various folks then throw around totally subjective *opinions* on "what it *looks* like."

After over 30 years, and probably a hundred (yes, a hundred ...) repeated imagings of "The Face" since 1998—curiously acquired by an Agency which repeatedly, simultaneously, professes with each new image "this is an object of NO scientific interest ..."—you'd think people would pretty much be past repeating the same tired "instant reactions" to each retaking of "new" images.

* * *

They're not.

Of course, after the original first "new" imaging of the Face, by NASA's *Mars Surveyor* spacecraft in 1998, we at *Enterprise* immediately realized that trying to evaluate the potential *artificial* possibilities for this object (and its surrounding structures)—based on "what does it *look* like?"—were not only unscientific ... they were pointless. There is no possibility of determining the *objective* reality of a subjective "work of art"—and on another planet!—via such unscientific, "opinion-based" criteria, let alone "proving" it ... to anyone.

Thus, we have waited patiently (for more than 25 years ...) for NASA's space technology to evolve (or, be "allowed" to evolve ... à la "Brookings") to the point where an unmanned robotic mission could be sent to Mars with a powerful enough telescope/camera to actually see the *structural elements* that, if the Face is artificial, *have* to be there:

Things like "rooms," and "walls" with exposed "beams," "girders," etc. etc.

With the new MRO Face image—that has now occurred.

So, what do we finally *see* close-up in the new Face image? With a resolution per pixel of about *a foot* (!), and a file size of over 300 megabytes, the resolution is finally good enough to detect these critical, small-scale *architectural* elements from orbit!

In this case, to *prove* the Face is artificial, we *have* to see the "brush strokes"—if this is, in fact, an intelligent mile-sized work of art.

And, we do!

Proof of the scientific validity (and practicality) of this approach came in the months leading up to this unannounced April, 2007 acquisition of MRO's first image of the Face. Soon after the spacecraft began routine science operations (in September, 2006) MRO was commanded to take an image of one of NASA's unmanned "stars" still operating on the Martian surface, the *Opportunity* Mars Rover—perched on the edge of a half-mile-wide Martian impact crater in the "Meridianii" region of the planet. From its circular orbit of about 180 statute miles straight overhead, the MRO ~19.5-inch HiRISE telescope/camera combination looked down ... and snapped an astonishingly detailed picture of a *known* "man-made artifact" on the surface of Mars.

As can be seen in this comparative enlargement, not only were the ~5-inch Rover *wheel tracks* in the sand easily visible from orbit (!), the *shadow* of the 3-inch-wide, 3-foot-high camera mast [Fig. E-11]—stretching out across the Martian Meridianni desert in the late afternoon sun—was *also* clearly visible ... from 180 miles!

If there *were* "eroded walls and girders" on the Face (and on the other artificial structures all over Mars that we've identified), this stunning demonstration of the visual acuity of the MRO "HiRISE camera" *proved* that we would be able to detect them!

Provided—

The MRO images (unlike other NASA photographs discussed earlier ...) are NOT "tampered with."

Given the demonstrable hold "Brookings" has had on the "honest side" of NASA for so long, the relentless way the *dishonest* side has used this aging 1950's sociological Study to repeatedly justify its continuing *censorship* of "what is really out there"—to instill an almost palpable *fear* in those scientists and engineers who might be moved to openly discuss what NASA's really found in the way of "ET artifacts" on other planets in the solar system—the fate of such a powerful new tool orbiting Mars, and the stunning images it can obviously now acquire, was (and is ..) still quite uncertain, even at this writing ....

So, when the first NASA MRO "Face image" was suddenly released, *just before we closed out the final sections of this book*, we (along with everyone else) truly did not know what to expect. However, after spending many days enlarging and analyzing different sections of this enormous object (the Face possesses almost two *square miles of exposed surface area* for such analyses ...), we have came to three critical conclusions:

A) This NASA image is NOT up to the technical standards of previous MRO images of Mars—such as the Odyssey Rover photo just discussed; for some inexplicable reason, the MRO "Face" image possesses a significant amount of what imaging scientists call "noise"—both as "random noise" across the entire image, and in the form of *rhythmic, banded patterns*. The latter appear as equally-spaced bars, both vertical and on a diagonal. This peculiar and intrusive "banding" tends to obscure the real, geometrically designed surface patterns we are searching for underneath (which may explain their presence ...) that represent the signature of actual artificial Martian *ruins*.

B) There are portions of this image, on the Face itself and on nearby formations, which show the tell-tale signs of image tampering with a "blur-tool," a device used to obscure detail in a digital CCD image such as this one.

But 2) despite this handicap, there are now clearly-defined remnants of *artificial constructions—ruins!—easily visible all across this extraordinary object* ... as you can see [Fig. E-12].

In this composite MRO Face image, the small area to the right of the "nose area" (outlined) is the "footprint" of the enlarged section (inset). In this enlargement one can easily discern row upon row of obviously collapsed *geometric ruins*. The striking, orderly arrangement includes a blatant, stair-stepped "wall" descending through the center of the image, as well as a host of other, equally rectilinear ruins below this blatantly geometric configuration.

The key to proper interpretation of aerial or satellite imagery of man-made ruins on Earth lies in noting the multiple examples of "parallel walls" and redundant *rectilinear geometry*; natural geologic features *cannot* present—except in very restricted contexts (where other geologic clues must also be present)—these *repeating* demonstrations of redundant geometric regularities and 90-degree relationships: right angle, enclosed rooms; repeating linear wall alignments; and redundant examples of geometrically organized "uniform-width" features.

Intelligently-designed ruins *always* do.

So, in comparing this same MRO sectional enlargement [Fig. E-13]with an aerial photograph of a 5th century terrestrial middle eastern ruin, note the number of strikingly parallel, regular geometric features that *both* images present, including those which even look strikingly like "avenues" and "roads" in the MRO Face image!

And ... all those parallel-width *walls*.

Further, in spite of obvious evidence of substantial erosion and decay, redundant examples are present in this MRO enlargement of entire "enclosed courtyards," "deliberately aligned constructions," and evidence of a "large-scale, organizational *plan*."

It is the presence of these multiple examples of "recognizable, *architectural* geometries"—in even this small region of the Face—that now confirms that this extraordinary object is, indeed, host to multiple *artificial structures*.

But, there is more.

Examining another region of the Face—this one approximately a mile from the first location, further down on the flat "platform section" at the base of the "chin area"—reveals another type of equally obvious constructions [Fig. E-14].

Sand and debris-filled, *geometric* "cavities"—strikingly similar to ancient Anasazi ruins in the American Southwest [Fig. E-15].

The fact that these ruined structures look somewhat different from the previous examples high above, near the Face's "nose," is due to two critical conditions:

1) The original architectural geometry was truly different—composed of larger (and deeper) individual "cells," consistent with a structural foundation for the vast mass of the entire "Face"; and 2), the ruins' current physical location—on the flat "platform area," at the base of the "chin" slope—allows all the eroded debris from higher on that slope to cascade down ... into these deep geometric cavities of former "rooms."

A third striking example of "artificiality" is on the forehead of the Face [Fig. E-16].

Captured in this region of the MRO Face image (outline-above), this geometric structure is a ~ 800-foot-long, *multi-storied ruin*—now almost totally obliterated by time and erosion. It is located just below the boundary between the heavily-eroded "cubicle transition zone" of a former surface coating above the "forehead," and an even more eroded, flat, "depression area," located above the Face's "eyebrow ridge."

The western end of this multi-leveled structure is the most recognizable as artificial—revealing complex, 3-D rectilinear geometry and parallel aligned shadows. There are additional ruins to the east, casting additional tell-tale geometric shadows on the patterned "ground." A similar-scaled set of "Iranian hilltop forts" [Fig. E-17] reveal comparable manmade ruined structures here on Earth.

Our next example is even more extraordinary—once you get past the fact that the sheer existence of ANY "ruins" on the Face is *totally* "extraordinary"— because NONE of these obviously artificial structures should even be there ....

Here [Fig. E-18] is where this striking example lurks—at the base of the Face's eastern "platform."

As can be seen from the detailed image comparison (below—top and bottom), what we are seeing is a "small," strikingly rectangular *collapsed* section of the Martian surface at the base of the much larger "Face platform"—an in-fallen section, measuring approximately 1000 feet long by several hundred feet wide ... and several *hundred* feet deep. Fascinating, *vertical* striations can be seen all along the right hand (eastern) edge of this deep "chasm," indicative of surviving *structural columns*—still holding up a multi-layered section of exposed, three-dimensional, distinctly *honey-combed artificial surface structure* [Fig. E-19]. Additional three-dimensional geometric patterns, carefully aligned with the major rectangular axis of this "collapse feature," can be seen further to the east (above—top right)—presenting an overall pattern of "a massive, three-dimensional, inexorably deteriorating artificial *complex* ...."

The impression of "looking down through multiple levels ... of *a vast, three-dimensional, highly-battered architectural framework ...*" is now *inescapable* [Fig. E-20].

Our last example is obviously (since, I've saved the best for last ...) the most amazing of these enlarged "snapshots" I've been able to capture from the full resolution MRO image: a series of obviously high-tech, obviously collapsed entire 3-D *structures* ... located on the "chin" of the Face—specifically, on the now highly eroded "bottom lip" [Fig. E-21].

An enlarged comparison—between this magnified section of the MRO image, and an entire, tilted (from a recent local earthquake) modern apartment house in South America [Fig. E-22]—illustrates the *crucial* point; if you look carefully at the top image, you will notice a bewildering number of straight lines, sharp edges, 90-degree angles, flat sides and more of those "parallel-width walls." These are all *unnatural* features, *never* seen on any ordinary "geological" formation, and certainly not in such extraordinary numbers and close, repetitive association.

These are the inevitable hallmark of closely-associated, shattered and eroding *high-tech structures*, whose sheer presence is an overwhelming confirmation of the completely artificial nature of the Face.

As can be seen in the lower image [Fig. E-22], these same striking *geometric* relationships are found in any modern city—the inevitable consequence of the construction of *repetitive, multi-storied structures* formed from basic *geometric units*.

So, what happens if you partially destroy (tilt, or even knock down) one or more of these constructed, closely adjacent units?

The result is vividly illustrated in [Fig. E-22]: you then see *multiple* sets of mutually *conflicting* rectilinear geometry ... *exactly* like what we see, over and over again, in this astonishing small section of the Face!

And, if you look carefully at an annotated enlargement [Fig. E-23], you can even see the infamous "beams and girders" we've been predicting for over a decade now, that *had to be confirmed* in any sufficiently high-resolution image of this ancient "high-tech structure" ....

Because—

Nothing in this image is natural. Let me repeat that: *nothing* that you see in this MRO enlargement [Fig. E-23] is natural.

Natural geology *doesn't* come with "parallel walls," "multiple, 3-D planes," "twisted beams"—or repetitive examples of obvious "thin girders."

High-tech structures always *do* ... regardless of their specific composition.

*   *   *

We have been proposing for over 15 years that the Face is, in fact, just such a massive assemblage of ancient, high-tech *buildings*—a literal "headquarters." MRO's stunning first image of The Face on Mars—and the multiple examples we've presented and discussed in this brief analysis of that first image—now *totally* support that view.

We see the ubiquitous, redundant presence—across all sections of this *mile-wide, upturned "statute"*—of striking "geometric patterns," "parallel walls," "rectilinear collapsed, sand-filled rooms" and even the elevated remains of a few, still clearly-recognizable, entire, tilted building fragments!

After more than 20 years of investigating Cydonia, after proposing test after scientific test for "archaeology," to essentially "deaf ears" in the National Aeronautics and Space Administration, and even (at times) not just apathy but against heavy *official opposition* from this same Agency ... it's finally:

Game ... set ... match!

The "Face on Mars"—after over a generation of ambiguous evidence, rancorous debate and even outright NASA-manipulated data (certainly, the first *MGS* Face image in 1998)—turns out to be *exactly* what we said it was at the United Nations in 1992.

The most unique example of an extraordinary *honeycombed arcology* in the solar system—because it looks like *us*, or, something we once *were*.

Still standing on the planet *Mars*.

Testimony to an extraordinary solar-system *full of ancient, ineffable ruins* ... left by an incredible civilization that (for reasons we must now figure out, and *soon*), completely, mysteriously ... disappeared.

Leaving only *us*.

* * *

The most serendipitous *Enterprise* discovery in all our years, of course, is "C3-PO," lying at the bottom of "Shorty Crater" on the Moon. His discovery, more than any other aspect of this long investigation, changes *everything*.

If he is a genuine thinking *robot*, endowed with bona fide "artificial intelligence" (AI), he represents an entirely new way of looking at this unknown, now-vanished brilliant civilization ... which left countless wonders abandoned, not only across the Moon, but *all across the solar system*.

The implications of that one fact—if it is a "fact"—and *if* the astronauts found and brought "him" home (or another robot *like* him—from any of the other Apollo landing sites)—obviously would have sent shockwaves throughout Washington ... if not (certainly among the "in crowd") all around the world. This discovery, not "economics," or "lack of interest in the Moon" by the American people during the end of the Apollo Program, in our opinion was quite likely *the real reason* why the entire Program was so suddenly and so unceremoniously terminated.

And, why no one has gone near the Moon ... for almost forty years.

Because—

His stunning presence on the Moon compellingly indicates that "Man" may have been preceded in this solar system, in addition to his own ancestors, by a vast integrated *population* of "other intelligences." The overriding question then becomes, in light of "Brookings," what if there are others like him "out there"—

But—*still functioning?!*

What if this was NASA's *real* "Dark Mission" all along ... to find and bring to Earth a functioning member of this projected AI population—a robot that the in-crowd could ultimately interrogate about the literal "secrets of the Universe" ... *firsthand?*

* * *

As we go to press, there is much that could not be covered in this volume, simply because of space and time considerations. We have demonstrated the "what" behind NASA's peculiar behavior vis-à-vis the "artifacts question" over the years. We have shown that their original assertions about the Face and Cydonia—that it was just a trick of light and shadow—are fallacious. We have demonstrated their duplicitous behavior with the Catbox image and the THEMIS infrared data. We have shown that NASA and its appendant bodies, like JPL, are willing to go to extreme lengths to disabuse the public from the notion that Mars once harbored life. They've even gone as far as grinding a possible fossil to dust rather than study it openly. Can one seriously doubt at this point that NASA has been manipulative— at the least— around the question of past or present life on Mars?

We have further shown that there is an undeniable "Orion-Osiris" Egyptian connection to both the Apollo program and our new space initiative. We have shown that all of the major power brokers inside NASA at the time of the Apollo program had connections to one of three secretive societies. And each of these societies has as its core faith a reverence for the same three ancient Egyptian gods; Isis Osiris and Horus. We have shown, over and over again, that key moments in NASA's exploration of the solar system have been planned around "stellar rituals" that pay homage to these long forgotten "gods" of ancient Egypt.

What we recognize we have yet to prove is the "why" of this strange behavior. The Brookings report alone seems insufficient to explain the artifacts cover-up and even less adequate to explain the occult naming rituals and curious affinity for stellar alignments. After forty years of *Star Trek* and

Star Wars, we would seem to have a populace that is not simply conditioned, but *eager* to find extraterrestrial life, or its remnants. Yet still, NASA hesitates, ignoring the obvious evidence and cloaking their true objectives in posters and paintings, spacecraft names and odd rituals. To find this truth, we will have to look back in time once again, not only to the occult history of NASA itself, but to our nation's founding and our own esoteric history as a people.

Finally, we must ask the question that NASA fears the most; that it must never allow to be asked: "If there are ruins on Mars, and ruins on the Moon, what happened to the builders of them?"

If this *was* Apollo's ultimate "deep black mission"—a Mission that John F. Kennedy was somehow convinced to undertake at the beginnings of his Administration—is this also, as we've asked earlier, the *real* reason for his murder?

Was his discovery of NASA's potential secret motivation for Apollo the hidden reason he quietly decided to turn around and *share* our lunar program with our *"arch enemies"* only a few months after its announcement? Did those protracted, behind-the-scenes negotiations with Nikita Khrushchev in the end, get him killed?

Based on the disturbing, compelling mosaic this combined scientific and documentary evidence now paints of NASA—and its unquestionably treasonous actions over these last ~40 years—we believe we have finally grasped the outlines—and have attempted to accurately reflect them here—of the staggering truth that has silently, simultaneously, both impelled and cowed the Space Agency for all these years:

That, the Human Race lives in a solar system *surrounded* by a vast array of "silent, ancient ghosts" ... countless extinct "extraterrestrial humans," just like ourselves—in fact, *our own* "great, great, great, great, great, great ... ancestors"—who once lived ... and built ... and walked amid what are still the almost incomprehensible remains of an awesome *extraterrestrial civilization*, spread across more than one adjoining world.

That, *this* is the ultimate meaning of "the Face on Mars"–

That, *this* is the real, *hidden* meaning of Neil Armstrong's infamous (and still deliberately "spun") "misstatement"—as he stepped for the first time onto the surface of the Moon that unforgettable July 20th night in 1969. With almost a billion people watching live, Armstrong uttered those still endlessly debated words:

"That's one small step for Man ... one giant leap for Mankind ...."

What Armstrong was actually acknowledging—to a "hidden audience" *in code* that night—as he, representing "Man," stepped onto the lunar surface for perhaps the first time in tens of thousands of years ... was that the human race ("Man") is only a *subset* of a vastly larger, vastly more ancient, vastly more knowledgeable, but *genetically related "Mankind."* And that we had somehow crawled up from the ashes of that once grand civilization, to do the same as they had done, to travel to another world, and touch the face of the gods...

That—*this* is the True History of the Human Race, the clandestine reason for the creation and naming of "Apollo" ... if not the ultimate "Dark Mission" for the existence of NASA itself? Was this the "protective layer of truth" that Armstrong implored the young people of America to remove at the 25th anniversary celebration of Apollo 11?

We may never know. After all, as one of our intelligence sources admitted about NASA some years ago:

"The lie is different at every level."

Epilogue Images

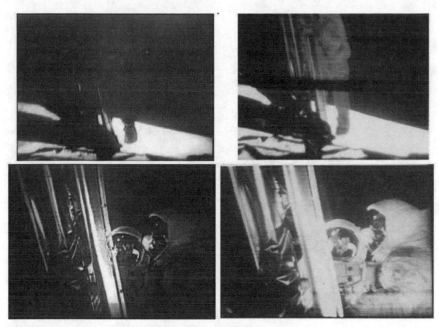

Fig. E-1 - Comparison of scan-converted TV broadcast from Apollo 11 (left) vs. Polaroid photos of SSTV monitors at Parkes tracking station in Australia (right).

Fig. E-2 – Comparison of Ken Johnston's version of AS14-66-9301 and current version from NASA archives. The glassy ruins still persist, but much of the detail has been lost in the intervening 35 years since Johnston obtained his early generation version.

Fig. E-3 – Another example from the current NASA archive showing "hints" of the amazing structures found on Ken Johnston's original prints.

Fig. E-4 – Astonishing "airglow limb" from Apollo 15 "lookback" image of the Moon. See also color figure 16.

Fig. E-5 – Enhancement of Apollo 17 photograph from Taurus Littrow landing site showing towering glass structures complete with meteor damage and a collapsed "spar" resting on Mons Vitruvius.

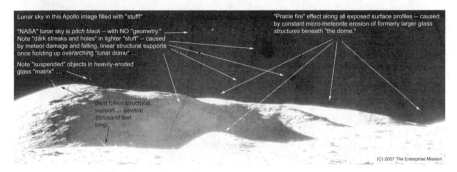

Fig. E-6 – Annotated version of AS17-136-20767 obtained from Italian website http://www.spacearchive.net/

Fig. E-7 – Close up of the Spar from AS17-136-20767.

Fig. E-8 – Hand assembled panorama showing the Mons Vitruvius dome.

Fig. E-9 – Close-up showing battered remnants of glass structures above Mons Vitruvius.

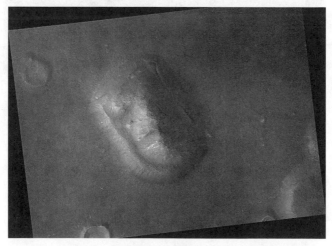

Fig. E-10 – Mars Reconnaissance Orbiter Image of the Face on Mars.

Fig. E-11 – Artists rendition of mast on Opportunity rover and visible shadow recorded by MRO. Rover tracks visible in orbital image are less than 3 inches across.

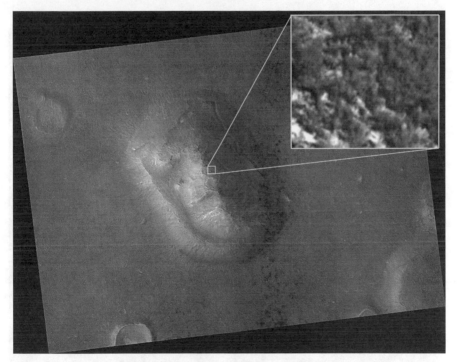

Fig. E-12 – Enhancement of wall like "Roman ruins" from the nose area of the Face on Mars

Fig. E-13 – Comparison of "nose ruins" from MRO image with 5th century Middle Eastern ruin.

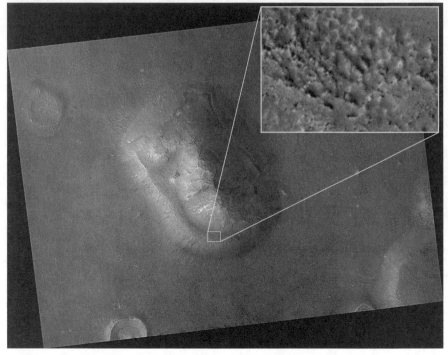

Fig. E-14 – Sand filled ruins from the Chin area of the Face

Fig. E-15 – Sand filled ruins compared to Anasazi ruins in New Mexico.

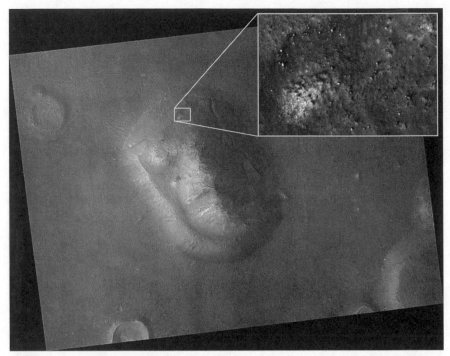

Fig. E-16 – Underlying ruins from the forehead of the Face.

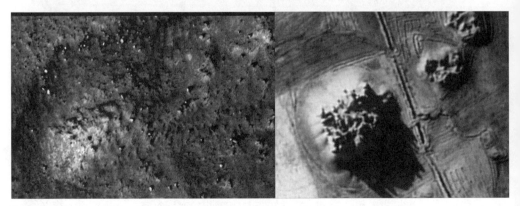

Fig. E-17 – The Forehead ruins compared to eroding fortresses in the Iranian desert.

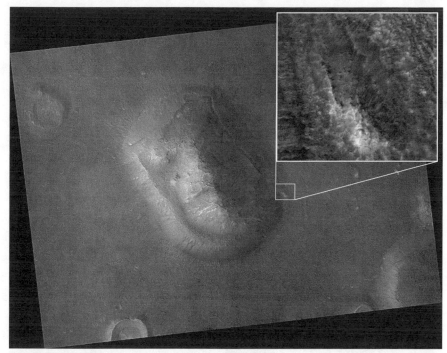

Fig. E-18 – Collapsed trench next to the Face.

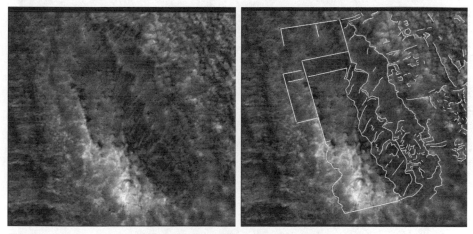

Fig. E-19 – Close-up of the collapsed trench with mark-up (Hoagland)

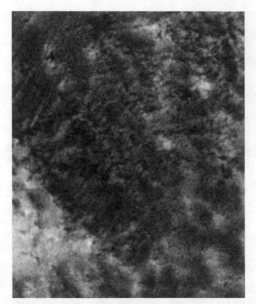

Fig. E-20 – Ultra close-up of geometric ruins near
the collapse area.

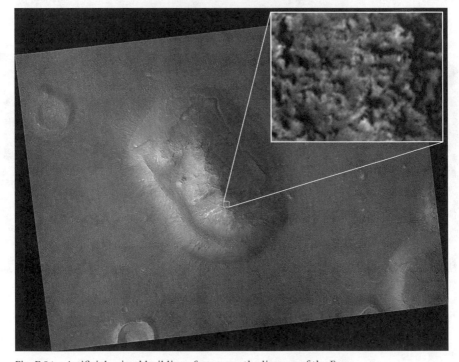

Fig. E-21 – Artificial ruined buildings from near the lip area of the Face

Fig. E-22 – Enlarged comparison with terrestrial ruins.

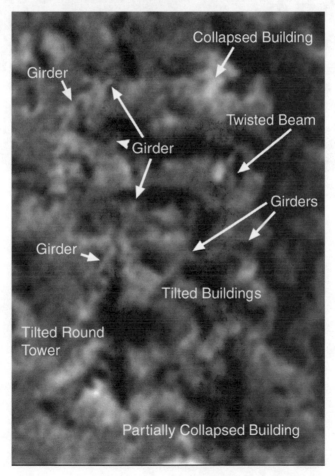

Fig. E-23 – Marked up ultra close-up of "chin ruins" on the Face

Fig. E-24 – "One small step for Man, one giant leap for Mankind."

Endnotes

Introduction

1   http://www.enterprisemission.com/images/act-1.gif
2   http://www.enterprisemission.com/images/act-7.gif
3   http://www.enterprisemission.com/images/act-1.gif
4   http://www.enterprisemission.com/images/brooking.gif
5   http://www.enterprisemission.com/images/brook-7.gif
6   http://www.thespacereview.com/article/735/1
7   http://history.nasa.gov/SP-4209/ch2-4.htm
8   http://www.enterprisemission.com/images/brook-9.gif
9   http://www.enterprisemission.com/tides.htm
10 http://www.enterprisemission.com/images/nytimes.gif
11 http://www.usatoday.com/tech/science/space/2005-09-27-nasa-griffin-interview_x.htm

Chapter One—The Monuments of Mars

12 Hoagland, Richard C. The Monuments of Mars—A City on the Edge of Forever. Fourth edition. p. 5.
13 Sagan, Carl. Cosmos. p. 140.
14 The Face on Mars. p. 68.
15 http://www.mcdanielreport.com/teeth.htm
16 http://www.planetarymysteries.com/mars/fiertek.html
17 "The McDaniel Report: On the Failure of Executive, Congressional and Scientific Responsibility in Investigating Possible Evidence of Artificial Structures on the Surface of Mars and in Setting Mission Priorities for NASA's Mars Exploration Program." p. 148-156.
18 The Monuments of Mars. p. 185.
19 "The McDaniel Report: On the Failure of Executive, Congressional and Scientific Responsibility in Investigating Possible Evidence of Artificial Structures on the Surface of Mars and in Setting Mission Priorities for NASA's Mars Exploration Program."
20 http://www.astrosurf.com/lunascan/blair.htm
21 http://www.vgl.org/webfiles/lan/cuspids/cuspids.htm
22 The Monuments of Mars. p. 325.
23 "The McDaniel Report." p. 98-101.
24 http://www.mcdanielreport.com/tvalues.htm, http://www.mcdanielreport.com/pntdchrt.htm, http://www.mcdanielreport.com/mounds.htm, http://www.mcdanielreport.com/geometry.htm http://www.mcdanielreport.com/sirvent.htm, http://www.mcdanielreport.com/flmnds2.htm
25 http://www.enterprisemission.com/message.htm
26 "The McDaniel Report." p. 126.
27 Posting from Ralph Greenberg to the Art Bell BBS, 12:40 p.m., February 23, 1999.
28 The Nephilim and the pyramid of the Apocalypse, Patrick Heron p. 2
29 The Nephilim and the pyramid of the Apocalypse, Patrick Heron p. 3

Chapter Two—Hyperdimensional Physics

30 Hubbard, W. B. Geophys. Space Phys. 18 (1980) 1.
31 Icarus, vol. 112, no. 2. p. 337-353.
32 "An experimental test of non-local realism" by S. Gröblacher et. al., Nature 446, 871, April 2007 | "To be or not to be local" by Alain Aspect, Nature 446, 866, April 2007
33 "My soul is an entangled knot, Upon a liquid vortex wrought. By Intellect in the Unseen residing. And thine doth like a convict sit, With marlinspike untwisting it, Only to find its knottiness abiding; Since all the tool for its untying." – James Clerk Maxwell, "A Paradoxical Ode."
34 Bulletin of the American Mathematical Society. [4 (1887), 54-7]
35 Bulletin of the Calcutta Mathematical Society, Vol. 20, 1928-29, p.202. "Oliver Heaviside: Sage in Solitude" (IEEE Press, New York, 1988, p.9, note 3.
36 "On the Partial Differential Equations of Mathematical Physics." (Mathematische Annalen, vol. 57, 1903. p. 333-335); "On an Expression of the Electromagnetic Field Due to Electrons by Means of Two Scalar Potential Functions." (Proceedings of the London Mathematical Society, vol. 1, 1904. p. 367-372.); Nikola Tesla, Colorado Springs Notes 1899-1900, Nolit, Beograd, Yugoslavia, 1978. p. 61-62.
37 http://www.apfn.org/Free_Energy/electromagnetic.pdf

38   http://jnaudin.free.fr/meg/meg.htm

39   Scientific American, September 1975. p. 29.

40   http://www.enterprisemission.com/images/pulsar-Wash-post1.gif

41   http://www.msnbc.com/news/320182.asp?cp1=1

42   Cook, Alan H. Interiors of the Planets. Cambridge University Press. p. 261.

43   "High-Resolution Maps of Jupiter at Five Microns," Astrophysics Journal, vol. 183. p. 1063-1073;

44   "Summary of Historical Data: Interpretation of the Pioneer and Voyager Cloud Configurations in a Time Dependent Framework." Science, vol. 204. p. 948-951.; "Infrared Images of Jupiter at 5-Micrometer Wavelength During the Voyager 1 Encounter." Science, vol. 204. p. 1007-8. Flaser, et. al. "Prospecting Jupiter in the Thermal Infrared with Cassini CIRS: Atmospheric Temperatures and Dynamics." American Astronomical Society, DPS meeting #33, #03.01

45   Nelson, J. H. "Planetary Position Effect on Short-Wave Signal Quality." Electrical Engineering, May 1952.

46   http://amasci.com/freenrg/tors/doc17.html

47   http://www.americanantigravity.com/search?articlelive=5641a08aaa928ff0619bb3a11e6f4b12&Query=Shipov&fromSmall=true&searchWhat=searchAll&Categories=0&searchField=searchContentBody

48   http://www.americanantigravity.com/documents/Shipov-Interview.pdf

49   http://www.bibliotecapleyades.net/esp_divinecosmos_1.htm

Chapter Three—Political Developments

50   Light Years – By Gary Kinder – Viking Press ISBN-10: 0670818860

51   http://www-mipl.jpl.nasa.gov/iplhistory.html

52   The Monuments of Mars. p. 405.

53   http://www.jpl.nasa.gov/jplhistory/the80/mars-observer-t.php

54   The Monuments of Mars. p. 423.

55   After its launch in 1991, NASA engineers discovered that a basic error had been made in the grinding of th Hubble's key reflective mirror. It was only years later that the optics were repaired on a shuttle mission, at the cost of hundreds of millions of dollars.

56   http://www.jfklibrary.org/Historical+Resources/Archives/Reference+Desk/Speeches/JFK/003POF03NewspaperPublishers04271961.htm

57   http://www.pbs.org/redfiles/moon/deep/interv/m_int_sergei_khrushchev.htm

58   SP-4209 The Partnership: A History of the Apollo-Soyuz Test Project http://history.nasa.gov/SP-4209/ch2-4.htm

59   http://www.jfklibrary.org/Historical+Resources/Archives/Reference+Desk/Speeches/JFK/003POF03_18thGeneralAssembly09201963.htm

60   http://history.nasa.gov/SP-4209/ch2-4.htm#source72

61   Public Papers of the Presidents of the United States, Lyndon B. Johnson, 1963-1964 I (Washington, 1964), pp. 72-73

62   SP-4209 The Partnership: A History of the Apollo-Soyuz Test Project http://history.nasa.gov/SP-4209/ch2-4.htm

63   http://www.jfklibrary.org/Asset+Tree/Asset+Viewers/Image+Asset+Viewer.htm?guid=%7BBFF5BEE4-D3FC-422D-9D39-946104F2B845%7D&type=lgmpd&num=1

64   http://209.132.68.98/pdf/kennedy_cia.pdf

65   http://www.spacewar.com/news/russia-97h.html

66   http://www.pbs.org/redfiles/moon/deep/interv/m_int_sergei_khrushchev.htm

67   http://mcadams.posc.mu.edu/russ/jfkinfo/jfk8/sound1.htm

68   Houston Chronicle coverage, Nov. 22, 1963 Edition: Blue Streak, By STAN REDDING and WALTERMANSELL, Chronicle Reporters

69   Public Law 88-215, An act making appropriations. . . for the fiscal year ending June 30, 1964, . . . , 88th Cong., 1st sess., 1963, p. 16

Chapter Four—The Crystal Towers of the Moon

70   http://www.mufor.org/tlp/lunar.html

71   E-mail communication from Ken Johnston Jr. to the authors. July 2, 2004.

72   http://www.hq.nasa.gov/office/pao/History/ap15fj/15solo_ops3.htm#proclus2

73   http://discovermagazine.com/1994/jul/rememberingapoll39

74   http://www.hq.nasa.gov/alsj/emj.html

75   http://www.hq.nasa.gov/alsj/kipp.html

76   http://www.apolloarchive.com/

77   http://www.alanbeangallery.com/

78   http://www.lunaranomalies.com/fake-moon.htm; http://www.lunaranomalies.com/fake-moon2.htm; http://www.lunaranomalies.com/rad.htm; http://www.lunaranomalies.com/c-rock.htm; http://www.lunaranomalies.com/coffin.htm

79   Mechanical Properties of Lunar Materials Under Anhydrous, Hard Vacuum Conditions: Applications of Lunar Glass Structural Components. Blacic, J. D. In: Lunar Bases and Space Activities of the 21st Century. Houston, TX, Lunar and Planetary Institute, edited by W. W. Mendell, 1985, p.487 1985lbsa.conf..487B

80   The Gold Bulletin

81   http://www.salon.com/news/feature/1999/07/20/aldrin/

82   http://books.guardian.co.uk/reviews/biography/0,6121,1468768,00.html

Chapter Five–A Conspiracy Unfolds

83   All We Did Was Fly to the Moon. p. 41.

84   Wallis Budge, E. A. Osiris and the Egyptian Resurrection, 1911.

85   Bauval and Gilbert. The Orion Mystery, 1994.

86   Temple, Robert K. G. The Sirius Mystery, 1976.

87   Bauval and Hancock. The Keeper of Genesis, 1996.

88   All We Did Was Fly to the Moon. p. 77.

89   Men From Earth. p. 248.

90   http://en.wikipedia.org/wiki/Osiris

91   Czarnik, Marvin. "The 'Where' and 'When' of Each Apollo Landing was Carefully Planned."

92   http://www.bu.edu/remotesensing/Faculty/El-Baz/FEBbio.html

93   Knight, Christopher and Robert Lomas. The Hiram Key.

94   "The New Age," Scottish Rite Journal, Volume LXXVII. Number 12.

95   http://www.tranquilitylodge2000.org/

96   Hunt, Linda. Secret Agenda.

97   http://www.majesticdocuments.com/personnel/vonkarman.php

98   http://www.enterprisemission.com/spaceact.html

99   http://www.spacepolitics.com/2006/08/23/griffin-fires-back-at-advisors

100 http://history.nasa.gov/SP-4209/ch1-4.htm

101 "CNN Breaking News," live broadcast, July 4, 1997.

102 http://www.enterprisemission.com/planet.htm

103 Bauval and Gilbert. The Orion Mystery, 1994.

104 Graham Hancock, The Mars Mystery. p. 50.

Chapter Six–New Mars Global Surveyor Images of Cydonia

105 http://www.anomalies.net/archive/cni-news/CNI.0843.html

106 http://www.virtuallystrange.net/ufo/updates/1998/mar/m06-016.shtml

107 http://www.metaresearch.org/solar%20system/cydonia/proof_files/proof.asp

Chapter Seven–An Eye for an Eye

108 http://www.cnn.com/TECH/space/9804/15/holliman/

109 http://marsweb.jpl.nasa.gov/msp98/news/mco990930.html

110 E-mail to Mike Bara.

111 http://nssdc.gsfc.nasa.gov/planetary/text/mpl_pr_20000322.txt

112 ftp://ftp.hq.nasa.gov/pub/pao/Goldin/2000/jpl_remarks.pdf

113 http://www.fas.org/irp/news/1993/931216i.htm

114 http://www.enterprisemission.com/empire.html

115 http://barsoom.msss.com/mars_images/moc/01_31_01_releases/cydonia/index.html

116 http://barsoom.msss.com/mars_images/moc/01_31_01_releases/cydonia/M18-00606d.gif

Chapter Eight–FACETS and the Face

117 http://www.enterprisemission.com/images/mars/lawnchair.jpg

118 http://www.enterprisemission.com/images/felinec2.jpg

119 http://science.nasa.gov/headlines/y2001/ast24may_1.htm?aol453399

120 http://www.enterprisemission.com/mola.htm

121 http://members.nbci.com/cydonia_institute/index.html

Chapter Nine–2001: A Mars Odyssey

122 http://www.sciencemag.org/feature/data/hottopics/se260002330p.pdf

123 http://www.msss.com/mars_images/moc/abs/lpsc2000/03_slopestreaks_1058.pdf

124 http://www.space.com/missionlaunches/odyssey_update_020121.html

125 http://www.space.com/scienceastronomy/solarsystem/odyssey_update_020226.html

127 http://msnbc.msn.com/news/577946.asp

128 http://www.summit-okinawa.gr.jp/tokusyu/ruins1.htm

Chapter Ten—Mars Heats Up

129 http://themis.la.asu.edu/zoom-20020724A.html
130 http://www.aas.org/publications/baas/v34n3/dps2002/302.htm
131 http://www.enterprisemission.com/ir_analysis.html
132 http://www.enterprisemission.com/THEMISSDPSIS.pdf
133 Algorithm Theoretical Basis Document for Decorrelation Stretch, version 2.2, August 15, 1996, Ronald E. Alley, Jet Propulsion Laboratory.
134 http://www.jargon.net/jargonfile/b/bamf.html
135 Phone conversation between Noel Gorelick and Michael Bara, September 6th, 2002.
136 http://www.newfrontiersinscience.com/martianenigmas/Papers/JBIS1990.pdf
137 http://www.enterprisemission.com/IRLiesfromASU.htm
138 http://mars.complete-isp.com/time/zubrin.html
139 http://themis.la.asu.edu/zoom-20021031A.html

Chapter Eleven - The True Colors of NASA

140 http://themis-data.asu.edu/img/V03814003.html
141 http://www.lpi.usra.edu/meetings/programs/mesurwa.txt
142 http://meted.ucar.edu/nwp/pcu2/gemsnow.htm
143 DiGregorio, B., G. Levin and P. Straat. Mars: The Living Planet. Frog Ltd., 1997.
144 http://oposite.stsci.edu/pubinfo/pr/97/23/PR.html
145 http://mars.spherix.com/color/color.htm
146 http://www.maasdigital.com/gallery.html
147 http://marsrovers.jpl.nasa.gov/spotlight/airbags01.html
148 http://www.jpl.nasa.gov/mer2004/rover-images/jan-06-2004/captions/image-7.html, http://www.jpl.nasa.gov/mer2004/rover-images/jan-06-2004/captions/image-6.html
149 http://marsrovers.jpl.nasa.gov/spotlight/airbags01.html
150 http://www.jpl.nasa.gov/mer2004/rover-images/jan-06-2004/captions/image-6.html
151 http://www.snapon.com/tool-storage/tool_storage_kra.asp
152 http://marsrovers.jpl.nasa.gov/gallery/press/spirit/20040119a.html
153 http://www.earthfiles.com/news/news.cfm?ID=650&category=Science
154 http://www.dlr.de/mars-express/images/230104
155 http://www.enterprisemission.com/colors.htm
156 http://faculty.weber.edu/sharley/AIFT/GSL-Life.htm
157 http://sps.k12.ar.us/massengale/protist_unrevised_notes_b1.htm

Chapter Twelve—Where the Titans Slept...

158 http://www.enterprisemission.com/millenn.htm
159 http://www.thespacereview.com/article/106/1
160 Mark Dwane Audio recording.
161 http://www.space.com/news/060728_cev_gao.html
162 http://www.keithlaney.net/Ahiddenmission/A17HMp1.html
163 http://www.hq.nasa.gov/office/pao/History/SP-4214/ch13-8.html
164 http://www.hq.nasa.gov/alsj/a17/a17j.html
165 http://www.hq.nasa.gov/alsj/a17/a17.html
166 Apollo Lunar Surface Journal
167 http://www.space.com/news/060728_cev_gao.html
168 http://www.collectspace.com/news/news-072006a.html

Epilogue - Richard C. Hoagland

169 http://www.space.com/news/060813_apollo11_tapes.html
170 http://www.hq.nasa.gov/alsj/ApolloTV-Acrobat5.pdf
171 http://www.hq.nasa.gov/office/pao/History/SP-4205/ch13-4.html#source40

Acknowledgments

Richard C. Hoagland, *first and foremost, would like to acknowledge all the folks he has met on this incomparable journey, who have contributed to where we are today. There are far too many of you, across far too many years, to thank individually. You know who you are.*

However, I must recognize some of the more outstanding contributions personally, starting with the talents of my friend and co-author in this daunting project, Mike Bara; put simply, without Mike this volume would have never reached your hands. Adam Parfrey, our esteemed publisher, also deserves considerable credit in that vein—not only for his unshakable faith in the merits of this work, but for inordinate patience in the face of almost interminable delays in seeing its completion.

I also want to send a special "thanks" to Nick and Dana, and Kynthia—as well as Ken J., Steve T., Keith L. ... the "other" Keith ("Scotty" R.) ... Stan T., Jay W., David L., David W., Tom VF., Rick S., Paola H., Paul D., Hollace D., Tim V., Bill A., Robin W., George G., Ted St. R., Arthur A., Alan C., Michael M., Doris L-M., Boris F., Patty M., David H., Charlie B., Bobby T., and Ron G.

Each of you know "why."

A special remembrance also to those colleagues whose contributions were invaluable, but who are no longer able to contribute—at least, not from this dimension: David L., Bruce D., Gene M and Gene R.

To three special furry friends: Luvcky, Shadow and Sasha—whose boundless loyalty and simple love are unmistakable examples of why family has never been a species definition; and, further in that regard, to the entire "Coast to Coast AM family," whose unwavering support has been a crucial reminder all these years of why we're doing this—together.

And to Art and George—for keeping us a "family."

Finally, to Robin—without whom, literally, I would not be here ....

I love you.

Michael Bara *would like to acknowledge the following persons, without which this project would never have been completed: as always, my friend, mentor and co-author Mr. Hoagland, the smartest man I have ever known, Leslie, who changed my life twice, Michael, of whom I am very proud, my brother Dave, Alyssa and Sherri, whose friendship has meant more to me than I can articulate, Zaphod, Cleo and Indy, who were constant companions as I worked on this, my mom and dad, and my sister Kelli. Steve Troy and Keith Laney, whose indefatigable work has led to many new discoveries.*

About the Authors

Richard C. Hoagland was science advisor to Walter Cronkite and CBS News during the Apollo program, former curator at the Hayden Planetarium and consultant to NASA. He is the co-originator, along with Eric Burgess, of the British Interplanetary Society, of the "Pioneer Plaque" currently carrying a message from mankind on the Pioneer 10 spacecraft. He is the author of The Europa Enigma, *the first extensive scientific article proposing the mechanism by which life might exist in the oceans of Jupiter's moon Europa. This article became the basis for Arthur C. Clarke's novel 2010. He is the principal investigator of the Enterprise Mission, an independent scientific research organization dedicated to the examination of more than 40 years of NASA data pointing to the possibility of archaeological ruins on Mars and the Moon. In 1993, the Angstrom Foundation, in Stockholm, Sweden, awarded the International Angstrom Medal of Excellence in Science to Hoagland for that continuing research. He is the author of the best selling* The Monuments of Mars—A City on the Edge of Forever, *now in its fifth edition.*

Michael Bara is an aerospace structural engineer with more than twenty-five years' experience in the field. He is currently a CAD/CAM consultant for one of the largest engineering software solution providers in North America. He is the author of numerous articles on the Enterprise Mission website, as well as on Art Bell's website. He is curator of the Lunar Anomalies Homepage, a website dedicated to the investigation of potentially artificial structures on the Moon.